MICROBIAL PRODUCTION OF NUCLEIC ACID-RELATED SUBSTANCES

205294

MICROBIAL PRODUCTION OF NUCLEIC ACID-RELATED SUBSTANCES

Edited by the Association of Amino Acid and Nucleic Acid:
Koichi OGATA Shukuo KINOSHITA
Toshinao TSUNODA Kô AIDA

Associate Editors:
Akira KUNINAKA Yoshio NAKAO Kiyoshi NAKAYAMA
Kunio OISHI Hiroshi OKADA Mitsuru SHIBUKAWA
Zenjiro SUGISAKI Yoshiki TANI

TP248.3
M5
1976

A HALSTED PRESS BOOK
KODANSHA LTD.
Tokyo
JOHN WILEY & SONS
New York–London–Sydney–Toronto

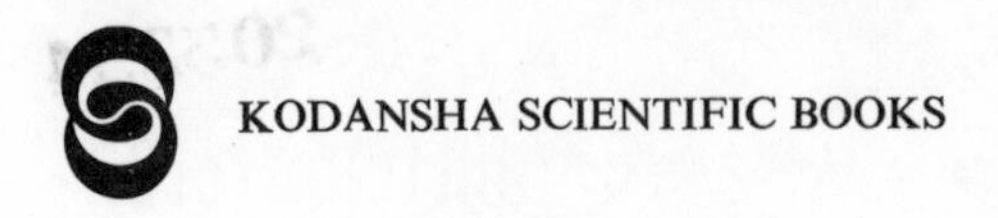

KODANSHA SCIENTIFIC BOOKS

Copyright © 1976 by Kodansha Ltd.

All rights reserved

No part of this book may be reproduced in any form, by photostat, microfilm, retrieval system, or any other means, without the written permission of Kodansha Ltd. (except in the case of brief quotation for criticism or review)

ISBN: 0 470-15167-6
LIBRARY OF CONGRESS CATALOG CARD NO.: 76-25309

Published in Japan by
KODANSHA LTD.
12-21 Otowa 2-chome, Bunkyo-ku, Tokyo 112, Japan

Published by
HALSTED PRESS
a Division of John Wiley & Sons, Inc.
605 Third Avenue, New York, N.Y. 10016, U.S.A.

PRINTED IN JAPAN

Contributors

Kô AIDA, Institute of Applied Microbiology, University of Tokyo, *Bunkyo-ku, Tokyo 113, Japan*

Ichiro CHIBATA, Research Laboratory of Applied Biochemistry, Tanabe Seiyaku Co. Ltd., *Yodogawa-ku, Osaka 532, Japan*

Akira FURUYA, Tokyo Research Laboratory, Kyowa Hakko Kogyo Co. Ltd., *Machida-shi 194, Japan*

Yoshio HIROSE, Central Research Laboratories, Ajinomoto Co. Inc., *Kawasaki-ku, Kawasaki 210, Japan*

Akira IMADA, Medicinal Research Laboratories, Takeda Chemical Industries Ltd., *Yodogawa-ku, Osaka 532, Japan*

Jiro ISHIYAMA, Central Research Laboratory, Kikkoman Shoyu Co. Ltd., *Noda-shi 278, Japan*

Kiyoshi ISONO, Institute of Physical and Chemical Research, *Wako-shi 351, Japan*

Joji KATO, Research Laboratory of Applied Biochemistry, Tanabe Seiyaku Co. Ltd., *Yodogawa-ku, Osaka 532, Japan*

Shukuo KINOSHITA, Kyowa Hakko Kogyo Co. Ltd., *Chiyoda-ku, Tokyo 100, Japan*

Yasushi KOMATA, Central Research Laboratories, Ajinomoto Co. Inc., *Kawasaki-ku, Kawasaki 210, Japan*

Akira KUNINAKA, Research Laboratory, Yamasa Shoyu Co. Ltd., *Choshi-shi 288, Japan*

Koji MITSUGI, Central Research Laboratories, Ajinomoto Co. Inc., *Kawasaki-ku, Kawasaki 210, Japan*

Sawao MURAO, Department of Agricultural Chemistry, Faculty of Agriculture, Osaka Metropolitan University, *Sakai-shi 591, Japan*

Yoshio NAKAO, Microbiological Research Laboratories, Takeda Chemical Industries Ltd., *Yodogawa-ku, Osaka 532, Japan*

Kiyoshi NAKAYAMA, Tokyo Research Laboratory, Kyowa Hakko Kogyo Co. Ltd., *Machida-shi 194, Japan*

Masao NARUSE, Pharmaceuticals Research, Development Center, Kyowa Hakko Kogyo Co. Ltd., *Chiyoda-ku, Tokyo 100, Japan*

Koichi OGATA, Department of Agricultural Chemistry, Faculty of Agriculture, Kyoto University, *Sakyo-ku, Kyoto 606, Japan*

Kunio OISHI, Institute of Applied Microbiology, University of Tokyo, *Bunkyo-ku, Tokyo 113, Japan*

Hiroshi OKADA, Central Research Laboratories, Ajinomoto Co. Inc., *Kawasaki-ku, Kawasaki 210, Japan*

Takuo SAKAI, Department of Agricultural Chemistry, College of Agriculture, University of Osaka Prefecture, *Sakai-shi 591, Japan*

Ken'ichi SASAJIMA, Institute for Fermentation, *Yodogawa-ku, Osaka 532, Japan*

Mitsuru SHIBUKAWA, Food and Fine Chemical Division, Asahi Chemical Industry Co. Ltd., *Chiyoda-ku, Tokyo 100, Japan*

Teruo SHIRO, Central Research Laboratories, Ajinomoto Co. Inc., *Kawasaki-ku, Kawasaki 210, Japan*

Tsuneo SOWA, Research Section, Foodstuff and Plant, Asahi Chemical Industry Co. Ltd., *Nobeoka-shi 882, Japan*

Zenjiro SUGISAKI, Noda Institute for Scientific Research, *Noda-shi 278, Japan*

Ikuo SUHARA, Biological Research Laboratories, Takeda Chemical Industries Ltd., *Yodogawa-ku, Osaka 532, Japan*

Yoshiki TANI, Department of Agricultural Chemistry, Faculty of Agriculture, Kyoto University, *Sakyo-ku, Kyoto 606, Japan*

Tatsurokuro TOCHIKURA, Department of Food Science and Technology, Faculty of Agriculture, Kyoto University, *Sakyo-ku, Kyoto 606, Japan*

Fusao TOMITA, Tokyo Research Laboratory, Kyowa Hakko Kogyo Co. Ltd., *Machida-shi 194, Japan*

Toshinao TSUNODA, Central Research Laboratories, Ajinomoto Co. Inc., *Kawasaki-ku, Kawasaki 210, Japan*

Kiyoshi WATANABE, Biochemical Research Laboratory, Kanegafuchi Chemical Industrial Co. Ltd., *Takasago-shi 676, Japan*

Masahiko YONEDA, Medicinal Research Laboratories, Takeda Chemical Industries Ltd., *Yodogawa-ku, Osaka 532, Japan*

Masaharu YOSHIKAWA, Central Research Laboratories, Ajinomoto Co. Inc., *Kawasaki-ku, Kawasaki 210, Japan*

Preface

In Japan, the microbial production of L-glutamic acid on a large scale began in 1957, and the possibility of producing 5′-inosinic acid and 5′-guanylic acid by the degradation of nucleic acid using microbial enzymes was demonstrated in 1959. Subsequently, various kinds of amino acids were produced on an industrial scale by fermentation, and in 1961 nucleic acid was produced commercially from yeast cells. This was followed by the development of a fermentation method for the direct accumulation of nucleic acid, and the practical coexistence of two-stage and single-stage fermentation methods became a distinctive feature of the industry. Furthermore, many kinds of studies on the microbial production of nucleotide derivatives have been undertaken by numerous researchers.

The "Association of Amino Acid Fermentation", founded in 1959, was renamed the "Association of Amino Acid and Nucleic Acid" from 1963 in view of the new results on the fermentative production of nucleic acid-related substances. The association has been managed by a number of committee members, secretaries and members under the leadership of former presidents, Drs. Kin'ichiro Sakaguchi, Hideo Katagiri, the late Toshinobu Asai, the late Teijiro Uemura, Koichi Yamada and Kei Arima, with the financial support of patron members. The association has already published 32 issues of a biannual bulletin, "*Amino Acid and Nucleic Acid*".

"*The Microbial Production of Amino Acids*" (Kodansha; Wiley) was published in 1972 and it was then decided to produce a companion volume dealing with research on the microbial production of nucleic acid-related substances. As a result, "*Microbial Production of Nucleic Acid-related Substances*" has now been prepared, with contributions by many leading researchers in Japan. In addition to the microbial production of nucleic acid-related substances, this book also covers the shift from two-stage fermentation to single-stage fermentation, the utilization of biochemical mutants, the release of metabolic control, increase in the permeability of the cell membrane, the linking of microbial and chemical processes, and the technological aspects of fermentation, all of which are important facets of modern applied microbiology. We believe that researchers engaged not only in the production of nucleic acid-related substances but also in the development of new fermentation processes will benefit from this book.

We are sincerely grateful to the editorial staff of Kodansha and to Mr. W.R.S. Steele for their invaluable assistance in the preparation of the English manuscripts which comprise this book.

Koichi OGATA Shukuo KINOSHITA
Toshinao TSUNODA Kô AIDA

March 1976

Contents

PART I: FUNDAMENTAL PRINCIPLES IN THE PRODUCTION OF NUCLEIC ACID-RELATED SUBSTANCES

PART II: PRODUCTION AND HYDROLYSIS OF NUCLEIC ACIDS

PART III: PRODUCTION OF NUCLEIC ACID-RELATED SUBSTANCES BY FERMENTATION

PART IV: PRODUCTION OF NUCLEIC ACID-RELATED SUBSTANCES FROM ADDED PRECURSORS

PART V: UTILIZATION OF NUCLEIC ACID-RELATED SUBSTANCES

List of Abbreviations

The following common abbreviations are used in this book without further explanation:

AMP, CMP, GMP, IMP, UMP, XMP—5′-monophosphates
(3′-monophosphates are prefixed "3′-")
ADP, ATP—adenosine 5′-di- and triphosphates
AICAR—5-amino-4-imidazole carboxamide riboside
AICARP—5-amino-4-imidazole carboxamide ribotide
CDP, CTP—cytidine 5′-di- and triphosphates
CoA—coenzyme A
cAMP—adenosine 3′,5′-monophosphate
DNA—deoxyribonucleic acid
DNase—deoxyribonuclease
FAD—flavin adenine dinucleotide
FMN—flavin mononucleotide
GDP, GTP—guanosine 5′-di- and triphosphates
mRNA—messenger RNA
NAD—nicotinamide adenine dinucleotide
NADP—nicotinamide adenine dinucleotide phosphate
Pi, PPi—inorganic phosphate and pyrophosphate
PRPP—phosphoribosyl pyrophosphate
RNA—ribonucleic acid
RNase—ribonuclease
rRNA—ribosomal RNA
SDS—sodium dodecyl sulfate
tRNA—transfer RNA
UDP, UTP—uridine 5′-di- and triphosphates

History and Prospects*

Investigations on the industrial production of nucleic acid-related substances, especially IMP and GMP, and on the formation of nucleoside derivatives by microorganisms have been done mostly in Japan. This work represents one of the most valuable achievements in the field of applied microbiology in Japan, together with the work on amino acid fermentation. However, the history of this research and industrial production is quite recent: in a period of less than 20 yr, a large number of reports have been published.

This introduction will offer a brief review of the background, progress and present state of the production and investigation of nucleic acid-related substances, with a discussion of future prospects. To avoid overlap with each chapter, references are omitted here, except for reviews. The citation of names is also limited to the principal researchers.

The Background

"5′-Nucleotide fermentation", was defined by Sakaguchi[1] to mean the microbial production of nucleic acid-related substances, centering on 5′-nucleotides. Since then, the development of the fermentation industry in Japan has been rapid as a result of the following advances in various fields.

* Koichi OGATA, Faculty of Agriculture, Kyoto University.

The development of industrial microbiology: Traditional fermentation industries in Japan have produced *sake* (rice wine), *miso* (soybean paste), soy sauce, etc. Extensive studies on microorganisms such as molds, yeasts and lactic acid bacteria which were used in these fermentations, were performed and continued by many researchers since the late 19th century. After World War II, the industrial production of antibiotics, yeasts, steroids, enzymes and amino acids has increased remarkably. These industries provided a basis for the development of 5′-nucleotide production in the following respects: screening of microorganisms, mutational improvement of productivity and large-scale culture in sealed tanks in antibiotics production; continuous culture in yeast production; the specific transformation of substrates in steroid production; the control of enzyme production; and research on the permeability of the cell membrane, the introduction of biochemical mutants for industrial use, fermentation conversion and the release of feedback inhibition or repression in amino acid production.
Progress in the biochemistry of nucleic acids and related substances: Since nucleic acid was first isolated from pus and named nuclein by Miescher in 1869, RNA and DNA have occupied a central position in biochemistry, especially since World War II. More recent work has dealt with the biosynthetic and biodegradative pathways of RNA, DNA and their component nucleotides, the mechanisms of metabolic control and assay methods. These reports are collected in "*Literatures on Nucleic Acid Fermentation*" published by the Laboratory of Applied Microbiology, Kyoto University,[2] and by Nakayama.[3–5]

Production of Nucleic Acid-related Substances

Several flavor-enhancing materials such as *konbu* (kelp-like seaweed), *katsuobushi* (dried bonito), *niboshi* (dried small sardine) and *shiitake* (a mushroom) have been used in Japan since ancient times. The taste of *konbu* was identified as being due to L-glutamate in 1908 by Ikeda, and subsequently glutamate was produced by the acid hydrolysis of plant protein. The fermentation method for the production of L-glutamate from carbohydrates using microorganisms was established in 1957, and since then various kinds of amino acid have been produced by fermentation. In 1913, Kodama found that the flavor of *katsuobushi* was due to the histidine salt of IMP, though the presence of the different isomers was not known at that time. Later, Kuninaka showed that the 5′-nucleotides, IMP, GMP and XMP, have flavor action, but 2′- and 3′-IMP do not.

A phosphodiesterase which releases AMP from RNA and an AMP deaminase which deaminates AMP to IMP are contained in snake venom and cow intestinal mucosa, and in animal muscle, respectively. However, little consideration was given to the use of such enzymes for the industrial production of nucleotides. A small amount of IMP, which originated from ATP in *niboshi*, the waste from fish canning or squid muscle, was produced by the extraction method in 1956–60.

In 1959, Kuninaka, Sakaguchi *et al.* demonstrated for the first time that an enzyme of *Penicillium citrinum* can decompose RNA to 5′-nucleotides. Kuninaka *et al.* also showed that GMP was a flavor component, in addition to IMP. Subsequently, Omura, Ogata *et al.* reported that a number of molds, actinomycetes and bacteria represented by *Streptomyces aureus* produced the same type of enzyme. On the other hand, the production of yeast cells had already been established using sulfite pulp waste as a raw material by Miwa *et al.* The culture of yeast for RNA production was then established and the industrial production of IMP and GMP began by degradation from yeast RNA using the enzymes produced by the two strains mentioned above. The degradation method is a two-stage fermentation. Single-stage fermentation to produce nucleotides by *de novo* synthesis became the next goal of researchers.

First, the screening of microorganisms to produce bases, purine nucleosides or purine nucleotides was carried out by the groups of Okabayashi, Arima, Furuya and Nakayama. However, no strain able to accumulate a large amount of nucleic acid-related substances extracellularly could be obtained. As for nucleotide excretion by cells as a result of autolysis, Higuchi and Uemura, and Nakao, Ogata *et al.* investigated *Saccharomyces cerevisiae* and a variety of yeasts, respectively. Ogata, Imada *et al.* detected the excretion of 5′-nucleotides by *Bacillus megaterium*.

After this transition period, the use of biochemical mutants was introduced for the production of nucleic acid-related substances. An adenine requiring strain of *B. subtilis* was found to accumulate IMP, inosine and hypoxanthine by Uchida, Kuninaka, Yoshino *et al.* in 1961. In 1962, the accumulation of a large amount of AICAR was achieved with a purine requiring strain of *B. subtilis* by Shiro, Takahashi *et al.* and it was used as a starting material for GMP synthesis by Yamazaki *et al.* This method is employed for the industrial production of GMP and is a typical example of the use of a combination of fermentation and chemical synthesis to avoid metabolic control by the end product in the fermentation process. In the same year, Hara, Kojima, Koaze *et al.* developed a process for the production of IMP from adenine, though it is not used practically. The chemically synthesized adenine was converted to adenosine by salvage synthesis, using a purine requiring strain of *B. subtilis*, phosphorylated to AMP by adenosine kinase of yeast, and then deaminated to IMP.

In 1963, Arima, Tamura *et al.* reported the accumulation of purine compounds by purine base requiring strains of several bacteria. In the same year, Nara, Kinoshita *et al.* and Aoki, Momose *et al.* established inosine fermentation with adenine requiring strains of *B. subtilis*, in which the amount of inosine reached 7 g/1. Several methods for the phosphorylation of inosine were attempted. These included selective phosphorylation using phosphotransferase by Katagiri, Yamada, Mitsugi *et al.*

The accumulation of purine compounds by various auxotrophs has been reported since 1964. The microorganisms used include *Brevibacterium ammoniagenes*, *Bacillus subtilis*, *B. megaterium*, *B. pumilus*, *Bacillus* sp., *Corynebacterium glutamicum*, *C. petrophilum*, *Arthrobacter paraffineus*, *A. aerogenes*, *Streptomyces showdoensis* and *Candida tropicalis*.[8] Details are given in the relevant chapters.

In 1968, Furuya, Kinoshita *et al.* found that a large amount of IMP was directly accumulated by an adenine requiring strain of *B. ammoniagenes* in the presence of a suitable Mn^{2+} concentration in the culture medium.

Although the accumulation of guanine compounds had been considered to be difficult, Demain *et al.* demonstrated the accumulation of GMP (1 g/l) by an adenine requiring strain of *C. glutamicum* in 1968. There are now several reports on the accumulation of guanine compounds.

The industrial production of IMP and GMP is at present carried out by both yeast RNA degradation and direct fermentation. Production in 1968–74 is shown in Table 1. About 3000 tons of IMP and GMP per annum is currently being produced. One reason for the continued development of the industry is the synergy of flavor between L-glutamate and IMP, found by Kuninaka *et al.* IMP and GMP are usually added to complex seasonings at levels of 8–12% and 1.5–2.0%, respectively, with respect to sodium glutamate.

TABLE 1

Monthly Output of IMP plus GMP (Estimated)

	No. of producers	Total monthly output (tons)
1968	5	210
1969	5	225
1970	5	235
1971	5	250
1972	5	258
1973	5	250
1974 (estimated)	5	250

(Nikkan Keizai News Agency)

In the course of establishment of the IMP and GMP industry, methods of preparing 3′-nucleotides by the degradation of RNA using an enzyme of *Rhodotorula glutinis*, 2′-nucleotides by the treatment of alkali-digested RNA with 3′-nucleotidase, and 5′-deoxynucleotides by the degradation of DNA with an enzyme of *Aspergillus querinus* were developed by Nakao, Ogata *et al.*, Igarashi and Kakinuma, and Nakao, Ogata *et al.*, respectively.

The practical production of orotic acid was achieved with a uracil requiring strain of *C. glutamicum* by Tanaka, Nakayama *et al.* D-Ribose production was achieved with an adenine, xanthine and shikimic acid requiring, and gluconate and ribose non-utilizing strain of *Bacillus* sp. by Sasajima, Yoneda *et al.*

Derivatives of Nucleic Acid-related Substances

The production of nucleoside and nucleotide derivatives was studied next. Methods for the production of ATP from adenosine with dried cells of bakers' yeast were developed by Tochikura, Ogata *et al.* and with cultures of *B. ammoniagenes* by Tanaka, Kinoshita *et al.* Murao *et al.* found a new enzyme, ATP pyrophosphotransferase, in *Streptomyces adephospholyticus* which could produce guanosine tetraphosphate from GDP. Okabayashi, Ide *et al.* and Ishiyama, Yokotsuka *et al.* observed the accumulation of cyclic AMP on addition of a purine base to cultures of *Brevibacterium liquefaciens*, and *Corynebacterium murisepticum* and *Microbacterium* sp., respectively.

Biosynthetic intermediates of polysaccharides and sugar nucleotides were prepared by Tochikura, Kawai, Kawaguchi *et al.*, and CDP-choline was prepared from CTP or CMP by Takeda *et al.*, Miyauchi, Yoshino *et al.*, and Tochikura, Kimura *et al.* with yeast.

Moreover, the production of FAD from AMP and FMN by *Sarcina lutea* was achieved by Sakai, Chibata *et al.*, and that of NAD with yeasts and *B. ammoniagenes* was investigated by Hayano, Kitahara *et al.* and Nakayama, Tanaka, Kinoshita *et al.* The practical production of CoA was achieved by a biosynthetic process from pantothenate, cysteine and AMP using *B. ammoniagenes* by Ogata, Tani, Shimizu *et al.* In addition, Kuninaka, Yoshino *et al.* reported the formation of *S*-adenosylmethionine by *Aspergillus tamarii.*

Various studies on nucleoside derivatives are still in progress.

Prospects

The nucleotide industry, which began with the production of the flavor components, IMP and GMP, is still based on the RNA degradation and direct fermentation methods. The RNA degradation method produces unutilized by-products, UMP and CMP. However, it may be possible to utilize fragments of these nucleotides (the pyrimidine ring or ribose moiety). The yield and cost of RNA might be restrictive when yeast cells are obtained by culture with pulp waste or molasses, but when the prospects for single-cell protein production from *n*-paraffin, methanol, ethanol, acetate, etc. are considered, it seems likely that good supplies of RNA as a by-product will be available.

In the direct fermentation method, which uses auxotrophs developed by many investigators, the main aim will be to increase the yield.

Research on nucleic acid-related subtances, which were originally of interest as food additives, is now focusing on medical applications of such substances as ATP, NAD, FAD, CoA, CDP-choline and orotic acid. Further, various kinds of physiologically active substances may become available using microbiological and chemical synthetic techniques; for instance, it is known that a number of antibiotics contain nucleoside moieties.

Considering that 10^4–10^7 microorganisms are present per gram of soil, there seems to be considerable scope for future research. For example, the physiological role of DNA formed extracellularly is obscure, and knowledge of the permeability of cell membranes and of metabolic pathways is still incomplete. RNA and DNA are essential for protein synthesis and the transfer of genetic information, so investigations on the production of these substances and their component nucleotides might be expected to produce interesting new data in this fundamental field.

References

1. K. Sakaguchi, *Shokuhin Kogyo, Rinji Zokan* (Japanese), p. 13, Korin Shoin, 1960.
2. Laboratory of Applied Microbiology, Kyoto University, "Literatures on Nucleic Acid Fermentation", *Amino Acid and Nucleic Acid*, no. 9, 114 (1964).
3. K. Nakayama, "Literatures on Nucleic Acid Fermentation (continued)", *ibid.*, no. 16, 155 (1967).
4. K. Nakayama, "Literatures on Nucleic Acid Fermentation", *ibid.*, no. 20, 163 (1969).
5. K. Nakayama, "Literatures on Nucleic Acid Fermentation (supplement)", *ibid.*, no. 26, 113 (1972); no. 32, 99 (1975).
6. S. Kodama, *Tokyo Kagaku Kaishi* (Japanese), **34**, 751 (1913).
7. A. Kuninaka, "Flavor Action and Production Process of 5'-Nucleotides", *Protein, Nucleic Acid and Enzyme*, **6**, 403 (1961).
8. K. Ogata, "The Microbial Production of Nucleic Acid-related Compounds", *Advan. Appl. Microbiol.*, **19**, 209 (1975).

PART I

FUNDAMENTAL PRINCIPLES IN THE PRODUCTION OF NUCLEIC ACID-RELATED SUBSTANCES

CHAPTER 1

Microbial Nucleolytic Enzymes
—Variety and Classification*1

In the standard systematic classification of enzymes (1972),[1] all well-known nucleolytic enzymes are classified in EC 3.1.4, phosphoric diester hydrolases (Table 1.1).*2 These nucleolytic enzymes can be divided into two classes based on the site of cleavage (Fig. 1.1): 5′-phosphomonoester-forming enzymes (enzymes producing mono- and/or oligonucleotides terminated by 5′-phosphate, hereafter referred to as 5′-P-producers), and 3′-phosphomonoester-forming enzymes (enzymes producing mono- and/or oligonucleotides terminated by 3′-phosphate or 2′:3′-cyclic phosphate, hereafter referred to as 3′-P-producers). This is still an absolute criterion:

*1 Akira KUNINAKA, Yamasa Shoyu Co. Ltd.

*2 RNases forming 3′-phospho-mono- and oligonucleotides via 2′:3′-cyclic phosphates were formerly classified in EC 2.7.7, nucleotidyltransferases.[2] They are now reclassified as follows: pancreatic RNase (EC 2.7.7.16 ⟶ 3.1.4.22), RNase T_2 (EC 2.7.7.17 ⟶ 3.1.4.23) and RNase T_1 (EC 2.7.7.26 ⟶ 3.1.4.8).

TABLE 1.1

List of Nucleolytic Enzymes Selected by the Commission on Biochemical Nomenclature[1)]

EC number	Name	Substrate	Product (terminal P)
3.1.4. 1	Phosphodiesterase I (5′-Exonuclease)	oligo[†1]	5′-P (mono)
3.1.4. 5	DNase I (Pancreatic DNase)	DNA	5′-P (oligo)
3.1.4. 6	DNase II	DNA	3′-P (oligo)
3.1.4. 7	Spleen endonuclease (Micrococcal nuclease)	RNA, DNA	3′-P (oligo)
3.1.4. 8	Guanyloribonuclease (RNase T_1, RNase N_1)	RNA	3′-P[†2] (Gp, oligo)
3.1.4. 9	Nucleate endonuclease (*Azotobacter* nuclease, Mung bean nuclease)	RNA, DNA	5′-P (oligo)
3.1.4.18	Phosphodiesterase II (3′-Exonuclease, Spleen phosphodiesterase)	oligo[†1]	3′-P (mono)
3.1.4.19	Oligonucleotidase	oligo[†1]	5′-P (mono)
3.1.4.20	Exoribonuclease (*E. coli* exo-RNase II)	RNA	5′-P (mono)
3.1.4.21	Single-stranded-nucleate endonuclease (*Neurospora crassa* endonuclease)	RNA, DNA (single-stranded)	5′-P (oligo)
3.1.4.22	RNase I (Pancreatic RNase)	RNA	3′-P[†2] (Up, Cp, oligo)
3.1.4.23	RNase II (Plant RNase, *E. coli* RNase I)	RNA	3′-P[†2] (oligo)
3.1.4.24	Endoribonuclease III	RNA (double-stranded)	5′-P[†3] (oligo)
3.1.4.25	Exodeoxyribonuclease I	DNA (single-stranded)	5′-P (mono, di)
3.1.4.26	Exodeoxyribonuclease II (Associated with *E. coli* DNA polymerase)	DNA	5′-P (mono)
3.1.4.27	Exodeoxyribonuclease III (Exonuclease III, DNA phosphatase-exonuclease)	DNA (double-stranded)	5′-P (mono)
3.1.4.28	Exodeoxyribonuclease IV (λ-Exonuclease)	DNA (double-stranded)	5′-P (mono)
3.1.4.29	Oligodeoxyribonucleate exonuclease (T_2 exonuclease)	Oligo-deoxyribo-nucleate	5′-P (mono)

TABLE 1.1—Continued

EC number	Name	Substrate	Product (terminal p)
3. 1. 4. 30	Endodeoxyribonuclease	DNA, alkylated DNA	5′-P (oligo)
3. 1. 4. 31	DNA 5′-dinucleotidohydrolase (Phage SP3 DNase)	DNA (single-stranded)	5′-P (di, tri)
3. 1. 4. 32	Endodeoxyribonuclease (ATP-and *S*-adenosyl-methionine-dependent, DNA restriction enzyme)	DNA (double-stranded)	5′-P (poly)
3. 1. 4. 33	Endodeoxyribonuclease (ATP-hydrolyzing)	DNA	5′-P (oligo)
3. 1. 4. 34	Hybrid nuclease (RNase H)	RNA–DNA hybrids	5′-P (oligo, ribo)

†1 Oligoribo- and oligodeoxyribonucleotides.
†2 2′:3′-cyclic phosphates are formed as intermediates.
†3 The product had been reported to be 3′-P,[1,5] but recently it was confirmed to be 5′-P.[61]

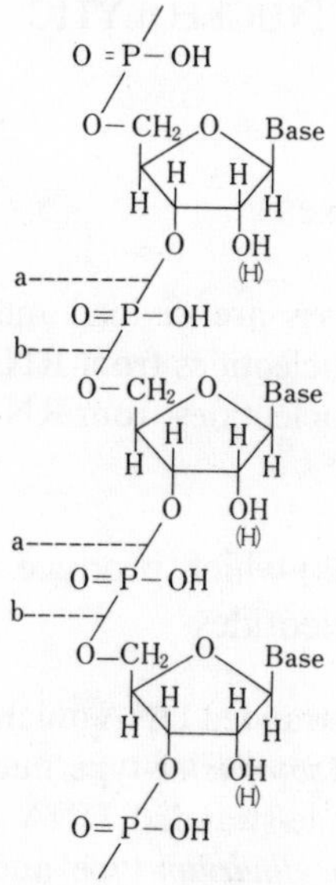

Fig. 1.1. Phosphodiester linkages in nucleic acids. 5′-P-Forming nucleases, RNases and DNases split linkage a, while 3′-P-forming nucleases, RNases and DNases split linkage b. Most 3′-P-forming RNases form 2′:3′-cyclic phosphates as intermediates.

no enzyme has been found that can split the internucleotide bond on both sides. 5′-P- and 3′-P-producers are further subdivided into three classes, non pentose-specific nucleases, RNases, and DNases.

Among nucleic acid-related compounds, the 5′-monoribonucleotides,[3] which cannot be produced directly by chemical decomposition of nucleic acids, are in the greatest demand for use as flavoring agents. In this chapter, microbial nucleolytic enzymes will be classified from the above viewpoint and reviewed, with particular emphasis on 5′-P-producers.

RNA nucleotidyltransferases (EC 2.7.7.6), DNA nucleotidyltransferases (EC 2.7.7.7), polynucleotide phosphorylases ((EC 2.7.7.8), see Chapter 19), autolysis-related enzymes (see Chapter 8), nonspecific phosphatases (EC 3.1.3.1 and EC 3.1.3.2), 5′-nucleotidases (EC 3.1.3.5), 3′-nucleotidases (EC 3.1.3.6), nucleosidases (EC 3.2.2) and nucleodeaminases (EC 3.5.4) will not be discussed in this chapter.

1.1 Microbial 5′-P-Forming Nucleolytic Enzymes

1.1.1 Microbial 5′-P-Forming Nucleases

Non pentose-specific, 5′-P-forming nucleases are divided into two classes: nucleases which produce mainly oligonucleotides from RNA and DNA (A) and nucleases which produce mononucleotides from RNA and DNA (B).

A. Microbial nucleases (endonucleases) which produce mainly 5′-phospho-oligonucleotides

Neurospora-type nucleases hydrolyze single-stranded DNA much faster than double-stranded DNA (see a, b and c). *Azotobacter*-type nucleases hydrolyze both single-stranded DNA and double-stranded DNA at approximately the same rate (see f and g). *Acrocylindrium*-type nucleases hydrolyze double-stranded DNA more rapidly than single-stranded DNA (see e).

a. *Aspergillus oryzae* (crystalline nuclease O) Uozumi *et al.*[49] crystallized nuclease O, which is activated and liberated from *A. oryzae* mycelia in the course of autolysis. The enzyme is specifically inhibited by an inhibitor (protein) found in the fresh mycelia. The mononucleotide contents in the degradation products of native DNA, heat-denatured DNA and RNA are 1.9%, 3.4%, and 13.6%, respectively. Nuclease O hydrolyzes heat-denatured DNA 19 times faster than native DNA.

b. *Neurospora crassa*[11,28) An endonuclease purified from conidia of *N. crassa* (EC 3.1.4.21) hydrolyzes heat-denatured DNA and rRNA at approximately the same rate. Native DNA is attacked at less than 0.1% of the rate found with denatured DNA. The enzyme is specific for polynucleotides lacking an ordered structure. The enzyme has a distinct preference, but not absolute specificity, for guanosine or deoxyguanosine residues within a polynucleotide. It is inhibited by EDTA, but this inhibition can be overcome by stoichiometric amounts of Co^{2+}. Another 5′-P-forming endonuclease was also purified from mitochondria of *N. crassa*.

c. Yeast Nakao *et al.*[52)] purified a nuclease from the supernatant fraction of a hybrid yeast (*Saccharomyces fragilis* × *S. dobzhanskii*). The enzyme hydrolyzes denatured DNA at approximately the same rate as yeast rRNA. Highly polymerized native *E. coli* DNA is almost inactive as a substrate.

d. *Streptomyces aureus*[54)] Culture filtrate of *S. aureus*, one of the microorganisms selected by Ogata *et al.*[36)] for the production of 5′-nucleotides (see 1.1.2.g), degrades RNA and DNA into 5′-mononucleotides.[25,39)] From the culture filtrate, Yoneda[54)] purified a nuclease forming 5′-phospho-oligonucleotides with chain lengths of more than three from RNA. This enzyme was first designated as RNase. Later, however, the enzyme was found to hydrolyze DNA more rapidly than RNA.[53)] The enzyme will be discussed in detail in Chapter 7.

e. *Acrocylindrium* sp. (*Fungi imperfecti*) Suhara *et al.*[53)] crystallized an endonuclease from culture filtrates of *Acrocylindrium* sp. The relative rates of hydrolysis of substrates were in the order, native DNA > heat-denatured DNA > RNA > poly(C) > poly(U) > poly(A) > poly(I). The enzyme hydrolyzes double-stranded polyribonucleotides more rapidly than single-stranded polyribonucleotides.

f. *Azotobacter agilis*[2,11,27)] Nucleate endonuclease purified from extracts of *A. agilis* (EC 3.1.4.9) hydrolyzes RNA, DNA (both native and denatured) and poly(A) to small 5′-phospho-oligonucleotides. Hydrolysis of shorter chains is much slower than that of longer chains. The enzyme does not act on poly(G) or poly(C).

g. *Serratia marcescens*[51)] An extracellular nuclease purified from *S. marcescens* culture fluid has the unusual property of hydrolyzing both single-stranded and double-stranded DNA and RNA at similar rates.

h. Main sources other than microorganisms 5′-P-Forming endonucleases have also been found in silkworm,[55)] chick pancreas,[56)] and rat liver.[57)]

Generally, 5′-P-forming endonucleases are activated by Mg^{2+} and Mn^{2+}. Suhara has summarized their properties.[53)]

B. Microbial nucleases which produce mainly 5′-mononucleotides

B-1. Nucleases with associated 3′-nucleotidase activity

Lately nucleases with associated 3′-nucleotidase activity have been reported in several molds and plants. These enzymes hydrolyze not only phosphodiester linkages in RNA and DNA but also phosphomonoester linkages in mono- and oligonucleotides terminated by 3′-phosphate*.

They are generally inactivated or inhibited by EDTA, and hydrolyze heat-denatured DNA faster than native DNA.

a. ***Penicillium citrinum* (nuclease P_1)** Kuninaka *et al.*[10,45] carried out screening for microorganisms degrading RNA into 5′-nucleotides for the first time, and selected *P. citrinum*. Nuclease P_1 purified from the culture of a pigmentless mutant of *P. citrinum* degrades RNA and DNA into 5′-mononucleotides, and liberates inorganic phosphate from ribo- and deoxyribonucleoside 3′-monophosphates and oligoribo- and oligodeoxyribonucleotides terminated by 3′-phosphate.[46–48] The optimum temperature is 70°C, and the optimum pH is usually 5 to 6. The enzyme is inactivated by EDTA and reactivated by Zn^{2+}. Nuclease P_1 will be discussed in detail in Chapter 6.

b. ***Monascus purpureus*** According to Saruno *et al.*,[40] a nuclease purified from culture filtrates of *M. purpureus* hydrolyzes RNA and heat-denatured DNA exonucleolytically into 5′-mononucleotides and also hydrolyzes 3′-nucleotides into nucleosides and inorganic phosphate.

c. ***Phoma cucurbitacearum*** Tone *et al.*[30] selected 48 strains capable of degrading RNA into 5′-nucleotides from about 960 strains of *Fungi imperfecti* isolated from plants. The selected strains belong to *Phyllosticta*, *Selenophoma*, *Phoma*, *Macrophomina*, *Phomopsis*, *Macrophoma*, *Ascochyta*, *Diplodia*, *Stagonospora*, *Septoria*, *Sphaceloma*, *Colletotrichum*, *Gloeosprium*, *Septogloeum*, *Pestalotia*, *Monilia*, *Cephalosporium*, *Acrocylindrium*, *Botrytis*, *Ovularia*, *Cylindrocladium*, *Arthrobotrys*, *Cephalothecium*, *Rhynchosporium*, *Fusoma*, *Piricularia*, *Nigrospora*, *Cladosporium*, *Fusicladium*, *Scolecotrichum*, *Brachysporium*, *Helminthosporium*, *Alternaria*, *Curvularia*, *Stemphylium*, *Macrosporium*, *Thyrospora*, *Cercospora*, *Corynespora*, *Cercosporina*, *Fusarium*, *Papularia*, *Epicoccum*, *Clathrococum*, *Isariopsis*, *Rhizoctonia* and *Sclerotium*. One of them, *P. cucurbitacearum* was grown on wheat bran, and a nuclease was extracted from the culture and purified. The enzyme preparation hydrolyzes RNA, DNA and 3′-AMP with an

---

* *E. coli* exonuclease III (EC 3.1.4.27, exodeoxyribonuclease III) specifically releases a terminal 3′-phosphate group from a DNA chain prior to exonucleolytic hydrolysis (see section 1.1.3.d.ii).

optimum temperature and pH of 70°C and 5.5, respectively. The enzyme is inactivated by EDTA. The rate of hydrolysis of DNA is about 50% that of RNA. *Phoma* nuclease, as well as the *Monascus* enzyme, seems to be quite similar to *Penicillium* nuclease P_1.

d. Main sources other than microorganisms 5′-P-Forming nucleases with associated 3′-nucleotidase activity have been found in mung bean,*[11,19,113,135] wheat seedling,[136] potato tubers,[114] ginkgo nuts,[115] tobacco,[116] and corn.[143] Unlike the microbial enzymes, however, these plant enzymes mainly produce 5′-phospho-oligonucleotides rather than 5′-mononucleotides.

B-2. Nucleases with associated 5′-nucleotidase activity

a. *Micrococcus sodonensis* An extracellular nuclease purified by Berry[58] from *M. sodonensis* hydrolyzes native and denatured DNA, RNA, poly (A), and adenylyl-3′,5′-adenosine. It is exonucleolytic in its action, requiring a free 3′-hydroxyl group. As the nuclease is associated with 5′-nucleotidase activity, the reaction products of the nuclease activity toward DNA and RNA are nucleosides. Mn^{2+} alone is required for monoesterase activity, while the diesterase activity requires Mg^{2+} in addition to Mn^{2+}.

B-3. Nucleases without phosphomonoesterase activity

a. *Aspergillus oryzae* (nuclease S_1)[2,31,49,137,144] Ando[31] isolated nuclease S_1 from "Takadiastase" (a product of *A. oryzae*). The enzyme specifically splits phosphodiester bonds of RNA and heat-denatured DNA. The digest of denatured DNA consists mainly of 5′-mononucleotides (93%). Nuclease S_1 is activated by Zn^{2+} (10^{-4}M) and inhibited by EDTA (2×10^{-2} M).

b. *Aspergillus quercinus* Ogata *et al.*[36] and Nakao *et al.*[20,25,26] reported that *A. quercinus* could degrade both RNA and DNA into 5′-nucleotides. Subsequently, Ohta *et al.*[59] purified an exonuclease degrading both RNA and DNA from the culture filtrate.

c. Black koji-molds According to Fujimoto *et al.*,[110] a phosphodiesterase isolated from cultures of black koji-molds, such as *Aspergillus niger*, hydrolyzes 3′-hydroxy-terminated dinucleotides or dinucleoside monophosphates as follows:

$$\text{pX-Y} \longrightarrow \text{pX} + \text{pY}$$
$$(2'\text{–}5')\text{pX-Y} \longrightarrow \text{pX} + \text{pY}$$

* According to the standard enzyme nomenclature,[1] mung bean nuclease, as well as *Azotobacter* nuclease, is classified in EC 3.1.4.9, nucleate endonuclease, although the mung bean enzyme, unlike the *Azotobacter* enzyme, is associated with 3′-nucleotidase activity and is not active against native DNA.

$$\text{X-Y} \longrightarrow \text{X} + \text{pY}$$
$$(2'\text{-}5')\text{X-Y} \longrightarrow \text{X} + \text{pY}$$

The enzyme splits 2′–5′ phosphodiester linkages and 3′–5′ phosphodiester linkages without specificity toward bases or sugars. In these respects, the enzyme is similar to snake venom phosphodiesterase (EC 3.1.4.1, phosphodiesterase I). However, the former seems to be much less active toward RNA and DNA than the latter.

d. *Streptomyces* sp. No. 41[2,29] An exonuclease isolated by Sugimoto *et al.*[29] from *Streptomyces* sp. No. 41 degrades RNA, RNase I core and DNA into 5′-nucleotides. The strain also produces a phosphodiesterase hydrolyzing RNA, DNA and bPNPP. This phosphodiesterase is suggested to be associated with phosphomonoesterase activity.

e. *Physarum polycephalum* (slime mold, nuclease P_{p-2})[16,41] Hiramaru *et al.*[41] isolated four RNases and two nucleases from cell extracts of *P. poly cephalum*. One of them, nuclease P_{p-2}, is activated by Zn^{2+}, and degrades RNA and denatured DNA. The main products from RNA are 5′-mononucleotides.

f. *Acrocylindrium* sp. Suhara[60] purified an exonuclease from *Acrocylindrium* sp. The enzyme is activated by Mg^{2+}, and has an optimum pH of 10 to 12. The yields of mononucleotides from yeast RNA, heat-denatured DNA and native DNA were 98%, 46% and 36%, respectively.

g. Main sources other than microorganisms Snake venom phosphodiesterase (EC 3.1.4.1, phosphodiesterase I)[6,11,13,19] and rat liver oligonucleotidase (EC 3.1.4.19)[141] are regarded as typical 5′-P-forming exonucleases.

1.1.2 Microbial 5′-P-Forming RNases[69]

a. *Pellicularia* Hasegawa *et al.*[35] investigated the RNA-degrading activity of about 300 strains of molds, and found strains belonging to *Pellicularia, Rhizoctonia,* and *Phaeoisariopsis* that could degrade RNA into 5′-nucleotides. They reported that a crude enzyme preparation obtained from culture filtrates of *Pellicularia* sp. H-11 formed 5′-mononucleotides from RNA in 74% yield. The enzyme, like nuclease P_1, is completely inactivated by dialysis against EDTA, and the activity is restored specifically by Zn^{2+}. If the enzyme is confirmed to act on DNA and 3′-nucleotides in the future, it should be regarded as a nuclease with associated 3′-nucleotidase activity (see section 1.1.1.B-1).

b. *Physarum polycephalum* (RNase P_{p-4})[16,41] RNase P_{p-4} isolated from *P. polycephalum* catalyzes the endonucleolytic hydrolysis of RNA, forming nucleotides bearing 5′-phosphate with a preference for purine as the 3′-terminal nucleotide. It is activated by Zn^{2+}.

c. *E. coli* Various types of nucleolytic enzymes have been described in *E. coli*. Among them, *E. coli* RNase II (EC 3.1.4.20, exoribonuclease),[11–13,16,33,43] RNase III (EC 3.1.4.24, endoribonuclease III),[61] RNase V,[9,16] and RNase H (EC 3.1.4.34, hybrid nuclease)[62] are regarded as 5′-P-forming RNases.

RNase II requires the presence of both a monovalent cation (K^+ or NH_4^+) and a divalent cation (Mg^{2+} or Mn^{2+}) for its activity. The enzyme catalyzes the progressive hydrolysis of a polyribonucleotide chain, beginning at the 3′-hydroxyl end and sequentially liberating 5′-mononucleotides from the chain until a small oligonucleotide ($n = 2$ to 4), which is itself comparatively resistant to hydrolysis, is formed. RNase III is highly specific for RNA:RNA duplex, and catalyzes endonucleolytic cleavage to yield 5′-phosphate and 3′-hydroxyl termini. RNase V is an exoribonuclease specifically attacking mRNA from the 5′- to the 3′-terminal, producing 5′-mononucleotides. RNase H specifically hydrolyzes the RNA strand of RNA-DNA hybrid. The degradation products were identified as oligonucleotides with 3′-hydroxyl and 5′-phosphate termini.

In addition, RNase IV[34] specifically cleaves R17 RNA at a site about one-third of the way from the 5′-end of the molecule, creating 15S and 21S fragments. The nature of the bonds cleaved by the enzyme is not known.

d. *Lactobacillus casei*[11,32] The enzyme extracted from the ribosomal fraction of *L. casei* cells degrades RNA to 5′-mononucleotides. The enzyme exhibits endonuclease and exonuclease activities, and is free from DNase and 3′-nucleotidase activities.

e. *Lactobacillus plantarum*[16,42] *L. plantarum* RNase II, like *E. coli* RNase II, catalyzes the progressive hydrolysis of polyribonucleotides, forming 5′-mononucleotides.

f. *Bacillus* AU2 Jacobsen *et al.*[63] purified an RNA phosphodiesterase 330-fold from cell-free culture fluid of *Bacillus* AU2 grown on RNA as the sole source of phosphorus. The enzyme is associated with 5′-nucleotidase activity. In the presence of Mg^{2+} or Mn^{2+}, this enzyme catalyzes the hydrolysis of RNA. While the initial products are 5′-mononucleotides, nucleosides are eventually formed due to the associated 5′-nucleotidase activity. Polymer cleavage appears to proceed from the 3′-hydroxyl terminus and to be exonucleolytic. The enzyme is similar to that from *Micrococcus sodonensis* (see 1.1.1.B-2.a). Phosphodiesterase activity in the former is, however, specific to RNA, while the latter is active against both RNA and DNA. In addition, ribo- and deoxyribonucleoside 5′-phosphates are good substrates for the 5′-nucleotidase activity of *Bacillus* AU2, as well as that of *M. sodonensis*.

g. Others Culture filtrates or cell extracts of many microorganisms, including a soil bacterium[7] and *Asterococcus*,[8] have been reported to degrade RNA into 5′-nucleotides, although the enzymes have not been purified. Ogata *et al.*[36] carried out systematic screening and found that culture filtrates of the following microorganisms could degrade RNA into 5′-mononucleotides:

Molds (including *Fungi imperfecti)*: *Acrocylindrium* sp.,[53,60] *Fusarium roseum*, *F. solani*, *Gliomastix convoluta*, *Helminthosporium sigmoideum* var. *irregulare*, *Verticillium niveostratosum*, *Anixiella reticulispora*, *Chaetomidium japonicum*, *Glomerella cingulata*, *Neurospora crasssa*,[28] *N. sitophila*, *Ophiobolus miyabeanus*, *Ophiostoma ulmi*, *Sordaria fimicola*, *Aspergillus elegans*, *A. fischeri*, *A. flavipes*, *A. melleus*, *A. nidulans*, *A. quercinus*,[20,25,59] *A. sulphureus*, and *A. ustus*.

Streptomyces: *Streptomyces coelicolor*, *S. albogriseolus*, *S. aureus*,[25,54] *S. gougerotii*, *S. griseoflavus*, *S. griseus*, *S. purpurascens*, *S. ruber*, and *S. viridochromogenes*.

Bacteria: *Bacillus brevis*,[22] *B. subtilis*,[22] and *Pseudomonas aeruginosa*.

Nakao *et al.*[21] found that an enzyme capable of degrading RNA into 5′-mononucleotides was produced in the cells at the early stage of the logarithmic growth phase of *Rhodotorula glutinis*. According to Sugimori,[64] dry cells of a new strain resembling *Pichia* degrade RNA into 5′- phosphomono- and oligonucleotides.

Nakao *et al.*[26] reported that most microorganisms producing ribonucleoside 5′-phosphates from RNA[36] also produced deoxyribonucleoside 5′-phosphates from DNA. Thus, there is a possibility that some of the above microorganisms contained non sugar-specific nucleases. In fact, four of these microorganisms, *Neurospora crassa*, *Streptomyces aureus*, *Acrocylindrium* sp. and *Aspergillus quercinus*, have been shown to have non sugar-specific 5′-P-forming nucleases (see 1.1.1.A.b.d.e, B-3.b.f). According to Mouri *et al*,[65] enzymes partially purified from mushrooms (*Lentinus edodes*, *Tricholoma matsutake* and *Psalliota bisporus*) also degrade RNA into four 5′-mononucleotides.

h. Main sources other than microorganisms In addition to barley rootlet (RNase I)[66] and mung bean sprouts,[67] cultured plant cells are particularly noteworthy. According to Ukita *et al.*,[68] cultured cells of *Vinca rosea* have more than 30 times as much 5′-P-forming RNase as the mother plant on a dry cell weight basis.

1.1.3 Microbial 5′-P-Forming DNases[17,78,137]

a. *Aspergillus oryzae* (DNase K_2) DNase K_2, purified by Kato *et al.*[70] from mycelia of *A. oryzae*, catalyzes the endonucleolytic hydrolysis of native and heat-denatured DNA.

b. *Rhizopus niveus* (DNase Rh) DNase Rh, purified by Kurono *et al.*[71]

from an *R. niveus* product "Gluczyme", hydrolyzes both native and heat-denatured DNA to 5′-phospho-oligonucleotides.

c. *Streptococcus* (streptodornase)[11,19] One of the most extensively examined bacterial DNases is the so-called "streptodornase" produced by streptococci. This 5′-P-forming endonuclease attacks double-stranded DNA at a greater rate than denatured DNA. Exhaustive hydrolysis of DNA leads to the formation of not more than 1.5% mononucleotides. The type of bond that is preferentially broken is pPy–pPu.

d. *E. coli*

i. *E. coli* $K_{12}\lambda$ DNases I through IV. Weissbach *et al.*[44] separated two endonucleases (DNases I and II) which preferentially attack native DNA and two exonucleases (DNases III and IV) which preferentially attack heat-denatured DNA from *E. coli* $K_{12}\lambda$.

ii. Exonucleases (EC 3.1.4.25–27, exodeoxyribonucleases I through III).[11,12,19] Exonuclease I (EC 3.1.4.25, exodeoxyribonuclease I)[72] hydrolyzes denatured single-stranded DNA to 5′-mononucleotides and terminal dinucleotides. Exonuclease II (EC 3.1.4.26, exodeoxyribonuclease II) continues its attack down to the mononucleotide stage. Unlike exonuclease I, the attack on native DNA is faster than on denatured DNA. The enzyme activity is associated with DNA polymerase. Exonucleases I and II both effect a purely exonucleolytic hydrolysis starting from the end carrying the 3′-terminal hydroxyl group, and successively liberating 5′-mononucleotides. Exonuclease III (DNA phosphatase-exonuclease, EC 3.1.4.27, exodeoxyribonuclease III)[73] carries out two types of reactions: the enzyme liberates the terminal 3′-phosphate group, if it is present, subsequently proceeds to exonucleolytic hydrolysis with the liberation of 5′-mononucleotides. The enzyme is specific for double-stranded DNA, and degradation stops when about 40–50% of the substrate has been hydrolyzed. 3′-Phosphomonoester linkages in oligodeoxyribonucleotides, mononucleotides and RNA and 5′-phosphomonoester linkage in DNA are not attacked by the enzyme (*cf.* 1.1.1B-1).

In addition, λ-exonuclease (EC 3.1.4.28, exodeoxyribonuclease IV) forms 5′-mononucleotides from the 5′-terminus of double-stranded DNA, and T_2 exonuclease (EC 3.1.4.29, oligodeoxyribonucleate exonuclease) catalyzes progressive hydrolysis from the 3′-hydroxyl terminus of oligodeoxyribonucleotides with chain lengths of less than 100 to give 5′-phospho-mononucleotides.

iii. Endonucleases I, II, V, etc. *E. coli* contains several characteristic endonucleases. Endonuclease I[17,78] requires Mg^{2+} or Mn^{2+} for activity. It degrades native DNA at a rate seven times greater than denatured DNA, yielding 5′-phospho-oligonucleotides of average chain length seven. The enzyme cleaves both strands of a DNA double helix at or near the same region. It is competitively inhibited by a variety of RNA's including tRNA. Endonuclease II (DNA-nicking enzyme, nickase)[74] isolated from phage T_4-infected and non-infected *E. coli* attacks double-stranded DNA to form single-strand breaks, creating 3′-OH and 5′-P termini at the sites of nicking. DNA-nicking enzymes belong to EC 3.1.4.30, endodeoxyribonuclease. According to Yasuda *et al.*,[75] Shimizu *et al.*[142] and Minton *et al.*,[77] endonuclease V (UV-endonuclease), which is specific for UV-irradiated double-stranded DNA, is induced by phage T_4. The enzyme produces single-strand incisions in UV-irradiated DNA on the 5′-side of pyrimidine dimers, and creates 3′-OH and 5′-P termini at the sites of nicking. Another UV-endonuclease has been found in *Micrococcus luteus*.[17] This enzyme, however, creates 3′-P and 5′-OH termini at the sites of nicking.

ATP-dependent DNases (EC 3.1.4.33, endodeoxyribonuclease (ATP-hydrolyzing))[76] have been reported in *E. coli*, *Micrococcus lysodeikticus*, *Diplococcus pneumoniae*, *Bacillus laterosporus*, *Mycobacterium smegmatis*, and *Haemophilus influenzae*. The *Micrococcus* enzyme cleaves DNA to give 5′-phospho-oligonucleotides of average chain length 5.5, with simultaneous hydrolysis of ATP to ADP. Native DNA with termini is hydrolyzed faster than heat-denatured DNA. In addition, *E. coli* restriction endonucleases[17,137,139] are highly specific endonucleases destroying foreign DNA from phages or cells of different strains. The restriction endonucleases catalyze endonucleolytic cleavage of one duplex strand of DNA, followed by cleavage of a second strand to give double-stranded fragments. The enzymes require Mg^{2+} for activity. In addition to Mg^{2+}, some restriction endonucleases specifically require ATP or both ATP and *S*-adenosylmethionine (EC 3.1.4.32, endodeoxyribonuclease (ATP- and *S*-adenosylmethionine-dependent)). The relationship of DNases to DNA synthesis has been reviewed by Lehman.[79]

e. Others As mentioned previously (section 1.1.2.g), Nakao *et al.*[26] found that most microorganisms capable of degrading RNA into 5′-mononucleotides[36] could also degrade DNA to 5′-mononucleotides. In particular, the extracellular enzymes of *Aspergillus quercinus* and *Streptomyces aureus* were suitable for the practical production of deoxynucleoside 5′-monophosphates.[25]

In addition, *Bacillus subtilis* phage SP3 DNase (EC 3.1.4.31, DNA 5′-dinucleotidohydrolase) catalyzes the progressive hydrolysis of denatured DNA from the 3′-terminus to give 90% 5′-phospho-dinucleotides and 10% 5′-phospho-trinucleotides.

f. Main sources other than microorganisms[78] 5′-P-Forming DNases are known to be present in bovine pancreas (EC 3.1.4.5, DNase I),[6] *Octopus vulgaris* hepatopancreas, lamb brain, etc.

1.2 Microbial 3′-P-Forming Nucleolytic Enzymes

1.2.1 Microbial 3′-P-Forming Nucleases

a. ***Staphylococcus aureus***[11–13,15,80] Staphylococcal nuclease (EC 3.1.4.7, micrococcal nuclease) was the first nuclease to be found that yielded 3′-nucleotides upon hydrolysis of polynucleotide chains.[15] Denatured DNA is hydrolyzed faster than native DNA. The enzyme functions as an endonuclease in the initial stages of its action and then functions as an exonuclease to the point where the resulting fragments are dinucleotides, at which point its action ceases.
b. Others Microbial 3′-P-forming nucleases are not widely distributed in microorganisms. They have been reported only in *Bacillus subtilis* Marburg strain[81] and SB 19 strain,[83] *Lactobacillus acidophilus*,[82] and *Physarum polycephalum* (nuclease P_{p-1}),[16,41] etc., in addition to *Staphylococcus.* As will be described later (section 1.2.3), 3′-P-forming DNases (EC 3.1.4.6, DNase II type) are not widely found in microorganisms, either. These facts are consistent with the finding[26] that most microorganisms capable of degrading RNA into 3′-nucleotides cannot degrade DNA into 3′-nucleotides.
c. Main sources other than microorganisms Bovine spleen enzyme (EC 3.1.4.7, spleen endonuclease; EC 3.1.4.18, phosphodiesterase II)[1,11,13,94] is the best known.*

1.2.2 Microbial 3′-P-Forming RNases

A. 3′-P-Forming, noncyclizing RNases

RNase purified by Otaka *et al.*[84] from bakers' yeast degrades yeast RNA completely into 3′-mononucleotides. It splits neither DNA nor nucleoside 2′:3′-cyclic phosphates. Thus, the enzyme is regarded as a 3′-P-forming, noncyclizing RNase which forms 3′-nucleotides directly from RNA without the formation of 2′:3′-cyclic phosphates as intermediates. No other noncyclizing RNases have been confirmed to exist, except for the possibility that *Ustilago* RNase U_4[2,16,105] may also be a 3′-P-forming, noncyclizing, nonspecific RNase (exonuclease).

---

* The relationship between staphylococcal nuclease and spleen phosphodiesterase is not clearly defined in *Enzyme Nomenclature.*[1] As will be shown in Chapter 6 (Fig. 6.1), however, the basic substrate structural elements for these 3′-P producers appear to be quite different.

B. 3′-P-Forming, cyclizing RNases

3′-P-Forming, cyclizing RNases can be classified into four groups:

Pyrimidine-specific RNases (EC. 3.1.4.22, RNase I) These enzymes catalyze the endonucleolytic cleavage of RNA at the 3′-position of a pyrimidine nucleotide residue. 2′:3′-Cyclic phosphates are formed as intermediates.* Formerly EC 2.7.7.16.

Guanine-specific RNases (EC 3.1.4.8, guanyloribonuclease) These enzymes catalyze the endonucleolytic cleavage of RNA at the 3′-position of a guanylate residue. 2′:3′-Cyclic phosphates are formed as intermediates.* Formerly EC 2.7.7.26.

Purine-specific RNases These enzymes catalyze the endonucleolytic cleavage of RNA at the 3′-position of a purine nucleotide residue. 2′:3′-Cyclic phosphates are formed as intermediates.*

Non-specific RNases (EC 3.1.4.23, RNase II) These enzymes catalyze the endonucleolytic cleavage of RNA at the 3′-position of purine and pyrimidine nucleotide residues. 2′:3′-Cyclic phosphates are formed as intermediates.* Formerly EC 2.7.7.17.

These cyclizing RNases can also be utilized for the synthesis of oligonucleotides from nucleoside 2′:3′-cyclic phosphates and a phosphate acceptor.[88,104,107,130)]

a. Pyrimidine-specific RNases (EC 3.1.4.22, RNase I) Although pancreatic RNase,[2,4,11)] which is the best known RNase, belongs to this group, pyrimidine-specific RNases have not been confirmed to exist in microorganisms. *Thiobacillus thioparus* RNase[2,16)] splits RNA preferentially at the 3′-position of a pyrimidine residue, but no cyclizing reaction has yet been demonstrated.

b. Guanine-specific RNases (EC 3.1.4.8, guanyloribonuclease) RNase T_1[1,2,11–13,16,85–93)] crystallized from *Aspergillus oryzae*, RNase N_1 (extracellular enzyme, crystallized[103)] and utilized for synthesizing oligonucleotides[104)]) and RNase N_3 (intracellular enzyme) from *Neurospora crassa*,[102)] *Ustilago sphaerogena* (smut fungus) RNase U_1,[105)] RNases from *Ustilago zeae*,[121)] *Actinomyces aureoventicillatus*,[16)] *Streptomyces albogriseolus*,[109)] and *S. erythreus*,[108)] *Physarum polycephalum* RNase P_{p-1}[16,41)] and *Acrocylindrium* sp. RNase[118)] are regarded as guanine-specific RNases.

c. Purine-specific RNases RNases U_2 and U_3 purified from *Ustilago sphaerogena*[2,16,105–107)] and *Pleospora* RNase[16)] are regarded as purine-specific RNases. It is interesting that they are complementary to RNase I in specificity. RNase U_2 is a useful tool for the synthesis of oligonucleo-

* The 2′:3′-cyclic phosphates have been described as the final products for a few RNases.

tides such as adenylyl-(3′,5′)-nucleosides, adenylyl-(3′,5′)-guanosine 2′:3′-cyclic phosphate and oligoadenylic acids.[107]

d. Non-specific RNases (EC 3.1.4.23, RNase II) *Aspergillus oryzae* RNase T_2[11,12,16,85,95] *A. saitoi* RNase M[97] and RNase Ms,[98] *Penicillium citrinum* RNase,[46] RNase (Rh)[99–101] crystallized from *Rhizopus niveus*, *Monascus purpureus* RNase,[125] *Neurospora crassa* RNase N_2 (intracellular enzyme),[102] *Rhodotorula glutinis* RNase,[38,111] *Candida lipolytica* RNase CL,[112] RNases of *Tricoderma koningi*[124] and *Tetrahymena pyriformis*,[123] *Physarum polycephalum* RNase P_{p-2} and RNase P_{p-3},[16,41] *Azotobacter agilis* RNase,[119] *E. coli* RNase I,[120,122] *Proteus mirabilis* RNase II,[126] *Salmonella typhimurium* RNase,[127] *Lactobacillus plantarum* RNase I,[42] extracellular RNase[12] crystallized from *Bacillus subtilis* H (*B. amyloliquefaciens*), *B. subtilis* Marburg strain RNase,[81] and intracellular ATP-inhibited RNases found in *B. subtilis*, *B. amyloliquefaciens*, *B. cereus*, *B. cereus* var. *mycoides* and *B. thuringiensis* are regarded as non-specific RNases. The ATP-inhibited RNase is a useful tool for the synthesis of diribonucleoside monophosphates.[130]

3′-P-Forming, non-specific RNases (cyclizing) are also distributed widely in plants.[11,66,115]

Among the above microbial RNases, enzymes from *Rhizopus*,[99–101] *Azotobacter*,[119] *Bacillus*,[128,129] and *Proteus*[126] have been shown to form 2′:3′-cyclic phosphates as the final products.

C. Others

It has not been confirmed whether the following 3′-P-forming RNases are cyclizing enzymes or not: RNases of *Aspergillus niger*,[96] *Mucor genevensis*,[117] *Monascus pilosus*,[117] *Endomyces* sp.,[131] *Bacillus pumilus*,[117] and *Lenzites tenuis*,[117] *Clostridium acetobutylicum* RNase II,[16] and RNases of *Mycobacterium avium*[132] and *Thiobacillus thioparus*.[2,16]

Nakao *et al.*[37] found 3′-P-forming RNase activity in culture filtrates of the following microorganisms: *Candida lipolytica*, *Endomycopsis fibuliger*, *Hansenula anomala*, *H. schneggii*, *Rhodotorula aurantiaca*, *R. glutinis*, *Spolobolomyces pararoseus*, *Torulopsis candida*, *T. miso*, *Zygosaccharomyces tikumaensis*, *Absidia coerulea*, *A. spinosa*, *Aspergillus fonsecaeus*, *A. flavus*, *Blackeslea circinaus*, *Cuninghamella echinulata*, *Gibberella fujikuroi*, *Gloeosporium laeticolor*, *Helicostylum piriforme*, *Merulius domesticus*, *Monascus ruber*, *M. anka*, *Penicillium citrinum*, *P. corymbiferum*, *P. baarnense* var. *beyma*, *P. asperum*, *P. crustosum*, *P. ehinulo-nalgiouense*, *Pestalotia diospyri*, *Pestalotispsis funerea*, *Poria vaporaria*, *Rhizopus chinensis*, and *R. niveus*.

1.2.3 Microbial 3′-P-Forming DNases (EC 3.1.4.6, DNase II)

Typical DNases of this type have been found in calf thymus, bovine spleen, and pig spleen.[11] DNase II and DNase II-like enzymes have been reviewed by Laskowski.[19] Koga *et al.*[50] purified an acid DNase from pupae of *Bombyx mori* (silkworm). The enzyme degrades both native and denatured DNA into 3′-phospho-mono-(16%) and oligonucleotides. Microbial 3′-P-forming DNases have been confirmed to exist only in *M. luteus* (UV-endonuclease)[17] and *A. oryzae* (DNase K_1).[70]

1.3 Microbial 2′-P-Forming Nucleolytic Enzymes

Enzymes purified by Kataya *et al.*[133] from millet, milo maize, and soybean sprouts form 2′-mononucleotides from RNA via nucleoside 2′:3′-cyclic phosphates. Such RNases have not been reported in microorganisms. In addition, Igarashi *et al.*[23,24,134] effectively prepared 2′-nucleotides by treating alkaline hydrolyzates of RNA (mixtures of 2′-nucleotides and 3′-nucleotides) with a partially purified *Bacillus subtilis* 3′-nucleotidase preparation.

1.4 Applications of Nucleolytic Enzymes

As described above, various types of nucleolytic enzymes are widely distributed in microorganisms. Furthermore, a microorganism generally produces several kinds of nucleolytic enzymes simultaneously. For example, *Aspergillus oryzae* has been shown to produce at least six kinds of nucleolytic enzyme (Table 1.2).

At present, only *Penicillium* and *Streptomyces* 5′-P-forming nucleases are being employed for the industrial production of 5′-nucleotides. In the future, however, other nucleolytic enzymes will be also applied for the industrial production of various nucleic acid-related compounds as demand for these compounds develops. As crude enzyme preparations are usually employed industrially, it is particularly important that the main reaction should proceed effectively without any interfering side reactions.

TABLE 1.2

Comparison of Nucleolytic Enzymes from *Aspergillus oryzae*†

Nucleases	Molecular weight	Optimum pH	Metal requirement	Substrate: RNA	Substrate: Denatured DNA	Substrate: Native DNA	Terminal phosphate	Main products (nucleotide)	Base specificity
Nuclease O	64,000	7.4～8.2	Mg^{2+}, Mn^{2+}	+	+	+	5′	oligo	no
Nuclease S_1	70,000～100,000	4.4	Zn^{2+}	+	+	−	5′	mono	no
DNase K_1		8.5～9.0	Mg^{2+}, Mn^{2+}	−	+	+	3′	oligo	G–G, G–A
DNase K_2		8.0	Mg^{2+}, Mn^{2+}	−	+	+	5′	oligo	G–G, G–A
RNase T_1	11,000	7.5	none	+	−	−	3′	oligo	G–X
RNase T_2	36,000	4.5	none	+	−	−	3′	mono	no

† From Uozumi *et al.* (1972).[49]

Several nucleolytic enzymes are commercially available for biochemical research. Enzymes developed in Japan, such as RNase T_1, RNase T_2, nuclease S_1 and nuclease P_1, have been used mainly for studying the structures of nucleic acids.

Excellent reviews[1,2,11–14,19,69,78,137–140,145] and invaluable lists of references[18] on nucleolytic enzymes have been published.

REFERENCES

1. Commission on Biochemical Nomenclature, *Enzyme Nomenclature*, Elsevier, 1972.
2. T. E. Barman, *Enzyme Handbook* I, II, Springer-Verlag, 1969; Supplement I, 1974; F. Egami and K. Nakamura, *Microbial Ribonucleases*, Springer-Verlag, 1969.
3. A. Kuninaka, *Nippon Nogeikagaku Kaishi* (Japanese), **34**, 489 (1960).
4. F. M. Richards and H. W. Wyckoff, *The Enzymes* IV (ed. P. D. Boyer), p. 647, Academic, 1971.
5. H. D. Robertson, R. E. Webster and N. D. Zinder, *J. Biol. Chem.*, **243**, 82 (1968).
6. M. Laskowski, Sr., *The Enzymes* IV (ed. P.D. Boyer), p. 289, p. 313, Academic, 1971.
7. T. P. Wang, *J. Bact.*, **68**, 128 (1954).
8. P. Plackett, *Biochim. Biophys. Acta*, **26**, 664 (1957).
9. M. Kuwano, C. N. Kwan, D. Apirion and D. Schlessinger, *Proc. Natl. Acad. Sci. U.S.A.*, **64**, 693 (1969).
10. A. Kuninaka, *Hakko Kyokaishi* (Japanese), **20**, 311 (1962); *Tanpakushitsu Kakusan Koso* (Japanese), **6**, 403 (1961); *Baioteku* (Japanese), **1**, 65, 140 (1970); *Shokuhin Kogyo* (Japanese), **4** (11), 9 (1961).
11. M. P. De Garilhe, *Enzymes in Nucleic Acid Research*, Holden-Day, 1967.
12. G. L. Cantoni *et al.*, *Procedures in Nucleic Acid Research*, Harper & Row, 1966.
13. J. N. Davidson, *The Biochemistry of the Nucleic Acids*, Methuen, 1969.
14. K. Ogata, *Amino Acid and Nucleic Acid*, no. 8, 1 (1963); *Seikagaku* (Japanese), **36**, 1 (1964).

15. C. B. Anfinsen, P. Cuatrecasas and H. Taniuchi, *The Enzymes* IV (ed. P. D. Boyer), p. 177, Academic, 1971.
16. T. Uchida and F. Egami, *The Enzymes* IV (ed. P. D. Boyer), p. 205, Academic, 1971.
17. I. R. Lehman, *The Enzymes* IV (ed. P. D. Boyer), p. 251, Academic, 1971.
18. Laboratory of Fermentation of Kyoto University, *Amino Acid and Nucleic Acid*, no. 9, 113 (1964); K. Nakayama, *ibid.*, no. 16, 155 (1967); no. 20, 163 (1969); no. 26, 113 (1972).
19. M. Laskowsky, Sr., *Advan. Enzymol.*, **29**, 165, (1967).
20. Y. Nakao and K. Ogata, *Agr. Biol. Chem.*, **27**, 291 (1963).
21. Y. Nakao, I. Nogami and K. Ogata, *ibid.*, **27**, 507 (1963).
22. A. Kakinuma, S. Igarashi and K. Ogata, *ibid.*, **26**, 213 (1962).
23. S. Igarashi and A. Kakinuma, *ibid.*, **26**, 218 (1962).
24. S. Igarashi, *ibid.*, **26**, 221 (1962).
25. Y. Nakao and K. Ogata, *ibid.*, **27**, 491 (1963).
26. Y. Nakao and K. Ogata, *ibid.*, **27**, 199 (1963).
27. A. Stevens and R. J. Hilmoe, *J. Biol. Chem.*, **235**, 3016, 3023 (1960).
28. S. Linn and I. R. Lehman, *ibid*,. **240**, 1287, 1294 (1965); **241**, 2694 (1966).
29. H. Sugimoto, T. Iwasa, and T. Yokotsuka, *Nippon Nogeikagaku Kaishi* (Japanese), **38**, 135, 567 (1964).
30. H. Tone, K. Sasaki, Y. Sayama, M. Takata and A. Ozaki, *Amino Acid and Nucleic Acid*, no. 10, 135 (1964); H. Tone, M. Umeda, T. Ishikura and N. Miyachi, *ibid.*, no. 10, 142 (1964); H. Tone and A. Ozaki, *Enzymologia*, **34**, 101 (1968).
31. T. Ando, *Seikagaku* (Japanese), **37**, 456 (1965); *Biochim. Biophys. Acta*, **114**, 158 (1966); K. Shishido and Y. Ikeda, *J. Biochem.* (Tokyo), **67**, 759 (1970); *J. Mol. Biol.*, **55**, 287 (1971); *Biochem. Biophys. Res. Commun.*, **42**, 482 (1971); K. Shishido and T. Ando, *Biochim. Biophys. Acta*, **287**, 477 (1972).
32. H. M. Keir, R. H. Mathog and C. E. Carter, *Biochemistry*, **3**, 1188 (1964).
33. P. F. Spahr, *J. Biol. Chem.*, **239**, 3716 (1964).
34. P. F. Spahr and R. F. Gesteland, *Proc. Natl. Acad. Sci. U.S.A.*, **59**, 876 (1968).
35. Y. Hasegawa, T. Nakai, Y. Fujimura, Y. Kaneko and S. Doi, *Nippon Nogeikagaku Kaishi* (Japanese), **38**, 461 (1964); Y. Fujimura, Y. Hasegawa, Y. Kaneko and S. Doi, *ibid.*, **38**, 467 (1964); *Agr. Biol. Chem.*, **31**, 92 (1957).
36. K. Ogata, Y. Nakao, S. Igarashi, E. Omura, Y. Sugino, M. Yoneda and I. Suhara, *Agr. Biol. Chem.*, **27**, 110 (1963).
37. Y. Nakao and K. Ogata, *ibid.*, **27**, 116 (1963).
38. Y. Nakao and K. Ogata, *ibid.*, **27**, 499 (1963).
39. A. Kakinuma, Y. Nakao, S. Igarashi and K. Ogata, *ibid.*, **28**, 300 (1964).
40. R. Saruno, H. Takahira and M. Fujimoto, *Hakkokogaku Zasshi* (Japanese), **42**, 475 (1964); E. Soeda, A. Murata, N. Ube and R. Saruno, *Seikagaku* (Japanese), **40**, 688 (1968); E. Soeda, S. Abe, A. Murata and R. Saruno, *Nippon Nogeikagaku Kaishi* (Japanese), **43**, N86 (1969).
41. M. Hiramaru, T. Uchida and F. Egami, *J. Biochem.* (Tokyo), **65**, 693, 701 (1969).
42. D. M. Logan and M. F. Singer, *J. Biol. Chem.*, **243**, 6161 (1968).
43. N. G. Nossal and M. F. Singer, *ibid.*, **243**, 913 (1968).
44. A. Weissbach and D. Korn, *ibid.*, **238**, 3383 (1963).
45. A. Kuninaka, S. Otsuka, Y. Kobayashi and K. Sakaguchi, *Bull. Agr. Chem. Soc. Japan*, **23**, 239 (1959); A. Kuninaka, M. Kibi, H. Yoshino and K. Sakaguchi, *Agr. Biol. Chem.*, **25**, 693 (1961).
46. A. Kuninaka, T. Fujishima and M. Fujimoto, *Amino Acid and Nucleic Acid*, no. 16, 28, (1967)
47. M. Fujimoto, A. Kuninaka and H. Yoshino, *Agr. Biol. Chem.*, **10**, 1517 (1969); **38**, 777, 785, 1555 (1974); **39**, 1991, 2145 (1975); M. Fujimoto, K. Fujiyama, A. Kuninaka and H. Yoshino, *ibid.*, **38**, 2141 (1974).
48. A. Kuninaka, M. Fujimoto and H. Yoshino, *ibid.*, **39**, 597, 603 (1975).
49. T. Uozumi, G. Tamura and K. Arima, *ibid.*, **32**, 963, 969, 1409 (1968); **33**, 25, 636, 645 (1969); **35**, 1109 (1971); T. Uozumi, T. Hino, G. Tamura and K. Arima, *ibid.*,

36, 434 (1972); T. Uozumi and K. Arima, *ibid.*, **38**, 1739 (1974).
50. K. Koga and S. Akune, *ibid.*, **36**, 1903 (1972).
51. M. Nestle and W. K. Roberts, *J. Biol. Chem.*, **244**, 5213, 5219 (1969).
52. Y. Nakao, S. Y. Lee, H. O. Halvorson and R. M. Bock, *Biochim. Biophys. Acta*, **151**, 114 (1968); S. Y. Lee, Y. Nakao and R. M. Bock, *ibid.*, **151**, 126 (1968).
53. I. Suhara and M. Yoneda, *J. Biochem.* (Tokyo), **73**, 647 (1973); I. Suhara, *ibid.*, **73**, 1023 (1973).
54. M. Yoneda, *ibid.*, **55**, 475, 481 (1964).
55. J. Mukai, *Biochem. Biophys. Res. Commun.*, **21**, 562 (1965); J. Mukai and E. Soeta, *Biochim. Biophys. Acta*, **138**, 1 (1967); E. Soeda, J. Mukai and S. Akune, *J. Biochem.* (Tokyo), **63**, 14 (1968); J. Mukai, *Anal. Biochem.*, **49**, 576 (1972); T. Funaguma and J. Mukai, *Comp. Biochem. Physiol.*, **44B**, 633 (1973); M. Himeno, I. Morishima, F. Sakai and K. Onodera, *Biochim. Biophys. Acta*, **167**, 575 (1968); T. Funaguma, H. Nomiyama and J. Mukai, *Amino Acid and Nucleic Acid*, no. 31, 130 (1975).
56. J. Eley and J. S. Roth, *J. Biol. Chem.*, **241**, 3063, 3070 (1966).
57. P. J. Curtis, M. G. Burdon and R. M. S. Smellie, *Biochem. J.*, **98**, 813 (1966); P. J. Curtis and R. M. S. Smellie, *ibid.*, **98**, 818 (1966).
58. S. A. Berry and J. N. Campbell, *Biochim. Biophys. Acta*, **132**, 78, 84 (1967).
59. Y. Ohta and S. Ueda, *Appl. Microbiol.*, **16**, 1293 (1968).
60. I. Suhara, *J. Biochem.* (Tokyo), **75**, 1135 (1974).
61. H. D. Robertson and J. J. Dunn, *J. Biol. Chem.*, **250**, 3050 (1975).
62. J. P. Leis, I. Berkower and J. Hurwitz, *Proc. Natl. Acad. Sci. U.S.A.*, **70**, 466 (1973); I. Berkower, J. Leis and J. Hurwitz, *J. Biol. Chem.*, **248**, 5914 (1973).
63. G. B. Jacobsen and V. W. Rodwell, *J. Biol. Chem.*, **247**, 5811 (1972).
64. T. Sugimori, *Amino Acid and Nucleic Acid*, no. 10, 129 (1964).
65. T. Mouri, W. Hashida, I. Shiga and S. Teramoto, *ibid.*, no. 17, 178 (1968).
66. Y. Nakagiri, Y. Maekawa, R. Kihara and M. Miwa, *Hakkokogaku Zasshi* (Japanese), **46**, 605, 610 (1968).
67. S. Ishii, H. Sugimoto and T. Yokotsuka, *Nippon Nogeikagaku Kaishi* (Japanese), **41**, 340, 348 (1967).
68. M. Ukita, A. Furuya, H. Tanaka and M. Misawa, *Agr. Biol. Chem.*, **37**, 2849 (1973).
69. M. Irie, *Tanpakushitsu Kakusan Koso* (Japanese), **12**, 386 (1967).
70. M. Kato and Y. Ikeda, *J. Biochem.* (Tokyo), **64**, 321, 329 (1968); **65**, 43 (1969); K. Shishido, M. Kato and Y. Ikeda, *ibid.*, **65**, 49 (1969); M. Kato, S. Ikawa, Y. Ikeda and M. Fuke, *ibid.*, **65**, 185 (1969).
71. Y. Kurono, K. Horikoshi and Y. Ikeda, *Agr. Biol. Chem.*, **35**, 1436 (1971).
72. Y. Takagi, *Kakusan Jikkenho Ge* (Bessatsu Tanpakushitsu Kakusan Koso (Japanese)), p. 91, Kyoritsu Shuppan, 1973.
73. S. Hirose, *ibid.*, p. 94.
74. T. Ando, *ibid.*, p. 108.
75. S. Yasuda and M. Sekiguchi, *ibid.*, p. 97.
76. Y. Takagi, *ibid.*, p. 103; *Tanpakushitsu Kakusan Koso* (Japanese), **16**, 1109 (1971).
77. K. Minton, M. Durphy, R. Taylor and E. C. Friedberg, *J. Biol. Chem.*, **250**, 2823 (1975).
78. R. Okazaki and K. Sugimoto, *Tanpakushitsu Kakusan Koso* (Japanese), **12**, 371 (1967).
79. I. R. Lehman, *Ann. Rev. Biochem.*, **36**, 645 (1967).
80. M. Ohno and N. Izumiya, *Tanpakushitsu Kakusan Koso* (Japanese), **15**, 101 (1970).
81. M. Nakai, Z. Minami, T. Yamazaki and A. Tsugita, *J. Biochem.* (Tokyo), **57**, 96 (1965).
82. W. Fiers and H. G. Khorana, *J. Biol. Chem.*, **238**, 2780, 2789 (1963).
83. J. M. Kerr, E. A. Pratt and I. R. Lehman, *Biochem. Biophys. Res. Commun.*, **20**, 154 (1965).
84. Y. Ohtaka, K. Uchida and T. Sakai, *J. Biochem.* (Tokyo), **54**, 322 (1963).
85. F. Egami, K. Takahashi and T. Uchida, *Progress in Nucleic Acid Research and Molecular Biology*, vol. 3, p. 59, Academic, 1964.

86. K. Takahashi, *Tanpakushitsu Kakusan Koso* (Japanese), **14**, 1127 (1969).
87. S. Minato, T. Tagawa and K. Nakanishi, *J. Biochem.* (Tokyo), **59**, 443 (1966).
88. T. Sekiya, Y. Furuichi, M. Yoshida and T. Ukita, *ibid.*, **63**, 514 (1968).
89. C. Shiozawa, *ibid.*, **66**, 733 (1969).
90. S. Irie, T. Itoh, T. Ueda and F. Egami, *ibid.*, **68**, 163 (1970).
91. T. Uchida, *ibid.*, **68**, 255 (1970).
92. T. Oshima and K. Imahori, *ibid.*, **70**, 197 (1971).
93. K. Takahashi, *ibid.*, **49**, 1 (1961); **51**, 95 (1962); **52**, 72 (1962); **60**, 239 (1966); **68**, 517, 659 (1970); **69**, 331 (1971); **70**, 477, 603, 617, 803, 945 (1971); **72**, 1469 (1972); **74**, 1279 (1973); **75**, 201 (1974).
94. A. Bernardi and G. Bernardi, *The Enzymes* IV (ed. P.D. Boyer), p. 329, Academic, 1971.
95. T. Uchida, *J. Biochem.* (Tokyo), **60**, 115 (1966); T. Uchida and F. Egami, *ibid.*, **61**, 44 (1967).
96. H. Horitsu, Y. Higashi and M. Tomoeda, *Agr. Biol. Chem.*, **38**, 933 (1974).
97. M. Imazawa, M. Irie and T. Ukita, *J. Biochem.* (Tokyo), **64**, 595 (1968); M. Irie, *ibid.*, **65**, 133, 907 (1969); **66**, 569 (1969); **69**, 965 (1971); M. Irie, M. Harada, T. Negi and T. Samejima, *ibid.*, **69**, 881 (1971); M. Harada and M. Irie, *ibid.*, **73**, 705 (1973).
98. K. Ohgi and M. Irie, *ibid.*, **77**, 1085 (1975).
99. S. Mantani, T. Yamamoto and J. Fukumoto, *Amino Acid and Nucleic Acid*, no. 18, 64 (1968); S. Mantani, J. Fukumoto and T. Yamamoto, *Agr. Biol. Chem.*, **36**, 242 (1972).
100. M. Tomoeda, Y. Eto and T. Yoshino, *Arch. Biochem. Biophys.*, **131**, 191 (1969).
101. T. Komiyama and M. Irie, *J. Biochem.* (Tokyo), **70**, 765 (1971); **71**, 973 (1972); **75**, 419 (1974).
102. N. Takai, T. Uchida and F. Egami, *Biochim. Biophys. Acta*, **128**, 218 (1966); *Seikagaku* (Japanese), **39**, 285, 473 (1967).
103. K. Kasai, T. Uchida, F. Egami, K. Yoshida and M. Nomoto, *J. Biochem.* (Tokyo), **66**, 389 (1969).
104. T. Koike, T. Uchida and F. Egami, *ibid.*, **70**, 55 (1971).
105. T. Arima, T. Uchida and F. Egami, *Biochem. J.*, **106**, 601, 609 (1968); J. Hashimoto, T. Uchida and F. Egami, *J. Biochem.* (Tokyo), **70**, 903 (1971); J. Hashimoto, K. Takahashi and T. Uchida, *ibid.*, **73**, 13 (1973); J. Hashimoto and K. Takahashi, *ibid.*, **76**, 1359 (1974).
106. T. Uchida, T. Arima and F. Egami, *J. Biochem.* (Tokyo), **67**, 91 (1970).
107. T. Koike, T. Uchida and F. Egami, *ibid.*, **69**, 111 (1971).
108. K. Tanaka, *ibid.*, **50**, 62 (1961); K. Tanaka and G. L. Cantoni, *Biochim. Biophys. Acta*, **72**, 641 (1963).
109. M. Yoneda, *J. Biochem.* (Tokyo), **55**, 469 (1964).
110. M. Fujimoto, K. Fujiyama, Y. Midorikawa, T. Fujishima, A. Kuninaka and H. Yoshino, *Ann. Mtg. Agr. Chem. Soc. Japan* (Japanese), 1975. Abstracts, p. 167.
111. K. Ogata, S. Song and H. Yamada, *Agr. Biol. Chem.*, **35**, 58 (1971); S. Song, H. Yamada and K. Ogata, *Ann. Mtg. Agr. Chem. Soc. Japan* (Japanese), 1973. Abstracts, p. 169.
112. A. Imada, A. J. Sinskey and S. R. Tannenbaum, *Proc. IV IFS: Ferment. Technol. Today*, p. 455, 1972; A. Imada, J. W. Hunt, H. V. DeSande, A. J. Sinskey and S. R. Tannenbaum, *Biochim. Biophys. Acta*, **395**, 490 (1975).
113. T. L. Walters and H. S. Loring, *J. Biol. Chem.*, **241**, 2870, 2881 (1966); H. S. Loring, J. E. McLennan and T. L. Walters, *ibid.*, **241**, 2876 (1966).
114. A. Nomura, M. Suno and Y. Mizuno, *J. Biochem.* (Tokyo), **70**, 993 (1971); M. Suno, A. Nomura and Y. Mizuno, *ibid.*, **73**, 1291 (1973).
115. A. Hara, H. Yano and T. Watanabe, *Nippon Nogeikagaku Kaishi* (Japanese), **43**, 13 (1969); A. Hara, K. Yoshihara and T. Watanabe, *ibid.*, **44**, 385 (1970).
116. A. E. Oleson, A. M. Janski and E. T. Clark, *Biochim. Biophys. Acta*, **366**, 89 (1974).
117. G. W. Rushizki, A. E. Greco, R. W. Hartley and H. A. Sober, *J. Biol. Chem.*, **239**, 2165 (1964).

118. I. Suhara, F. Kawashima-Kusaba, Y. Nakao, M. Yoneda and E. Ohmura, *J. Biochem.* (Tokyo), **71**, 941 (1972).
119. I. Shiio, K. Ishii and S. Shimizu, *ibid.*, **59**, 363 (1966).
120. Y. Anraku and D. Mizuno, *Biochem. Biophys. Res. Commun.*,**18**, 462 (1965); *J. Biochem.* (Tokyo), **61**, 70, 81 (1967).
121. M. Yanagida, T. Uchida and F. Egami, *Nippon Nogeikagaku Kaishi* (Japanese), **38**, 531 (1964).
122. P. F. Spahr and B. R. Hollingworth, *J. Biol. Chem.*, **236**, 823 (1961).
123. L. H. Lazarus and O. H. Scherbaum, *Biochim. Biophys. Acta*, **142**, 368 (1967).
124. M. Harada and M. Irie, *Seikagaku* (Japanese), **41**, 587 (1969).
125. E. Soeda, M. Deguchi, A. Murata and R. Saruno, *Ann. Mtg. Agr. Chem. Soc. Japan* (Japanese), 1973. Abstracts, p. 168.
126. M. S. Center and F. J. Bahal, *Biochim. Biophys. Acta*, **151**, 698 (1968).
127. K. Chakraburtty and D. P. Burma, *J. Biol. Chem.*, **243**, 1133 (1968).
128. S. Nishimura and B. Maruo, *Biochim. Biophys. Acta*, **40**, 355 (1960).
129. M. Yamasaki and K. Arima, *ibid.*, **139**, 202 (1967); **209**, 475 (1970); M. Yamasaki, K. Yoshida and K. Arima, *ibid.*, **209**, 463 (1970); M. Yamasaki, M. Sugi and G. Tamura, *Agr. Biol. Chem.*, **37**, 2595 (1973).
130. M. Saito, Y. Furuichi, K. Takeishi, M. Yoshida, M. Yamasaki, K. Arima, H. Hayatsu and T. Ukita, *Biochim. Biophys. Acta*, **195**, 299 (1969).
131. T. Hattori and S. Nakamura, *Seikagaku* (Japanese), **38**, 563 (1966).
132. Y. Tsugita and K. Matsui, *ibid.*, **41**, 588 (1968).
133. K. Kataya, K. Nishimoto, M. Akimoto and H. Kawata, *Nippon Nogeikagaku Kaishi* (Japanese), **43**, 345 (1969).
134. A. Kakinuma and S. Igarashi, *Agr. Biol. Chem.*, **28**, 131 (1964).
135. P. H. Johnson and M. Laskowski, Sr., *J. Biol. Chem.*, **243**, 3421 (1968); **245**, 891 (1970); A. J. Mikulski and M. Laskowski, Sr., *ibid.*, **245**, 5026 (1970).
136. D. M. Hanson and J. L. Fairley, *ibid.*, **244**, 2440 (1969); W. D. Kroeker, D. M. Hanson and J. L. Fairley, *ibid.*, **250**, 3767 (1975); W. D. Kroeker and J. L. Fairley, *ibid.*, **250**, 3773 (1975).
137. T. Ando, *Kagaku to Seibutsu* (Japanese), **13**, 342 (1975).
138. K. Imahori *et al.*, *Seikagaku Jikken Koza* (Japanese), 2-II, Tokyo Kagaku Dojin, 1976.
139. H. Sugisaki and M. Takanami, *Tanpakushitsu Kakusan Koso* (Japanese), **20**, 915 (1975).
140. E. A. Barnard, *Ann. Rev. Biochem.*, **38**, 677 (1969).
141. M. Futai and D. Mizuno, *J. Biol. Chem.*, **242**, 5301 (1967).
142. K. Shimizu, S. Yasuda and M. Sekiguchi, *Seikagaku* (Japanese), **43**, 495 (1971).
143. C. M. Wilson, *Plant Physiol.*, **43**, 1332, 1339 (1968).
144. K. Shishido and T. Ando, *Kagaku to Seibutsu* (Japanese), **13**, 605 (1975).
145. M. Yamasaki, *ibid.*, **13**, 555 (1975).

CHAPTER **2**

Nucleotide Biosynthesis and Its Regulation*

Several excellent reviews have been published on nucleotide biosynthesis, such as those by Buchanan and Hartman,[1] Most,[2] Magasanik[3] and Hartman.[4] There are two routes for the biosynthesis of purine nucleotides: the *de novo* route and the salvage route. The latter is also important in the microbial production of nucleic acid-related substances.

2.1 THE DE NOVO BIOSYNTHESIS OF PURINE NUCLEOTIDES

2.1.1 Precursors of the Purine Ring

The precursors of the purine ring were detetermined by Buchanan

* KÔ AIDA, Institute of Applied Microbiology, University of Tokyo.

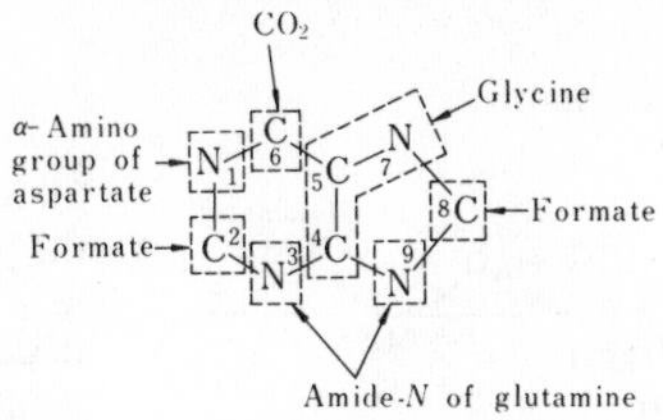

Fig. 2.1. Precursors of the purine ring of nucleotides.

and Hartman,[1] Greenberg[5] and others with intact animals and pigeon liver homogenates using isotopic tracers and ^{15}N-labeled compounds, as shown in Fig. 2.1. The origin of N-1 of the purine ring is the α-amino group of aspartate and N-3 and N-9 are from amide nitrogen of glutamine. Glycine enters positions 4, 5 and 7 as an intact molecule. Formate is the source of C-2 and C-8, and carbon dioxide labels position 6.

For the biosynthesis of one molecule of IMP, one molecule of PRPP, one aspartate, one glycine, two glutamine, two formate and one CO_2 are required, and eight molecules of ATP are also necessary as an energy source.

2.1.2 The *de novo* Biosynthetic Process

Greenberg[6] succeeded in the enzymatic biosynthesis of purine nucleotides using a homogenate of pigeon liver, and a series of experiments by Buchanan's group[1] clarified the outline of this pathway. The pathway and related enzymes thus established are shown in Fig. 2.2. This pathway has been confirmed to operate in microorganisms such as *Bacillus subtilis*[7] and others. The pathway consists of two parts. One involves the formation of IMP from ribose-5-phosphate (formed from glucose by the Warburg-Dickens pathway), and the other consists of AMP and GMP formation from IMP.

The first step of IMP biosynthesis is the formation of PRPP from ribose-5-phosphate and ATP. PRPP is aminated by glutamine to form 5-phosphoribsylamine (PRA). Glycine is incorporated into PRA to form glycinamide ribotide (GAR). Formylglycinamide ribotide (FGAR) is formed from GAR by transfer of the formyl group from a formylated derivative of tetrahydrafolate. FGAR is aminated by glutamine to form formylglycinamidine ribotide (FGAM). By the cyclization of FGAM, the imidazole ring is formed as aminoimidazole ribotide (AIRP). The next steps are the incorporation of CO_2 and aspartate to form 5-amino-4-imidazolecarboxylic acid ribotide (CAIRP), then SAICARP. By the elimination of fumarate

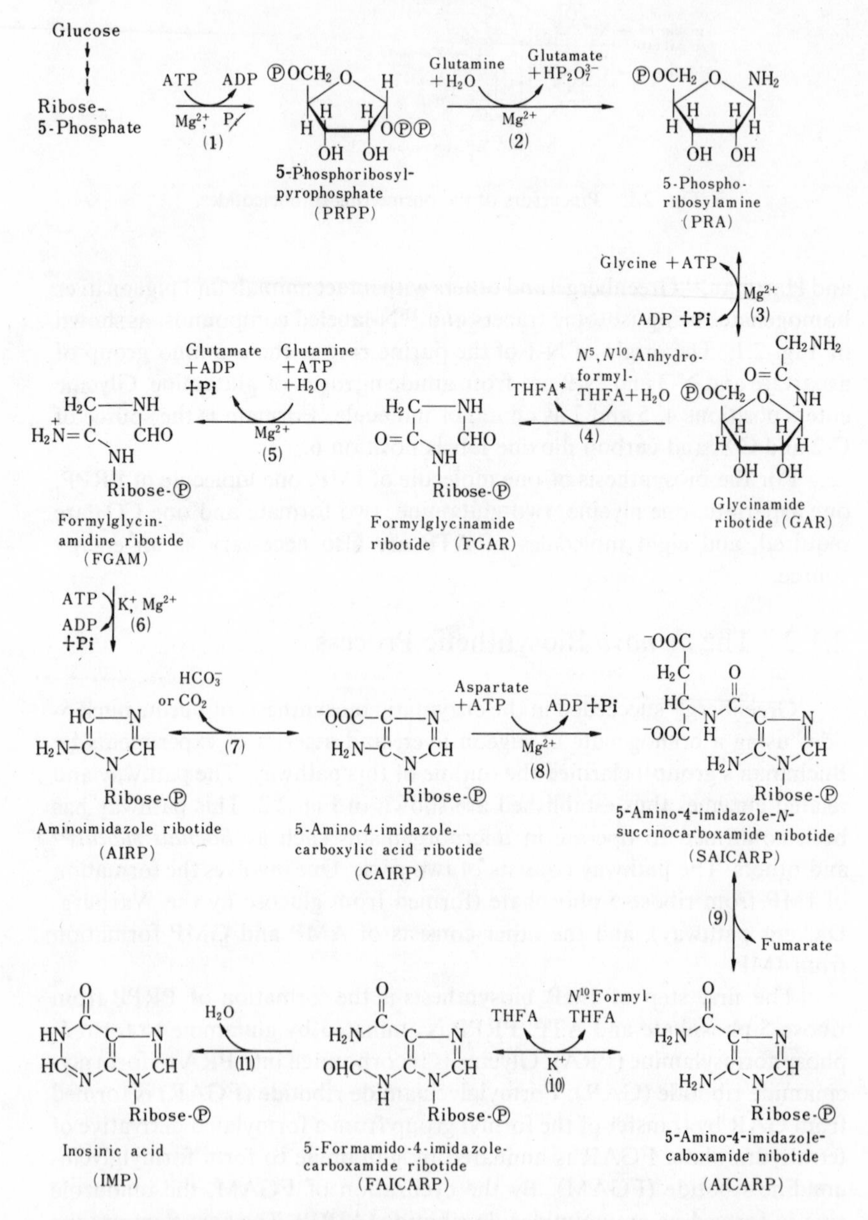

Fig. 2.2. Biosynthetic pathway of purine nucleotides, showing related enzymes. (* THFA: tetrahydrofolic acid)

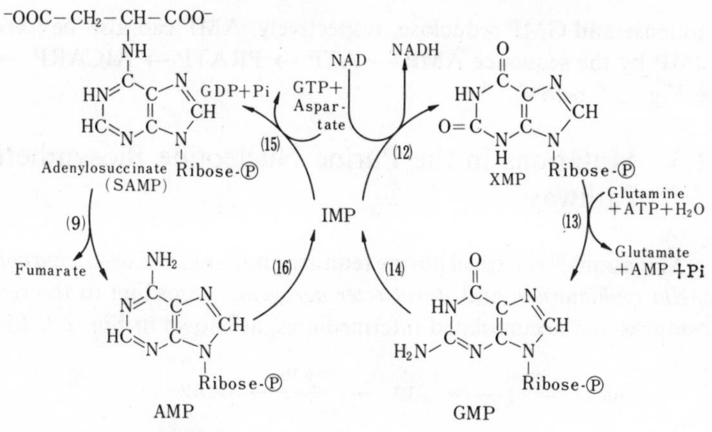

$^{-}OOC-CH_2-CH-COO^{-}$

Adenylosuccinate (SAMP) Ribose-Ⓟ

GDP+Pi (15) GTP+ Aspar- tate

NAD NADH (12)

XMP Ribose-Ⓟ

(9) Fumarate

IMP

(13) Glutamine $+ATP+H_2O$ Glutamate +AMP+Pi

(16) (14)

AMP Ribose-Ⓟ

GMP Ribose-Ⓟ

Fig. 2.2.—*Continued*

(1) PRPP synthetase
(2) PRPP amidotransferase
(3) Phosphoribosyl glycinamide synthetase
(4) Phosphoribosylglycinamide formyltransferase
(5) Phosphoribosylformylglycinamidine synthetase
(6) Phosphoribosylaminoimidazole synthetase
(7) Phosphoribosylaminoimidazole carboxylase
(8) Phosphoribosylaminoimidazole succinocarboxamide synthetase
(9) Adenylosuccinate lyase
(10) Phosphoribosylaminoimidadole carboxamide formyltransferase
(11) IMP cyclohydrolase
(12) IMP dehydrogenase
(13) XMP aminase
(14) GMP reductase
(15) Adenylosuccinate synthetase
(16) AMP deaminase

from SAICARP, AICARP is formed. By transfer of the formyl group from formyltetrahydrofolate to AICARP, 5-formamido-4-imidazole-carboxamide ribotide (FAICARP) is formed. Finally, IMP is produced by the dehydration of FAICARP.

Two branch points to form AMP and GMP occur from IMP. AMP is formed by way of adenylosuccinate (SAMP) and GMP is formed by way of XMP. Adenylosuccinate lyase, which catalyzes the formation of AMP from SAMP, also accomplishes the conversion of SAICARP to AICARP.

AMP and GMP thus formed are incorporated into DNA and RNA by way of ATP and GTP, and AMP is also involved in the formation of histidine by way of ATP and 1-(5′-phosphoribosyl)adenosine triphosphate (PRATP). To control the supply of AMP and GMP, AMP and GMP are interconvertible. Both AMP and GMP can be converted to IMP, by AMP deaminase and GMP reductase, respectively. AMP can also be converted to IMP by the sequence AMP → ATP → PRATP → AICARP → IMP (see Fig. 2.5 below).

2.1.3 Mutations in the Purine Nucleotide Biosynthetic Pathway

Magasanik[8] classified purine requiring mutants of *Escherichia coli, Salmonella typhimurium* and *Aerobacter aerogenes* according to the required substances and accumulated intermediates, as shown in Fig. 2.3. Mutants

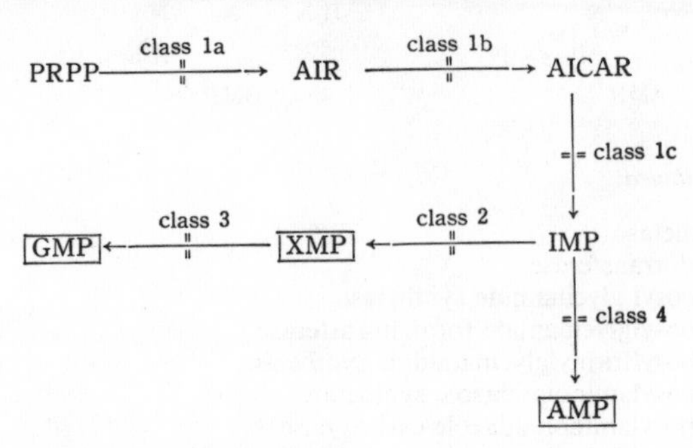

Fig. 2.3. Classification of mutants blocked in the purine nucleotide biosynthetic pathway.[8]

belonging to class I are nonexacting purineless mutants, and require adenine, guanine, xanthine and hypoxanthine. They have the ability to interconvert AMP and GMP, but cannot form IMP by the *de novo* route. This group of mutants was further classified into three groups according to the blocked position on the IMP biosynthetic pathway. Class 1 a requires purines, their precursors and vitamin B_1, and class 1 b requires purines and AICARP, accumulating AIRP. Class 1 c accumulates AICAP or its riboside. Class 1 c mutants are blocked at the reaction of AICARP to form IMP. Mutants of class 2 lack IMP dehydrogenase and require xanthine or guanine for growth. Mutants of class 3 lack XMP deaminase, require guanine, and accumulate xanthosine in the culture broth. Mutants of class 2 and class 3 are able to synthesize AMP, but not GMP. Mutants of class 4 have the ability to synthesize GMP, but not AMP. They specifically require adenine for growth. There are two types in this group: one accumulates SAICARP due to the loss of adenylosuccinate lyase, and the other accumulates inosine due to the loss of adenylosuccinate synthetase. These findings provided a starting point for the screening of mutants for the microbial production of nucleic acid-related substances.

2.2 THE *DE NOVO* BIOSYNTHESIS OF PYRIMIDINE NUCLEOTIDES

The biosynthesis of pyrimidine nucleotides begins with the formation of carbamylaspartate from aspartate and carbamylphosphate, which is formed from NH_3 and CO_2 in the presence of ATP, as shown in Fig. 2.4. The pyrimidine ring is formed by the dehydration of carbamylaspartate to form dihydroorotic acid. Orotic acid, formed by the oxidation of dihydroorotic acid, then reacts with PRPP to form orotidine-5′-phosphate. The latter is transformed to UMP by decarboxylation. UMP is phosphorylated by ATP to form UTP. CTP can be formed either by the incorporation of the amide

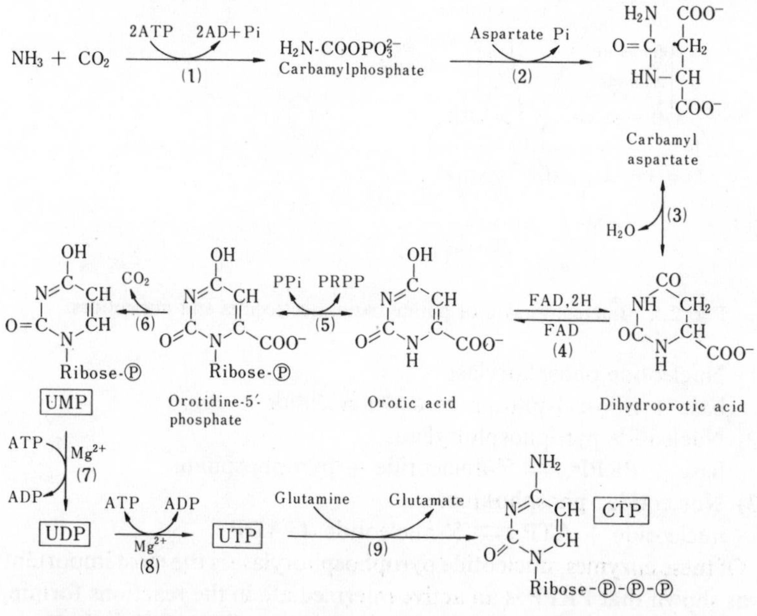

Fig. 2.4. Biosynthetic pathway of pyrimidine nucleotides, showing related enzymes.
(1) carbamylphosphate synthetase,
(2) aspartate transcarbamylase,
(3) dihydroorotase,
(4) dihydroorotate dehydrogenase,
(5) orotate phosphoribosyltransferase,
(6) orotidylic decarboxylase,
(7) nucleoside monophospho kinase,
(8) nucleoside diphosphokinase,
(9) cytidylate synthetase.

nitrogen of glutamine into UTP or from UTP and NH_3 in the presence of ATP.

2.3 Salvage Synthesis of Nucleotides

In addition to the *de novo* biosynthetic route, purine and pyrimidine nucleotides can be formed from purine and pyrimidine bases by processes known as salvage biosynthesis. The fact that auxotrophs blocked in the pathway of IMP biosynthesis can grow in media supplemented with any one of the four common purine bases shows that microorganisms can interconvert these compounds and synthesize IMP from purine bases (Fig. 2.5). The following three enzymes are important in salvage biosynthesis:

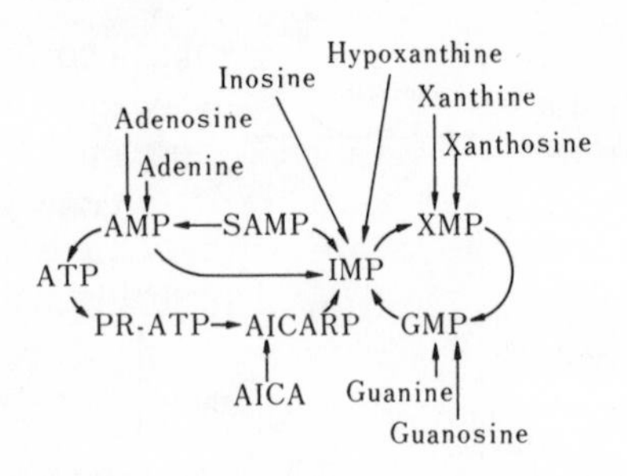

Fig. 2.5. Interconversion of purine bases, nucleosides and nucleotides.

(1) Nucleotide phosphorylase
base + ribose-1-phosphate $\rightleftharpoons$ nucleotide + Pi
(2) Nucleotide pyrophosphorylase
base + PRPP $\rightleftharpoons$ 5′-nucleotide + pyrophosphate
(3) Nucleotide phosphokinase
nucleoside + ATP $\rightleftharpoons$ 5′-nucleotide + ADP

Of these enzymes, nucleotide pyrophosphorylase is the most important. It was shown that PRPP is an active intermediate in the reactions forming nucleotides such as IMP, GMP and AMP from ribose-5-phosphate and purine bases such as hypoxanthine, guanine and adenine, respectively, in the presence of ATP.

Enzyme 1 Ribosephosphate pyrophosphokinase:

$$\text{ribose-5-P} + \text{ATP} \xrightarrow[\text{Pi}]{\text{Mg}^{2+}} \text{PRPP} + \text{ADP}$$

Enzyme 2 Nucleotide phosphorylase:

$$\text{purine base} + \text{PRPP} \xrightarrow[\text{Pi}]{Mg^{2+}} \text{5}'\text{-purine nucleotide} + \text{PPi}$$

(hypoxanthine, guanine, adenine, xanthine) (IMP, GMP, AMP, XMP)

Overall:

$$\text{ribose-5-P} + \text{ATP} + \text{purine base} \xrightarrow[\text{Pi}]{Mg^{2+}} \text{5}'\text{-purine nucleotide} + \text{ADP} + \text{PPi}$$

These reactions have been applied industrially to produce purine nucleotides from the corresponding purine bases using microorganisms such as *Brevibacterium ammoniages*, as will be described elsewhere in this book.

2.4 Regulation of Purine and Pyrimidine Nucleotide Biosynthesis

Nucleotides are components of nucleic acids and precursors of RNA and DNA. They are also involved in the biosynthesis of various nucleotide-containing coenzymes.

For instance, cAMP acts on various metabolic reactions as an effector. Therefore, the mechanism of regulation of nucleotide biosynthesis is not simple. In considering the microbial production of nucleic acid-related substances, the existence of interconversion reactions must also be taken into account.

Details of the regulation mechanisms of nucleotide biosynthesis have been published in reviews such as those of Blakley[9)] and Momose.[10)] Demain[11)] also discussed this problem in relation to the production of purine nucleotides by fermentation.

2.4.1 Regulation of IMP Biosynthesis

Switzer[12)] reported that PRPP synthetase was inhibited by ADP, ATP, GTP, UTP and tryptophan, which is partially biosynthesized from PRPP. Atkinson *et al.*[13,14)] reported that PRPP synthetase was inhibited more strongly by ADP than by GDP, CDP and tryptophan. Under energy-limited conditions, the ratio of ADP to ATP increases and the formation of PRPP and the rate of subsequent biosynthetic reactions decrease. Atkinson defined a parameter, "energy charge", to describe the ratio of ADP to ATP and showed the importance of "energy charge" in metabolic regulation. Details of this work have been published in a monograph.[15)]

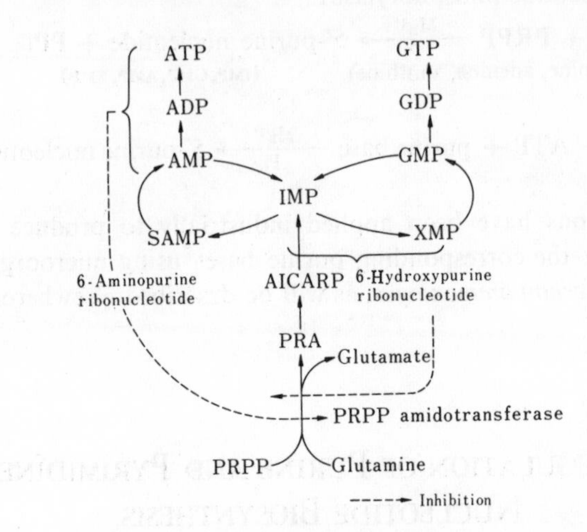

Fig. 2.6. Regulation of IMP biosynthesis.

PRPP amidotransferase catalyzes the first reaction of purine nucleotide biosynthesis, forming PRA from PRPP and glutamine (Fig. 2.6). Wyngaarden[16] found that the enzyme purified from pigeon liver was inhibited by ATP, GMP, AMP, IMP and GDP. The apparent k_i for AMP is 2 mM. Normal Michaelis-Menten kinetics were observed and inhibition by AMP was competitive with respect to PRPP. Purine nucleotides which show inhibition could be divided into two groups. One consists of 6-hydroxypurine nucleotides such as GMP and IMP, and the other consists of 6-aminopurine nucleotides such as ADP and AMP. Mixtures of nucleotides belonging to the same group do not show inhibition greater than the sum of the inhibition due to each nucleotide alone. However, when two nucleotides belonging to different groups (GMP + AMP or IMP + ADP) are present, inhibition increases cooperatively. This phenomenon is called cooperative end product inhibition. Each inhibitor of the two groups binds to a different allosteric site of the enzyme. This type of inhibition is rational considering that the end products of purine metabolism are interconvertible. The purified enzyme from *A. aerogenes* is inhibited most effectively by GMP (apparent k_i, 0.4 mM) and AMP (apparent k_i, 0.55 mM). Inhibition by GMP and AMP is also competitive with respect to PRPP. GTP, ADP and IMP are less effective, while ATP and adenine show no inhibition. As with the pigeon liver enzyme, cooperativity of GMP and AMP is observed. Further, this enzyme is repressed by GMP, AMP and other substances. When adenine and guanine are present in excess, the enzyme is repressed completely. When both adenine and guanine concentrations are limited, repression is released, and when only the guanine concentration is limited, repression is partially released. This situation is appropriate for the conversion of *de novo* biosynthesis to salvage biosynthesis.

2.4.2 Regulation of AMP and GMP Biosynthesis

As shown in Fig. 2.7, AMP and GMP are biosynthesized separately from IMP, and both reactions are practically irreversible. If there is no regulation mechanism, the sequence IMP ⟶ XMP ⟶ GMP ⟶ IMP involves simply loss of ATP. The sequence IMP ⟶ SAMP ⟶ AMP ⟶ IMP also involves only loss of energy due to the deamination of AMP. These cycles are prevented by the presence of the regulation mechanisms shown in Fig. 2.7. IMP dehydrogenase is inhibited by AMP and GMP reductase is inhibited by ATP. GMP also represses IMP dehydrogenase. Similarly, AMP and GTP inhibit SAMP synthetase and AMP deaminase, respectively. Further, it is important for metabolic regulation that the energy donor in the reaction of SAMP to AMP be GTP, while that in the reaction of XMP to GMP be ATP. Due to this regulation mechanism, when the GMP level in the cells increases, the metabolic flow from IMP changes to AMP and *vice versa*.

The metabolism of nucleotides is also related to histidine biosynthesis. The sequence AICARP ⟶ IMP ⟶ AMP ⟶ ATP ⟶ PR-ATP ⟶ AICARP forms a cycle and histidine is synthesized from PR-ATP by way of imidazole glycerophosphate. This cycle plays a catalytic role and during one operation of the cycle, one molecule of imidazole glycerophosphate

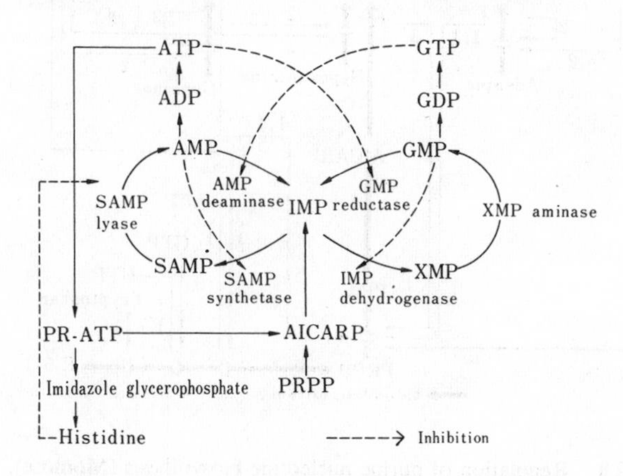

Fig. 2.7. Regulation of AMP and GMP biosynthesis.

is formed from PRPP, aspartate, N^{10}-formyltetrahydrofolate and ammonia without loss of purine nucleotide. Fumarate and tetrahydrofolate are reconverted to asparate and N^{10}-formyltetrahydrofolate, respectively, and are utilized for the synthesis of histidine. Histidine inhibits the conversion of ATP to PR-ATP. If an excess of histidine is present in the medium, the sequence AMP⟶ATP⟶PR-ATP⟶AICARP⟶IMP does not occur.

Regulation mechanisms also operate in salvage biosynthesis. All purine mononucleotide pyrophosphorylases in *B. subtilis* are inhibited by guanine nucleotides, dGTP and GTP. In the presence of an excess of either, the accumulation of the direct precursor of guanine nucleotides, XMP, is inhibited. XMP formation from IMP is also inhibited by GMP, as mentioned previously. Therefore, GMP regulates its own synthesis both by *de novo* and salvage biosynthesis.

Momose *et al.*[17–20] studied the regulation of purine nucleotide biosynthesis in *B. subtilis*. The results on feedback inhibition are summarized in Fig. 2.8. Feedback repression in the same organism is as follows: IMP dehydrogenase is not repressed by adenine, hypoxanthine or xanthinine

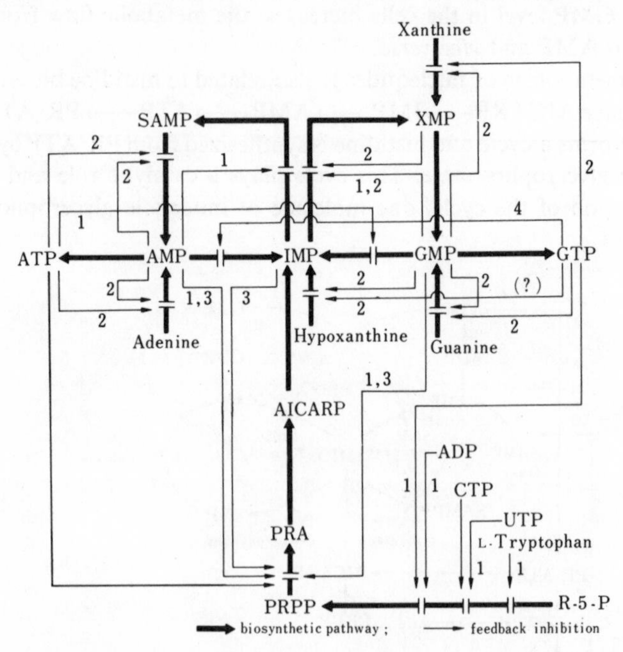

Fig. 2.8. Regulation of purine nucleotide biosynthesis (Momose).

1, Enterobacteriaceae ; 2, B. subtilis ; 3, pigeon liver ; 4, calf brain.

derivatives, but is repressed by guanine derivatives. SAMP synthetase is specifically repressed by adenine derivatives. PRPP amidotransferase, IMP transformylase and adenylosuccinase are repressed strongly by adenine and guanine derivatives, slightly by hypoxanthine derivatives and not at all by xanthine derivatives.

2.4.3 Regulation of Pyrimidine Nucleotide Biosynthesis

Carbamylphosphate synthetase, which catalyzes the formation of carbamylphosphate from NH_3, CO_2 and ATP, is involved not only in the biosynthesis of pyrimidine nucleotides but also in the biosynthesis of arginine and urea, as shown in Fig. 2.9. This enzyme exists in a single form in *E.*

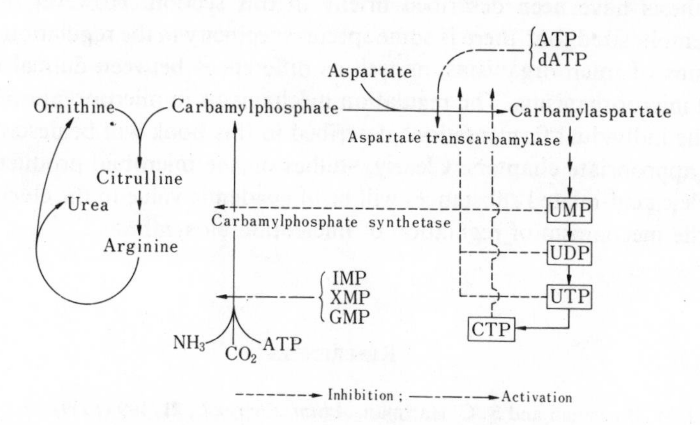

Fig. 2.9. Regulation of pyrimidine nucleotide biosynthesis.

coli but as two isozymes in a strain of *Neurospora.*[21] It is controlled by both arginine and uracil in *E. coli*, but in *Neurospora* sp. arginine and uracil each control only one of the isozymes. The enzyme from *E. coli* is inhibited by uridine nucleotides in the order UMP > UDP > UTP, and is activated by purine nucleotides such as IMP, XMP and GMP. Thus, it is controlled by both negative and positive feedback mechanisms. The enzyme is repressed by orginine and uracil. The percent repression by arginine or uracil is the same whether the second repressor is present or not, i.e., repression is cumulative.

The next reaction in pyrimidine nucleotide biosynthesis is the formation of carbamylaspartate from carbamylphosphate and aspartate. Aspartate transcarbamylase, which catatalyzes this reaction, is historically well known since it led to the discovery of feedback inhibition and repression by Yates and Pardee.[23] Extremely detailed enzymatic studies have been done on this enzyme as an allosteric protein.[24] It is inhibited by CTP and uridine nucleotides such as UTP, UDP and UMP. CTP inhibits the enzyme most strongly. In the case of purine nucleotides, GTP and dGTP weakly inhibit its activity, but ATP and dATP activate it. ATP and dATP eliminate the inhibition by CTP. Gerhart *et al.*[24] separated 11.2S aspartate transcarbamylase into a 5.8S catalytic subunit and a 2.8S regulatory subunit by *p*-hydroxymercuribenzoate treatment, and regenerated native enzyme on warming the subunits together in the presence of mercaptethanol.

2.4.4 Conclusion

The mechanisms of regulation of purine and pyrimidine nucleotide biosynthesis have been described briefly in this section. However, it must be emphasized that there is some species specificity in the regulation mechanisms of microorganisms, as well as differences between animal tissues and microorganisms. The regulation mechanisms in microorganisms used in the individual fermentations described in this book will be described in the appropriate chapters. Clearly, studies on the microbial production of nucleic acid-related substances will be of academic value in the elucidation of the mechanism of regulation of nucleotide biosynthesis.

REFERENCES

1. J. M. Buchanan and S. C. Hartman, *Advan. Enzymol.*, **21**, 199 (1959).
2. A. G. Most, *Bact. Rev.*, **24**, 309 (1960).
3. B. Magasanik, *The Bacteria* (ed. I. G. Gunsalus and R. Y. Stanier), vol. III, p. 295, Academic, 1960.
4. S. C. Hartman, *Metabolic Pathways*, 3rd ed. (ed. D. M. Greenberg), vol. IV, p. 1, Academic, 1970.
5. M. R. Greenberg, *J. Biol. Chem.*, **190**, 611 (1951).
6. G. R. Greenberg, *Arch. Biochem.*, **19**, 337 (1948).
7. H. Nishikawa, H. Momose and I. Shiio, *J. Biochem.* (Tokyo), **63**, 150 (1968).
8. B. Magasanik, *Ann. Rev. Microbiol.*, **11**, 221 (1957).
9. R. L. Blakley and E. Vitol, *Ann. Rev. Biochem.*, **37**, 201 (1968).
10. H. Momose, *Tanpakushitsu Kakusan Koso* (Japanese), **13**, 781 (1968).
11. A. L. Demain, *Progress in Industrial Microbiology*, vol. 8, p. 35, Churchill, 1968.
12. R. L. Switzer, *Fed. Proc.*, **26**, 560 (1697).
13. D. E. Atkinson and L. Fall, *J. Biol. Chem.*, **242**, 3241 (1967).
14. L. Klungsøyr, J. H. Hagerman, L. Fall and D. E. Atkinson, *Biochemistry*, **7**, 4035 (1968).
15. K. Aida, *Hakko to Biseibutsu I* (Japanese), (ed. T. Uemura and K. Aida), p. 32, Kyoritsu Shuppan, 1971.
16. R. J. McCollister, W. R. Gilbert, D. M. Ashton and J. B. Wyngaarden, *J. Biol. Chem.*, **240**, 358 (1964).
17. H. Momose, H. Nishikawa and N. Katsuya, *J. Gen. Appl. Microbiol.*, **11**, 211 (1965).

18. H. Momose, *ibid.*, **13**, 39 (1967).
19. H. Momose, H. Nishikawa and I. Shiio, *J. Biochem.* (Tokyo), **59**, 325 (1966).
20. H. Nishikawa, H. Momose and I. Shiio, *ibid.*, **62**, 92 (1967).
21. R. H. Davis, *Science*, **142**, 1652 (1963).
22. A. Piérad, N. Glansdorff, M. Merglay and J. M. Wisme, *J. Mol. Biol.*, **14**, 23 (1965).
23. R. A. Yates and A. B. Pardee, *J. Biol. Chem.*, **227**, 677 (1957).
24. J. C. Gerhart and A .B. Pardee, *ibid.*, **237**, 891 (1962).
25. J. C. Gerhart and H. K. Schachaman, *Biochemistry*, **4**, 1054 (1965).

CHAPTER 3

Separation and Determination of Nucleic Acid-related Substances*

The methods available for the separation and determination of nucleic acid components have been described in various reviews and books.[1,2] Each method has different advantages and disadvantages: for instance, high-speed liquid chromatography permits the rapid analysis of very small samples, but the equipment is expensive. On the other hand, paper chroma-

* Ikuo SUHARA, Takeda Chemical Industries Ltd.

tography, paper electrophoresis and thin layer chromatography are very simple, and under optimum conditions are very effective.

In this chapter, various methods for the separation and determination of nucleic acid-related compounds will be described.

3.1 Ion Exchange Column Chromatography

Ion exchange chromatography of nucleic acid components has been developed and reviewed by Cohn *et al.*[3–5] and widely used as a reliable and standard method of separation and determination, particularly for the isolation and identification of unknown compounds.

3.1.1 Separation and Determination of Nucleotides

Cohn *et al.*[6] separated 3′(2′)-mononucleotides in alkaline digests of RNA and 5′-mononucleotides in enzymic digests of RNA using a column of Dowex-1 ($HCOO^-$) with formic acid-ammonium formate as an eluant. Bergkvist *et al.*[7,8] modified this method and applied it to the analysis of acid-soluble nucleotides. These methods have been used for the separation and determination of 5′-nucleotides in the enzymic digests of nucleic acids,[9,10] in the culture filtrates of microorganisms,[11,12] in various foods[13] and others.[14] Subsequently, Hurlbert *et al.*[15,16] separated nucleotides by gradient elution with formic acid and ammonium formate; this method has been used for the separation of coenzyme nucleotides[17] and unknown nucleotides. Cohn *et al.* reported the separation of ribonucleotides[18] and deoxyribonucleotides[19] using Dowex-1 or -2 (Cl^-). Although a chloride-form anion exchanger gives poorer separation than formate-form resin, nucleotides can be eluted with dilute hydrochloric acid (0.003 N HCl), and this system is suitable for large-scale separation. Gradient elution with chloride ions has been used for the separation of acid-soluble nucleotides.[17,20]

On the other hand, separation by cation exchange resin has been reported by Cohn,[3] Blattner,[21] and Katz.[22] In this method, 5′-mononucleotides cannot be separated from 2′- or 3′-isomers, but this method has the advantage of being rapid, and has been used for the separation of 5′-mononucleotides,[23] and for analysis of the base composition of RNA.

3.1.2 Separation and Determination of Nucleosides

Under certain conditions, nucleosides form complexes with boric acid to give anions, and are separable on a column of Dowex-1.[24,25] This method was used for the separation of inosine from hypoxanthine in culture filtrates of *Bacillus subtilis*.[26]

3.1.3 Separation and Determination of Bases

Four bases were separated on a Dowex-50 column using 2 N HCl[3] or gradient elution with ammonium formate (pH 4.0),[27] or on a column of Dowex-1 (Cl^-) using NH_4Cl-NH_4OH (pH 10.6).[18]

3.1.4 Separation and Determination of Bases, Nucleosides, and Nucleotides

Anderson *et al.* reported the separation of five bases, five nucleosides and eight nucleotides on a column of Dowex-1 (CH_3COO^-) (1 × 150 cm) by gradient elution with acetate buffer (pH 4.4) within 24 h (Fig. 3.1).[28] In this method, fine particles of resin (smaller than 400 mesh) were used, and a high-pressure mini flow pump was necessary.

This analytical system performed quantitative separation of nucleic acid components with good resolution and speed, and therefore seems to be very suitable for the analysis of nucleic acid-related substances in the culture filtrates of microorganisms and nuclease digests of nucleic acids.

Hori *et al.*[29,30] modified the conditions of gradient elution and were able to separate nucleosides with a higher degree of resolution. Furthermore, Anderson *et al.*[31,32] reported the separation of deoxyribonucleosides and ribonucleoside di- and triphosphates.

These methods have been used in the analysis of enzymic and alkaline digests[33] of RNA, and in the separation and quantitative determination of inosine and guanosine produced by microorganisms.[34]

3.1.5 Separation and Determination of Oligonucleotides

Oligonucleotides of low molecular weight are separable on a column of Dowex-1,[35] but those of high molecular weight are not, because of nonpolar and irreversible adsorption onto the resin. Staehelin *et al.*[36,37] used DEAE-cellulose for the separation of RNase digests of RNA, and Tomlinson *et al.*[38,39] added 7 M urea to the elution buffer to prevent secondary binding between oligonucleotides. Under these conditions, oligonucleotides are bound electrostatically via phosphate anions to the quaternary ammonium groups on the cellulose and their order of elution is a function of their net negative charge.

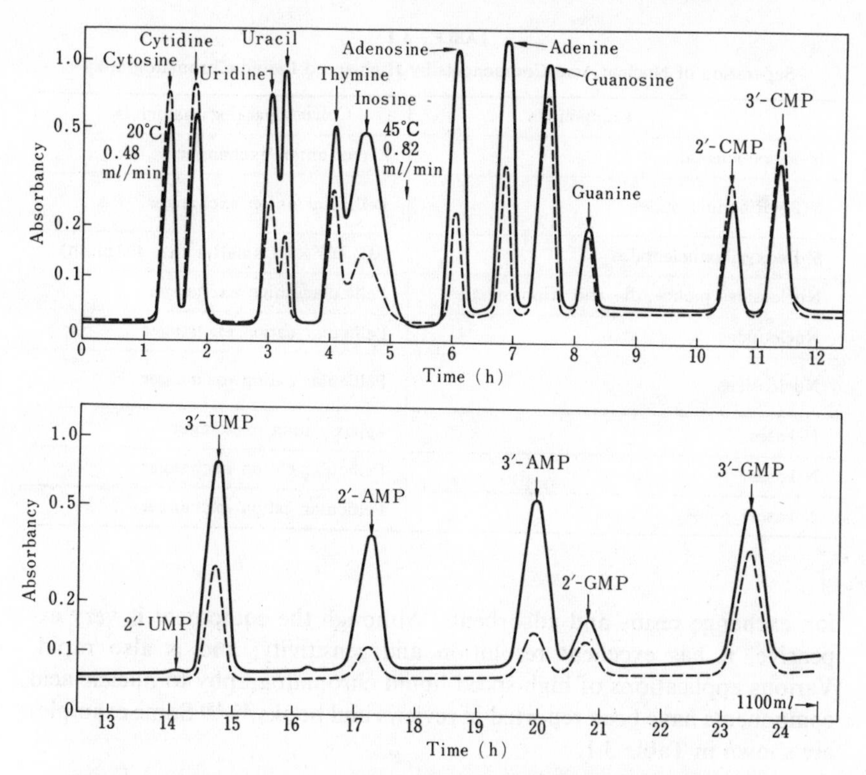

Fig. 3.1. Separation of a mixture of ribonucleotides, nucleosides, and bases by anion exchange. Exchanger: Dowex-1 × 8 (acetate), less than 400 mesh, 0.9 × 150 cm. Eluting solution: linear gradient from 0.6 to 2 M sodium acetate, pH 4.4. Flow and temperature: 0.48 ml/min at 20°C, changing to 0.82 ml/min at 45°C. Absorbancy is shown at 260 (solid line) and 280 nm (dashed line).[28]
(Source: ref. 28. Reproduced by kind permission of Elsevier Scientific Publishing Co., The Netherlands.)

This method has been applied to the separation of limit digests of nucleic acids by nuclease.[40–42]

3.2 HIGH-SPEED LIQUID CHROMATOGRAPHY

Since Horvath, Preiss, and Lipsky[43] described the high-pressure separation of nucleotides in 1967, high-speed liquid chromatography has made considerable progress with the development of instrumentation,

TABLE 3.1

Separation of Nucleic Acid Components by High-speed Liquid Chromatography

Compounds	Column packing materials
5′-Ribonucleotides	Zypax anion exchanger
3′(2′)-Ribonucleotides	Pellicular anion exchanger
5′-Deoxyribonucleotides	AG 50W×4 (smaller than 400 mesh)
Nucleosides mono-, di- and triphosphates	Pellicular anion exchanger
Nucleosides	Pellicular cation exchanger
Nucleosides	Pellicular cation exchanger
N-bases	Zypax cation exchanger
N-bases	Pellicular cation exchanger
N-bases	Pellicular cation exchanger

ion exchange resins and adsorbents. Although the equipment is very expensive, it has excellent resolution and sensitivity, and is also rapid. Various applications of high-speed liquid chromatography to nucleic acid components have been reported in reviews and books.[44,45] Some examples are shown in Table 3.1.

3.3 ENZYMATIC DETERMINATION OF NUCLEIC ACID-RELATED SUBSTANCES

3.3.1 Purine Derivatives

Kalckar reported methods of enzymatic determination of purine compounds based on ultraviolet spectrophotometry.[49–51] In these methods, purine compounds are determined specifically by utilizing the specificity of enzymes, and a small amount of sample may suffice for an analysis. Therefore, it is possible to determine purine compounds in culture filtrates of microorganisms or in nuclease digests of nucleic acids without purification or concentration. However, it is necessary to carry out a recovery experiment in order to ensure that the enzyme preparations employed are functioning properly under the test conditions.

Methods of enzymatic determination of adenine compounds are

TABLE 3.1—*Continued*

Column size	Carrier	Temperature, Pressure	Ref.
2.1mm×100cm	0.06M H_3PO_4–0.002M KH_2PO_4	60°C, 900 p.s.i.	46)
1mm×300cm	0.01M KH_2PO_4 (pH 3.35) 1M KH_2PO_4 (pH 4.3)	80°C, 500 p.s.i.	45)
8mm× 85cm	0.25M ammonium formate		21)
1mm×193cm	0.04M–1.5M ammonium formate	71°C, 51 atm	47)
1mm×152cm	0.02M $NH_4H_2PO_4$ (pH 5.6)	39°C, 131 atm	48)
2.4mm× 25cm	0.4M ammonium formate (pH 4.75)	55°C, 4400 p.s.i.	45)
2.1mm×100cm	0.01M HNO_3	63°C, 735 p.s.i.	46)
1mm×250cm	0.025M $NH_4H_2PO_4$ (pH 4.4)	70°C, 2800 p.s.i.	44)
1mm×300cm	0.05M $NH_4H_2PO_4$ (pH 2.5)	70°C, 1200 p.s.i.	44)

TABLE 3.2

Enzymatic Determination of Adenine Compounds

Compounds	Enzymes	Assay Method	Ref.
Adenosine, Deoxy-adenosine	Adenosine deaminase (calf intestinal mucosa)	Adenosine → Inosine ΔE_{265}nm	50, 51)
AMP	5′-Nucleotidase (bull semen) Adenosine deaminase	AMP → Adenosine → Inosine ΔE_{265}nm	50, 52)
	AMP deaminase (rabbit muscle)	AMP → IMP ΔE_{265}nm	50, 53)
ADP	Myokinase (muscle) Adenylic deaminase	2ADP ⇌ ATP+AMP AMP → IMP ΔE_{265}nm	50, 51)
	Pyruvate kinase (rabbit muscle) Lactate dehydrogenase	ADP+PEP ⇌ ATP+pyruvate Pyruvate+NADH+H^+ ⇌ Lactate+NAD^+ ΔE_{340}nm	54)
ATP	Luciferase (firefly)	ATP+Mg^{2+}+O_2+luciferin → Light	55, 56)
	Hexokinase (yeast) Glucose-6-phosphate dehydrogenase (yeast)	ATP+glucose → Glucose-6-phosphate +ADP Glucose-6-phosphate+NADP → 6-Phosphogluconate+NADPH+H^+ ΔE_{340}nm	57)
cAMP	Binding protein (protein kinase)	Competitive protein binding method	58, 59)

TABLE 3.3

Enzymatic Determination of Hydroxypurine Compounds

Compounds	Enzymes	Assay Method	Ref.
Hypo-xanthine Xanthine	Xanthine oxidase (raw cream)	Hypoxanthine → Xanthine → Uric acid ΔE_{290}nm	49, 51)
Inosine	Nucleoside phosphorylase (rat liver) Xanthine oxidase	Inosine → Hypoxanthine → Xanthine → Uric acid ΔE_{290}nm	49, 51)
IMP	5′-Nucleotidase Nucleoside phosphorylase Xanthine oxidase	IMP → Inosine → Hypoxanthine →→ Uric acid ΔE_{290}nm	49, 51)
Guanine	Guanase (rat liver) Xanthine oxidase	Guanine → Xanthine → Uric acid ΔE_{290}nm	49, 51)
Guanosine	Nucleoside phosphorylase	Guanosine → Guanine (phosphate and arsenate buffer) Color reaction with Folin phenol reagent	34)

shown in Table 3.2, and those for hypoxanthine and guanosine derivatives are shown in Table 3.3. These enzymatic analyses have been used for screening 5′-phosphodiesterase in the culture filtrates of microorganisms[60,61] and for the quantitative determination of inosine[26,62] or guanosine[34] derivatives produced by microorganisms.

Grassl[63] reported a method for the determination of GMP using GMP kinase, pyruvate kinase and lactate dehydrogenase. Cha *et al.*[64] reported a method for the differential determination of GMP, GDP and GTP using a succinate thiokinase-pyruvate kinase system.

3.3.2 Other Methods of Enzymatic Determination

5′-Mononucleotides can be determined by following the amount of inorganic phosphate liberated by the action of 5′-nucleotidase from bull semen.[52,65–67].

3′-Mononucleotides can be determined using nucleoside cyclic phosphate diesterase.[68–70] Nonspecific phosphomonoesterase[71] from human prostate gland has been used for determination of the terminal phosphate of oligonucleotides[72] and mononucleotides.

3.4 PAPER CHROMATOGRAPHY

Since the separation of nucleic acid components by paper chromatography was first demonstrated by Vischer and Chargaff,[73] progress has been rapid and this method has come into general use. It has the advantage of being rapid and simple, although it is less suitable for the isolation of materials in quantity.

The general procedures and solvent systems have been described in several reviews and books.[74,75] With low concentrations of nucleic acid components, or if high concentrations of salts are present in the samples, concentration or desalting of the samples by active charcoal treatment is necessary. Whatman No. 1 paper is suitable for the separation and determination of nucleic acid-related substances in the culture filtrate of microorganisms. If it is difficult to visualize a UV spot due to overlapping fluorescent substances, Lumicolor plates can be used (Wako Pure Chemical Ind. Co.). Guanosine derivatives can be discriminated from other compounds by intense blue fluorescense after exposure to conc. HCl vapor and by disappearance of the fluorescense after exposure to conc. NH_4OH vapor. For quantitative determination, the UV spot is cut out, eluted with 0.01 or 0.1 N HCl and its absorbance at 260 nm or at λ_{max} measured. The filter paper contains a small amount of UV-absorbing material which can be eluted with dilute acid and can thus be removed by prior washing of the paper with a chromatographic solvent and water or dilute HCl and water. Otherwise, appropriate blanks can be developed.

Vischer *et al.*[76] measured the absorbance at 260 nm (or 250 nm) and 290 nm and calculated the contents from the difference. When larger quantities of material are to be separated, Whatman No. 3MM paper is suitable.

For the separation of four bases, isopropanol-HCl[77] or methanol-HCl[78] has been used. Kirby described the solvent system methanol: ethanol:conc. HCl:H_2O (50:25:6:19) for the separation of four nucleotides.[78] This system was used for the separation of enzymatic digests[79] (5′-mononucleotides) of RNA. Lane developed a solvent system of 75% ethanol using paper dipped in 10% saturated ammonium sulfate for the separation of nucleosides and 2′(3′)-nucleotides.[80] The use of saturated ammonium sulfate, water and isopropanol[81] (79:19:2) or saturated ammonium sulfate, 1 N sodium acetate and isopropanol[82] (80:18:2) is suitable for the separation of 2′-, 3′- and cyclic nucleotides of adenine or guanine. This solvent system has been used for the separation of T_1 type RNase digests.[83,84]

Isobutyric acid-ammonia,[85,86] isobutyric acid-acetic acid-ammonia,[87] *n*-butanol-acetic acid-water,[88] or two-dimensional systems[89,90] have also been used as solvent systems for paper chromatography in the analysis of the microbial production of inosine.

3.5 Paper Electrophoresis

Nucleic acids and their components bear a variety of ionizable groups and it is therefore possible to separate and determine these derivatives based on the differences of net charge at a given pH. This manipulation can be carried out in a short time and using a very small amount of material. Various electrophoretic conditions have been reported in several reviews.[91,94] The separation of four nucleotides can be carried out at pH 3.5 in 0.1 M formate or acetate buffer for 20–30 min at 100 volt/cm.[94] The separation of IMP, GMP, UMP, and CMP can be carried out using 0.1 M formate buffer (pH 2.5) for 30–40 min at 50–100 volt/cm.[92,93] Four bases can be separated at pH 1.8 for 30 min at 100 volt/cm.[94]

Borate buffer (pH 9.2–9.5) is suitable for the separation of 3′(2′)-nucleotides from 5′-nucleotides, for the separation of ribonucleotides from deoxyribonucleotides[94] or for the separation of bases, nucleosides and nucleotides. For the separation of cyclic nucleotides from non-cyclic nucleotides, phosphate buffer (pH 7.4) has been used.[82,91,95] These methods were employed for the analysis[93] of enzymic and alkaline digests of yeast RNA, and for the analysis of inosine[26,62,96] and guanosine[34] in culture filtrates of microorganisms. Kuninaka *et al.*[97] separated four nucleotides in nuclease digests by paper electrophoresis in combination with paper chromatography during screening for 5′-phosphodiesterase produced by microorganisms.

3.6 Thin Layer Chromatography

Thin layer chromatography of nucleic acid components has been developed by Randerath.[98] In view of its simplicity, sensitivity, high resolution and reduced tailing, thin layer chromatography has become a powerful and widely used analytical tool. The analytical method and developing solvents have been described in several reviews.[99–102]

Coffey *et al.*[103] succeeded in the separation of ribonucleosides, deoxyribonucleosides and DNA bases on a thin layer of MN-cellulose 300/DEAE using 0.005 N HCl as a developing solvent. Furthermore, RNA bases can be separated using a mixture of saturated ammonium sulfate:1 N sodium acetate:isopropanol (80:18:2).

Randerath[99] separated ATP, ADP and AMP on a thin layer of DEAE-cellulose, using 0.02–0.03 N HCl as a developing solvent. This system is available for the separation of other nucleoside mono-, di, and triphosphates.

3′(2′)-Ribonucleotides can be separated on a thin layer of MN-cellulose 300/DEAE, using isobutyric acid:conc. NH_4OH:water[103] (66:2:32). Yoshimoto *et al.*[104] used a thin layer of DEAE-cellulose with ammonium carbonate:acetone:water for the separation and determination of 3′:5′-cyclic AMP in a culture filtrate of *Brevibacterium liquefaciens*.

3.7 Color Reactions

Dische[105] and Schmidt *et al.*[106] have reviewed the methods of determination of nucleic acid components by color reactions. Guanine and xanthine, but not adenine or hypoxanthine, react with the Folin phenol reagent to give a blue color.[107] This reaction has been applied to the quantitative determination[34] of guanosine in combination with nucleoside phosphorylase, which is specific for guanosine and inosine. Adenine[108] and hypoxanthine[109] react with *N*-1-naphthyl ethylenediamine hydrochloride after reduction with zinc dust and diazotization with $NaNO_2$ to give a red color in an analytical procedure.

Ribose reacts with orcinol reagent to give a green color[110,111] and deoxyribose reacts with diphenylamine to give a blue color.[111,112] Both reactions have been applied to the quantitative determination of RNA or DNA, but there is a difference in the reactivity of purine nucleotides and pyrimidine nucleotides with the reagent.

AICA and AICAR react with the Bratton-Marshall reagent to give a red color.[113,114] Buchanan *et al.*[115] applied this reaction to the quantitative determination of SAICARP. The reaction was carried out in the cold. As SAICAR and AICAR both react with the Bratton-Marshall reagent in the cold, the amount of AICAR should be substracted for the determination of SAICAR.

3.8 Separation by Active Charcoal

Nucleic acid components are readily adsorbed on active charcoal in acidic solution and can be eluted with acetone-water, alkaline methanol, alkaline ethanol or pyridine-water. However, various salts, amino acids, phosphoric esters of sugars and proteins are not adsorbed on active charcoal. Therefore, samples containing nucleic acid components can be desalted, concentrated and partially purified prior to column chromatography, paper chromatography or paper electrophoresis. This method has been used in the determination of nucleotides in milk,[116] various foods,[13, 65] culture filtrates[11] of microorganisms, and for the isolation of inosine[117] and xanthosine[118] from culture filtrates of microorganisms.

3.9 Miscellaneous Methods of Determination

Lee[119] and Pratt *et al.*[120,121] reported a computer method for the determination of the base compositions of oligo- and polynucleotides.

Hoff-Jørgensen[122,123] has developed a bioassay method for deoxyribonucleosides using *Thermobacterium acidophilus*, which requires deoxyribonucleoside as an essential growth factor.

References

1. Ed. E. Chargaff and J. N. Davidson, *The Nucleic Acids*, vol. 1, Academic, 1955.
2. Ed. S. P. Colowick and N. O. Kaplan, *Methods in Enzymology*, vol. 3, p. 671, Academic, 1957.
3. W. E. Cohn, *Science*, **109**, 377 (1949).
4. W. E. Cohn, *The Nucleic Acids* (ed. E. Chargaff *et al.*), vol. 1, p. 211, Academic, 1955.
5. W. E. Cohn, *Methods in Enzymology* (ed. S. P. Colowick *et al.*), vol. 3, p. 724, Academic, 1957.
6. W. E. Cohn and E. Volkin, *Nature*, **167**, 483 (1951).
7. R. Bergkvist and A. Deutsch, *Acta Chem. Scand.*, **8**, 1877 (1954).
8. R. Bergkvist, *ibid.*, **10**, 1303 (1956).
9. Y. Nakao and K. Ogata, *Agr. Biol. Chem.*, **27**, 499 (1963).
10. H. Tone, M. Umeda, T. Ishikura and N. Miyaji, *Amino Acid and Nucleic Acid*, no. 10, 142 (1964).
11. K. Ogata, A. Imada and Y. Nakao, *Agr. Biol. Chem.*, **26**, 586 (1962).
12. Y. Tsukada and T. Sugimori, *ibid.*, **28**, 471 (1964).
13. N. Nakajima, K. Ichikawa, M. Kamada and E. Fujita, *Nippon Nogeikagaku Kaishi* (Japanese), **35**, 797, 803 (1961).
14. A. Kakinuma and S. Igarasi, *Agr. Biol. Chem.*, **28**, 131 (1964).
15. R. B. Hurlbert, H. Schmitz, A. F. Brumm and V. R. Potter, *J. Biol. Chem.*, **209**, 23 (1954).
16. R. B. Hurlbert, *Methods in Enzymology*, (ed. S. P. Colowick *et al.*), vol. 3, p. 793, Academic, 1957.

17. S. Suzuki, *Seikagaku* (Japanese), **35**, 737 (1963).
18. W. E. Cohn, *J. Am. Chem. Soc.*, **72**, 1471 (1950).
19. E. Volkin, J. X. Khym and W. E. Cohn, *ibid.*, **73**, 1533 (1951).
20. J. L. Strominger, *Biochim. Biophys. Acta*, **17**, 283 (1955).
21. F. R. Blattner and H. P. Erickson, *Anal. Biochem.*, **18**, 220 (1967).
22. S. Katz and D. G. Comb, *J. Biol. Chem.*, **238**, 3065 (1963).
23. K. Arai, T. Tanaka and T. Saito, *Eiyo to Shokuryo* (Japanese), **17**, 278 (1964).
24. J. X. Khym and W. E. Cohn, *The Nucleic Acids* (ed. E. Chargaff *et al.*), vol. 1, p. 237, Academic, 1955.
25. K. S. McCully and G. L. Cantoni, *J. Biol. Chem.*, **237**, 3760 (1962).
26. I. Nogami and S. Igarasi, *J. Takeda Res. Lab.* (Japanese), **22**, 99 (1963).
27. C. F. Crampton, F. R. Frankel, A. M. Benson and A. Wade, *Anal. Biochem.*, **1**, 249 (1960).
28. N. G. Anderson and S.F.C. Ladd, *Biochim. Biophys. Acta*, **55**, 275 (1962).
29. M. Hori and E. Konishi, *J. Biochem.* (Tokyo), **56**, 375 (1964).
30. M. Hori, *Methods in Enzymology* (ed. L. Grossman *et al.*), vol. 12, p. 381, Academic, 1967.
31. N. G. Anderson, *Anal. Biochem.*, **4**, 269 (1962).
32. N. G. Anderson, J. G. Green, M. L. Barker, Sr. and F. C. Ladd, *ibid.*, **6**, 153 (1963).
33. I. Suhara, F. Kusaba-Kawashima, Y. Nakao, M. Yoneda and E. Ohmura, *J. Biochem.* (Tokyo), **71**, 941 (1972).
34. I. Nogami, M. Kida, T. Iijima and M. Yoneda, *Agr. Biol. Chem.*, **32**, 144 (1968).
35. E. Volkin and W. E. Cohn, *J. Biol. Chem.*, **205**, 767 (1953).
36. M. Staehelin, E. A. Peterson and H. A. Sorber, *Arch. Biochem. Biophys.*, **85**, 289 (1959).
37. M. Staehelin, *Biochim. Biophys. Acta*, **49**, 11 (1961).
38. R. V. Tomlinson and G. M. Tener, *J. Am. Chem. Soc.*, **84**, 2644 (1962).
39. G. M. Tenner, *Methods in Enzymology* (ed. L. Grossman *et al.*), vol. 12, p. 398, Academic, 1967.
40. S. Y. Lee, Y. Nakao and R. M. Bock, *Biochim. Biophys. Acta*, **151**, 126 (1968).
41. I. Suhara, *J. Biochem.* (Tokyo), **73**, 1023 (1973).
42. M. Fujimoto, K. Fujiyama, A. Kuninaka and H. Yoshino, *Agr. Biol. Chem.*, **38** 2141, (1974).
43. C. Horvath, B. Preiss and S. R. Lipsky, *Anal. Chem.*, **39**, 1422 (1967).
44. C. A. Burtis and D. R. Gere, *Nucleic Acid Constituents by Liquid Chromatography*, Varian Aerograph.
45. D. R. Gere, *Modern Practice of Liquid Chromatography* (ed. J. J. Kirkland), p. 417, Interscience, 1971.
46. J. J. Kirkland, *J. Chromatogr. Sci.*, **8**, 72 (1970).
47. C. G. Horvath, B. A. Preiss and S. R. Lipsky, *Anal. Chem.*, **39**, 1442 (1967).
48. C. G. Horvath and S. R. Lipsky, *ibid.*, **41**, 1227 (1969).
49. H. M. Kalckar, *J. Biol. Chem.*, **167**, 429 (1947).
50. H. M. Kalckar, *ibid.*, **167**, 445 (1947).
51. H. M. Kalckar, *ibid.*, **167**, 461 (1947).
52. L. A. Heppel and R. J. Hilmoe, *ibid.*, **188**, 655 (1951).
53. Y. P. Lee, *Methods in Enzymology* (ed. S. P. Colowick *et al.*), vol. 6, p. 102, Academic, 1963.
54. D. Jaworek, W. Gruber and H. U. Bergmeyer, *Methods of Enzymatic Analysis*, 2nd. ed. (ed. H. U. Bergmeyer), vol. 4, p. 2127, Academic, 1974.
55. B. L. Strehler and J. R. Totter, *Methods of Biochemical Analysis* (ed. D. Glick), vol. 1, p. 341, Interscience, 1954.
56. B. L. Strehler and W. D. McElroy, *Methods in Enzymology* (ed. S. P. Colowick *et al.*), vol. 3, p. 871, Academic, 1957.
57. A. Kornberg, *J. Biol. Chem.*, **182**, 779 (1950).
58. A. G. Gilman, *Proc. Natl. Acad. Sci. U.S.A.*, **67**, 367 (1970).

59. G. M. Walton and L. D. Garren, *Biochemistry*, **9**, 4223 (1970).
60. E. Ohmura, Y. Sugino, M. Yoneda, I. Suhara, K. Ogata, Y. Nakao, S. Igarasi, K. Tanabe, A. Kakinuma and A. Imada, *2nd. Mtg. Kanto Division of Agr. Chem. Soc. Japan*, Tokyo, 1960.
61. K. Ogata, Y. Nakao, S. Igarasi, E. Ohmura, Y. Sugino, M. Yoneda and I. Suhara, *Agr. Biol. Chem.*, **27**, 110 (1963).
62. T. Suzuki, I. Nogami, Y. Kitahara, M. Ishikawa and M. Yoneda, *J. Takeda Res. Lab.* (Japanese), **26**, 126 (1967).
63. M. Grassl, *Methods of Enzymatic Analysis*, 2nd. ed. (ed. H. U. Bergmeyer), vol. 4, p. 2162, Academic, 1974.
64. S. Cha and Ch. J. M. Cha, *Anal. Biochem.*, **33**, 174 (1970).
65. N. Nakajima, K. Ichikawa, I. Yoshimura, C. Kuriyama, M. Kamada and E. Fujita, *Nippon Nogeikagaku Kaishi* (Japanese), **37**, 558 (1963).
66. G. Schmidt, *Methods in Enzymology* (ed. S. P. Colowick *et al.*), vol. 3, p. 766, Academic, 1957.
67. L.A. Heppel and R. J. Hilmoe, *ibid.*, vol. 2, p. 546, Academic, 1955.
68. S. Igarasi and A. Kakinuma, *Agr. Biol. Chem.*, **26**, 218 (1962).
69. K. Shimada and Y. Sugino, *Biochim. Biophys. Acta*, **185**, 367 (1969).
70. Y. Anraku, *J. Biol. Chem.*, **239**, 3412, 3420 (1964).
71. S. E. Kerr and F. Chernigoy, *ibid.*, **228**, 495 (1957).
72. G. Schmidt, *Methods in Enzymology* (ed. S. P. Colowick *et al.*), vol. 3, p. 761, Academic, 1957.
73. E. Vischer and E. Chargaff, *J. Biol. Chem.*, **168**, 781 (1947).
74. G. R. Wyatt, *The Nucleic Acids* (ed. E. Chargaff *et al.*), vol. 1, p. 243, Academic, 1955.
75. R. Markham, *Methods in Enzymology* (ed. S. P. Colowick *et al.*), vol. 3, p. 743, Academic, 1957.
76. E. Vischer and E. Chargaff, *J. Biol. Chem.*, **176**, 703 (1948).
77. G. R. Wyatt, *Biochem. J.*, **48**, 584 (1951).
78. K. S. Kirby, *Biochim. Biophys. Acta*, **18**, 575 (1955).
79. Y. Fujimura, Y. Hasegawa, Y. Kaneko and S. Doi, *Agr. Biol. Chem.*, **31**, 92 (1967).
80. B. G. Lane, *Biochim. Biophys. Acta*, **72**, 110 (1963).
81. R. Markham and J. D. Smith, *Biochem. J.*, **49**, 401 (1951).
82. R. Markham and J. D. Smith, *ibid.*, **52**, 558 (1952).
83. K. Tanaka, *J. Biochem.* (Tokyo), **50**, 62 (1961).
84. K. Sato and F. Egami, *ibid.*, **44**, 753 (1957).
85. B. Magasanik, E. Vischer, R. Doniger, D. Elson and E. Chargaff, *J. Biol. Chem.*, **186**, 37 (1950).
86. K. Nakayama, T. Nara, T. Suzuki, Z. Sato, M. Misawa and S. Kinoshita, *Amino Acid and Nucleic Acid*, no. 8, 81 (1963).
87. K. Nakayama, T. Suzuki, Z. Sato and S. Kinoshita, *J. Gen. Appl. Microbiol.*, **10**, 133 (1964).
88. R. Aoki, H. Momose, Y. Kondo, N. Muramatsu and T. Tsuchiya, *ibid.*, **9**, 387 (1963).
89. K. Uchida, A. Kuninaka, H. Yoshino and M. Kibi, *Agr. Biol. Chem.*, **25**, 804 (1961).
90. G. D. Dorough and D. L. Seaton, *J. Am. Chem. Soc.*, **76**, 2873 (1954).
91. J. D. Smith, *The Nucleic Acids* (ed. E. Chargaff *et al.*), vol. 1, p. 267, Academic, 1955.
92. N. Nakajima, *Bull. Jap. Soc. Sci. Fisher.* (Japanese), **32**, 155 (1966).
93. Y. Nakao and K. Ogata, *Agr. Biol. Chem.*, **27**, 291 (1963).
94. T. Tsumita, *Tanpakushitsu Kakusan Koso* (Japanese), **9**, 213 (1964).
95. R. Markham and J. D. Smith, *Biochem. J.*, **52**, 552 (1952).
96. M. Fujimoto and K. Uchida, *Agr. Biol. Chem.*, **29**, 291 (1963).
97. A. Kuninaka, S. Otuka, Y. Kobayashi and K. Sakaguchi, *Bull. Agr. Chem. Soc.*, **23**, 239 (1959).
98. K. Randerath, *Angew. Chem.*, **73**, 436 (1961).

99. K. Randerath, *ibid.*, **74**, 484 (1962).
100. T. Hashizume, *Kagaku* (Japanese), **18**, 275 (1963).
101. M. Honjo, *Kagaku no Ryoiki* (Japanese), suppl. 64: Hakuso Chromatography, vol. 2, p. 1, Nankodo, 1964.
102. H. K. Mangold, *New Biochemical Separations* (ed. A. T. James *et al.*), p. 786, Van Nostrand, 1964.
103. R. G. Coffey and R. W. Newburgh, *J. Chromatogr.*, **11**, 376 (1963).
104. A. Yoshimoto, M. Ide and T. Okabayashi, *Amino Acid and Nucleic acid*, no. 10, 124 (1964).
105. Z. Dische, *The Nucleic Acids* (ed. E. Chargaff *et al.*), vol. 1, p. 285, Academic, 1955.
106. G. Schmidt, *Methods in Enzymology* (ed. S. P. Colowick *et al.*), vol. 3, p. 775, Academic, 1957.
107. G. H. Hitchings, *J. Biol. Chem.*, **139**, 843 (1941).
108. D. L. Woodhouse, *Arch. Biochem.*, **25**, 347 (1950).
109. S. Friedman and J. S. Gots, *Arch. Biochem. Biophys.*, **39**, 254 (1952).
110. W. Mejbaum, *Hoppe-Seyler's Z. Physiol. Chem.*, **258**, 117 (1939).
111. W. C. Schneider, *J. Biol. Chem.*, **161**, 293 (1945).
112. K. Burton, *Biochem. J.*, **62**, 315 (1956).
113. A. C. Bratton and E. K. Marshall, Jr., *J. Biol. Chem.*, **128**, 537 (1939).
114. J. M. Ravel, R. E. Eakin and W. Shive, *ibid.*, **172**, 67 (1948).
115. L. N. Lukens and J. M. Buchanan, *ibid.*, **234**, 1791 (1959).
116. A. Kobata, Z. Suzuoki and M. Kida, *J. Biochem.* (Tokyo), **51**, 277 (1962).
117. T. Nara, M. Misawa, K. Nakayama and S. Kinoshita, *Amino Acid and Nucleic Acid*, no. 8, 94 (1963).
118. K. Nakayama, T. Suzuki, Z. Sato and S. Kinoshita, *ibid.*, no. 8, 88 (1963).
119. S. Lee, D. McMullen, G. L. Brown and A. R. Stokes, *Biochem. J.*, **94**, 314 (1965).
120. A. W. Pratt, J. N. Total, G. W. Rushizky and H. A. Sober, *Biochemistry*, **3**, 1831 (1964).
121. W. Guschlbauer, E. G. Richards, K. Beurling, A. Adams and J. R. Fresco, *ibid.*, **4**, 964 (1965).
122. E. Hoff-Jørgensen, *Biochem. J.*, **50**, 400 (1952).
123. E. Hoff-Jørgensen, *Methods in Enzymology* (ed. S. P. Colowick *et al.*), vol. 3, p. 781, Academic, 1957.

PART II

PRODUCTION AND HYDROLYSIS OF NUCLEIC ACIDS

CHAPTER **4**

Production of RNA*

The production of flavor-enhancing ribonucleoside 5′-phosphates from RNA by enzymatic degradation was independently established by two groups, Kuninaka, Sakaguchi *et al.* and Omura, Ogata *et al.* This has encouraged the development of an efficient method for the production of microbial RNA by culture techniques. It was already known that microorganisms active in protein synthesis were rich in RNA and were better sources of RNA than plants and animals.

Microbial RNA[1–3] can be divided on the basis of its biochemical properties into three main groups, mRNA, tRNA and rRNA, each of which has a particular function in the biosynthesis of proteins. The existence of RNA in mitochondria has recently been reported.

The proportions of mRNA, tRNA and rRNA in total RNA are usually 5%, 10–15% and 75–80%, respectively. mRNA and rRNA exist as complexes with proteins, but tRNA exists in a free form.

Many methods for the extraction and purification of RNA have been devised in order to prepare bulk RNA as a reagent for biochemical research. Crestfield's method,[4] in which a surfactant is used, is thought to

---

* Kiyoshi WATANABE, Kanegafuchi Chemical Industrial Co. Ltd.

cause some degradation of the RNA in the case of yeasts. On the other hand, Osawa's method[5] is designed to avoid such degradation. Generally, bulk RNA is composed of 24S, 17S and 4S RNA as the main components; 24S, 17S particles derived from rRNA account for a large proportion of the total.

Many reports on the RNA content of microorganisms have been published since the 1940's and great variations have been observed: 0.3–51% in the case of bacteria, 2.7–11% in the case of yeasts, and 0.7–28% in the case of molds, as reviewed by Ogata.[6] A large amount of data on the biosynthesis of intracellular RNA has accumulated since 1960, when the industrial production of nucleotides by fermentation became feasible.

4.1 Formation and Accumulation of RNA in Bacteria

The biosynthesis of high molecular RNA in bacterial cells has been studied cytologically and biochemically in relation to the biosynthesis of protein and DNA. Cohen[7] reported that RNA biosynthesis began soon after the inoculation of microorganisms, and Belozersky[8] showed that the RNA content of growing cells was higher than that of resting cells. Malmgren *et al.*[9] reported that the maximum RNA content was obtained at the end of the lag phase. From the work of Morse *et al.*,[10] using *Escherichia coli* labeled with ^{32}P, it became clear that most RNA was synthesized in the lag phase. Wade *et al.*[11] examined 12 species of bacteria and showed that the RNA content in growing cells was 1.3 to 16.5 times that in resting cells, but no clear relationships between growth rate and RNA content were found among the different species (Table 4.1). The ratio of the four ribonucleotides was not affected by active cell division. Taking the AMP content as 1.0, the GMP was 1.31–1.36, UMP 0.93–0.99 and CMP 0.87–1.01.

Neidhardt *et al.*[12] studied the mechanisms of macromolecular biosynthesis by a strain of *Aerobacter aerogenes* which required guanidine and amino acid. The RNA content was high in rapidly growing cells, but the ratio of tRNA and high molecular RNA (particulate RNA) had no relationship to the growth rate. At the growth temperature, the ratio of RNA to protein was a direct function of the growth rate, and the formation of RNA and protein was proportional to the growth rate. On a change from high to low growth rate, there was a very marked decrease of RNA biosynthesis but the decreases of DNA and protein biosynthesis were slight.

In the opposite case, from low to high growth rate, the increase of

TABLE 4.1

RNA Contents of Various Bacteria[11]

Organism	Medium†	Doubling time (min)	RNA/dry wt (%) Resting cells (1)	RNA/dry wt (%) Dividing cells (2)	RNA ratio (2)/(1)
Aerobacter aerogenes	CCY	30.9	4.4	26.6	6.0
Bacillus anthracis	TMB	48	1.45	24	16.5
Bacillus cereus	TMB	23	3.5	31.5	9.0
Chromobacterium prodigiosum	CCY	28.5	7.8	32.1	4.1
Chromobacterium violaceum	TMB	60	7.2	30.3	4.2
Corynebacterium hoffmannii	CCY	128	25.4	51	2.0
E. coli	CCY	27.3	15.6	34.5	2.2
Proteus vulgaris	TMB	27.3	12.6	35	2.8
Staphylococcus aureus	CCY	30	5.2	10	1.9

† CCY, Casein-casein-yeast medium; TMB, tryptic-meat broth medium.

RNA biosynthesis was rapid but those of DNA and protein were rather slow. At any growth rate, the RNA consisted of 14% tRNA and 86% rRNA. The formation of RNA was strongly affected by changes of the physiological culture conditions, and varied by up to three times under some conditions. The biosynthesis of RNA decreased at the early stationary phase, and it was considered that the consumption of organic or inorganic growth factors, exhaustion of the main carbon or energy sources, limitation of oxygen, or accumulation of toxic metabolites might be the reasons.

Efficiency of RNA formation is expressed in terms of RNA content, cell growth rate and cell yield with respect to raw material. In the case of continuous culture, the efficiency is given by the yield of RNA (cell yield × RNA content) and speed of RNA production (growth rate × concentration of cells × RNA content). Generally bacteria have high growth rates and high RNA contents, but few studies of RNA production from natural raw materials have been made. This may be due to the relatively high cost of raw materials, low cell yield, ease of contamination, low recovery of cells and difficulty in the RNA extraction process. Yamada *et al.*[13] reported 5% RNA content in *Pseudomonas* sp. assimilating kerosene. Kuno *et al.*[14] isolated 9 strains of gram-negative bacteria (three genera, *Achromobacter*, *Alcaligenes* and *Pseudomonas*) from *n*-alkane assimilating bacteria; the growth rate was 0.65 h^{-1}. In jar fermenter culture, the RNA production was markedly affected by the concentration of dissolved oxygen, at pH 6.8 and 30° C. Under these conditions, a strain belonging to *Alcaligenes* sp. exhausted 7% (w/v) *n*-alkane within 5–6 h, and showed high RNA accumulation (10 g/1) in the culture broth.

4.2 FORMATION AND ACCUMULATION OF RNA IN YEASTS

Katchman *et al.*[15] and Chayen *et al.*[16] reported that high RNA contents are present in *Saccharomyces cerevisiae* and *Candida utilis*, respectively, at their logarithmic growth phase, as in bacteria.

Early in 1948, Di Carlo *et al.*[17] demonstrated that the RNA contents in various bakers' yeasts varied from 4.1 to 7.2%, depending on the composition of the media, and increased in higher concentrations of ammonium and phosphate ions. Analytical methods for RNA content in microorganisms were investigated by Kuroiwa *et al.*,[18] and it was consequently found that the Ogur-Rosen method was suitable for the determination of the RNA content in dried cells of microorganisms. They also demonstrated the need for denaturation of the cell wall for the determination of RNA in fresh yeast or in yeast dried at low temperatures. As yeast cell walls were only slightly damaged under these conditions, the cells should be frozen overnight in dry ice-ether as a pretreatment for RNA determination. RNA contents determined by this method in 10 samples of brewers' yeast, bakers' yeast and other yeasts were 2.6% to 6.8%. Further Kuroiwa *et al.*[19] proposed an improved Ogur-Rosen method involving the removal of acid-soluble compounds. The base compositions of RNA from several yeasts are listed in Table 4.2.

Characteristically, the RNA content in yeasts is very high, compared with the DNA content. Therefore, in the late 1950's, yeast RNA was recognized as a possible industrial source of nucleic acid-related flavor substances, and many reports appeared on the production of yeast RNA. Ogata *et al.*[20] reported that the RNA of microorganisms having high

TABLE 4.2

Base Compositions of Various Yeast RNA's

	adenine	uracil	guanine	cytosine	$\frac{\text{adenine+uracil}}{\text{guanine+cytosine}}$	Ref.
mRNA	28.6	28.7	21.2	21.7	1.34	4, 5)
tRNA	19.7	22.0	30.0	23.0	0.79	4, 5)
rRNA	25.6	26.8	29.0	18.6	1.10	4, 5)
Bulk RNA (1)	25.7	26.7	27.0	20.6	1.10	author's data
Bulk RNA (2)	24.8	28.1	27.9	19.3	1.12	author's data

RNA content, especially yeasts, might be the best raw material for the industrial production of 5′-nucleotides by their enzymatic method. They determined the RNA contents of 500 strains of yeast, and found that the RNA contents were 3.5% to 9.2% on cultivation for 18 h, and 2.3% to 5.8% on cultivation for 42 h. The RNA content in many yeasts was generally high in young cells, and high RNA productivity in the culture broth was obtained in the genera of *Candida*, *Pichia* and *Saccharomyces*. With cheap sulfite waste liquor as a raw material, some yeast strains which grew well were found among the genera of *Candida* and *Hansenula*, but on the whole the cell yields were rather low. It was demonstrated that the RNA content in yeast cells varied according to the nutrient composition, and was generally high at the early logarithmic growth phase and in media with low C/N ratios.

Yamaguchi *et al.*[21] carried out screening for RNA-rich yeasts among several tens of yeast strains using synthetic media containing 2% glucose, and found that *Candida* species, especially *C. utilis* and *C. mycoderma*, showed excellent RNA accumulation. It was also found that the RNA content reached a maximum in the course of the inductive growth phase leading to the logarithmic growth phase, and the RNA accumulation reached a maximum at the late logarithmic growth phase. RNA accumulation by yeasts was hardly affected by the kind of saccharide, but ammonium sulfate, peptone and yeast extract were excellent nitrogen sources. Yamaguchi *et al.*[22] investigated the effect of amino acids on RNA formation in *C. utilis* using a synthetic medium containing 2% glucose and ammonium sulfate. The results showed that RNA accumulation in the yeast was enhanced about 35% compared to the control by the addition of 5 mg/ml of L-leucine at the late logarithmic growth phase. It was also found that several amino acids, including leucine, depressed RNA accumulation in ammonium sulfate-deficient media.

Yamashita *et al.*[23] found that RNA content varied with the type of carbon source and other culture conditions, even with the same microorganism and growth phase. It thus appeared that the RNA content in microorganisms did not necessarily parallel the growth rate, but depended on the strain of microorganism, carbon source, composition of media and conditions of culture. During screening tests, it was found that the RNA productivity in the broth and the RNA content of *C. utilis* cells both increased markedly on the addition of metal ion mixtures containing Fe^{2+}, Zn^{2+} and Cu^{2+} to the basal medium (glucose 3.0%, $(NH_4)_2PO_4$ 1.0%, K_2HPO_4 0.06%, $MgSO_4 \cdot 7H_2O$ 0.05%), without markedly affecting the multiplication of the cells; remarkable effectiveness of Zn^{2+} levels above 0.25 ppm was observed, as shown in Table 4.3.

The effect on cultures of *C. utilis* and *S. cerevisiae* of about 200 select-

TABLE 4.3 Effect of Zn^{2+} on Yeast RNA Production

Concentration of Zn^{2+} in medium (ppm)	8 h			10 h		
	Amount of cells (%)	Amount of RNA (mg/l)	RNA content (%)	Amount of cells (%)	Amount of RNA (mg/l)	RNA content (%)
0	0.32	212	6.6	0.51	224	4.4
0.125	0.33	228	6.9	0.52	385	7.4
0.25	0.35	380	10.9	0.67	630	9.4
0.37	0.32	398	12.4	0.60	583	9.7
0.5	0.37	423	11.4	0.65	720	11.1
1.0	0.32	341	10.7	0.63	647	10.3
4.0	0.33	364	11.0	0.64	708	11.1
64.0	0.42	474	11.3	0.58	617	10.6
512.0	0.35	427	12.2	0.70	794	11.3
1024.0	0.42	530	12.6	0.65	733	11.3

TABLE 4.4 Effect of Anisomycin on RNA Production in *S. cerevisiae*

Level of anisomycin (ppm)	Amount of cells (%)	Amount of RNA (mg/cl)	RNA content (%)
0	1.22 (100)†	9.39 (100)	7.7 (100)
0.17	1.22 (100)	9.82 (105)	8.1 (105)
0.50	1.21 (99)	10.32 (110)	8.5 (110)
0.67	1.17 (96)	10.31 (110)	8.8 (114)
0.83	1.11 (91)	10.39 (111)	9.3 (121)
1.17	1.02 (84)	9.49 (101)	9.3 (121)
1.67	0.84 (69)	7.69 (82)	9.1 (118)
3.33	0.54 (44)	5.32 (57)	9.9 (129)

† ( ) = Percent of control.

TABLE 4.5 Composition of Nucleotides in Three Fractions of *S. cerevisiae*

	Control			With anisomycin (1.2 ppm)		
	RNA frac.	Low mol. wt. frac.	Extracellular frac.	RNA frac.	Low mol. wt. frac.	Extracellular frac.
Amount of nucleotides						
mg/dl of broth	79.86	10.26	3.22	94.14	16.50	2.88
% (to dried cells)	6.4	0.8	0.3	8.1	1.4	0.2
Molecular composition of nucleotides						
AMP (%)	25.9	53.9	47.1	25.7	30.5	31.5
GMP (%)	29.9	26.4	52.9	29.2	46.7	68.5
CMP (%)	19.9	10.7	0	20.5	7.3	0
UMP (%)	24.3	9.0	0	24.6	15.4	0

ed synthetic organic compounds and natural products was investigated, but no positive results were obtained. Further, they screened active substances promoting RNA formation in yeast from culture broths of 200 strains of actinomycetes; the culture filtrate of *Streptomyces* sp. S-22, isolated from a soil sample, was reproducibly effective, especially with *S. cerevisiae*. The effective substance in the culture broth of *Streptomyces* sp. S-22 was isolated as in crystal form and was identified as anisomycin, an antiprotozoal and antifungal antibiotic. As shown in Table 4.4, the addition of high concentrations of anisomycin depressed the growth of yeast cells, but the RNA content in the cells increased. With less than 1 ppm, both the RNA content and RNA productivity increased. The effect of anisomycin on RNA formation was remarkable in species of *Saccharomyces*, which exhibited a low minimum inhibitory concentration of the antibiotic. Anisomycin and cycloheximide, among known antifungal antibiotics, exhibit similar effects, i.e., these antibiotics are inhibitors of protein biosynthesis and inhibit the peptidyl transferase reaction on 80S ribosomes in fungal cells. They both show the same effectiveness on RNA formation in *S. cerevisiae*. These results are interesting in that inhibition of protein biosynthesis and promotion of RNA formation can occur independently and simultaneously. As shown in Table 4.5, anisomycin has almost no effect on the composition of nucleotides in intracellular high molecular RNA, low molecular RNA or in the extracellular nucleotide pool in the culture broth of *S. cerevisiae*.

Takahashi *et al.*[24] isolated two strains of *C. tropicalis* which could efficiently assimilate hydrocarbons, from soil; their RNA contents were 5.3% and 4.4%. Doi *et al.*[25] selected *C. lipolytica* from 800 strains of hydrocarbon assimilating yeasts isolated from nature, as a strain with a high RNA content. From this strain, a potassium chloride-sensitive mutant was obtained by treatment with *N*-methyl-*N'*-nitro-*N*-nitrosoguanidine. The RNA content of the mutant was enhanced about 1.5-fold. It has been suggested that the enhancement of the RNA content of the mutant did not represent an increase of a particular RNA, but that both tRNA and rRNA increased simultaneously. Ogata *et al.*[26] isolated a methanol assimilating yeast strain, *Kloeckera* sp., which contained 4.4% RNA on aerobic cultivation in a methanol medium.

4.3 PRODUCTION OF YEAST RNA

Economic mass production of dry yeast for human and animal use was established in Germany during World War II. So-called torula yeast (mainly *C. utilis*), which grows well in wood sugar or sulfite waste liquor containing both hexose and pentose, was preferred and used as the strain for the production of human food yeast (Kunst Fleisch) by a continuous culture method. Industrial production capacity for the dry yeast for human consumption had reached 15,000 tons per annum at the end of the War. As the sulfite waste liquor foamed strongly during aeration, the Waldhof-type fermenter was developed. It decreased the need for antifoaming agents and improved the culture productivity. An example of a Waldhof-type fermenter[27] is shown in Fig. 4.1. In the cylindrical culture tank (1), another cylinder (2) exists. The air from the shaft (6) is passed into the culture medium through six spinners (5) of a rapidly rotating aeration wheel (4). The culture broth passes down into the cylinder (2) from the outside and reduces the strong foaming with sulfite waste liquor. The temperature is adjusted by means of the water in the cooling coil (3). Feeding and the harvest of the culture are performed continuously from the top of the tank and the overflow pipe (8), respectively. After the War, it became desirable to remove hexose or pentose from sulfite waste liquor to reduce water pollution, and dry yeast production for animal feed or food use was industrialized in many countries. In Japan, Kihara *et al.* first worked on yeast

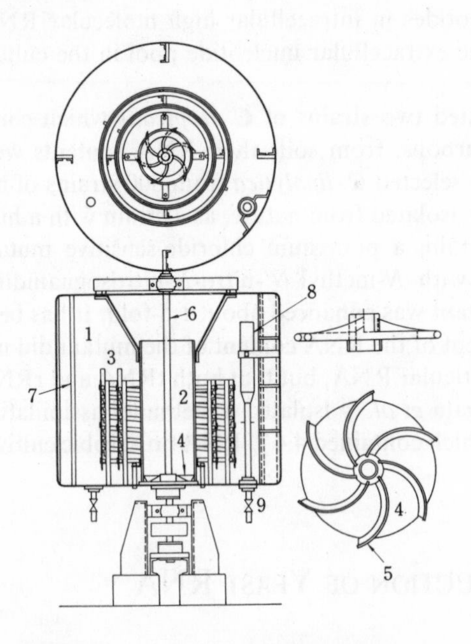

Fig. 4.1. A Waldhof-type fermenter.[27]
1, Fermenter; 2, draft tube; 3, cooling coil; 4, aeration wheel; 5, spinner; 6, shaft; 7, baffle plate; 8, overflow pipe; 9, outlet.

production from sulfite waste liquor during World War II. After the War, Yamada *et al.*[28] surveyed continuous culture methods, and Miwa *et al.*[29] worked on a pilot plant. Based on these results, a production plant for feed-grade dry yeast from sulfite waste liquor was constructed and operated at the Inuyama Plant of Toyo Spinning Co.

Yeast can be continuously cultivated in a cheap medium such as sulfite waste liquor or molasses; the cell yield based on sugar material is rather high. The DNA/RNA ratio is low in yeast, and the separation of yeast cells from the culture broth and extraction of yeast RNA are technologically easy. Residual yeast cells after extraction of the yeast RNA are also utilizable. In view of these advantages of yeast, Ogata *et al.*[6,20] concluded that the preferred RNA source as an industrial raw material for flavor enhancing 5′-nucleotides by the enzymatic method should be yeast RNA. Miwa *et al.*[30] demonstrated that when 2–4% phosphoric acid was added to the medium (with respect to sugar), the RNA content of the yeast reached a maximum, and in the case of continuous fermentation, the faster the charging speed of medium to the tank (within the possible range), the higher was the RNA content obtained. As for the culture of yeast RNA from several kinds of saccharide raw materials such as molasses and glucose, it was demonstrated by Watanabe[31] that the addition of zinc ions (above 0.25 ppm) and phosphoric acid (more than 0.15%) to the culture medium and an oxygen absorption rate (K_d) of more than 10×10^{-6} g mol. O_2/atm.min.ml were important. Nakamura *et al.*[32] demonstrated the continuous culture of yeast RNA in an air-lift type fermenter, in which air was well dispersed into the broth (the K_d was more than 20×10^{-6} g mol. O_2/atm.min.ml) and the cell concentration was 35 g/l. During continuous fermentation, a part of the culture broth was continuously flash-evaporated by an external evaporator, and the cooled and condensed broth was recycled into the fermenter. In order to cool the culture broth in an air-lift type fermenter, Sasa *et al.*[33] developed a recycle system for cooled broth using a vacuum jet and self-evaporation of partially discharged broth from the fermenter. An example of an air-lift type fermenter[34] is shown in Fig. 4.2. The culture broth is circulated at high speed by the air-lift effect of air blown in through the pipe. The air-foam is further dispersed during circulation and thus the oxygen-transfer rate is increased. Based on this and other research and engineering results, three producers of sulfite waste liquor and one producer of molasses in Japan have manufactured dry yeast with a high RNA content amounting to 10,000–20,000 tons per annum since 1962. Recently, a tendency to use glucose, acetic acid and ethyl alcohol as the main carbon sources has developed in order to reduce the COD or BOD value in the waste water. The yeast strain should not be pathogenic, and *C. utilis* has generally been used. The RNA content of

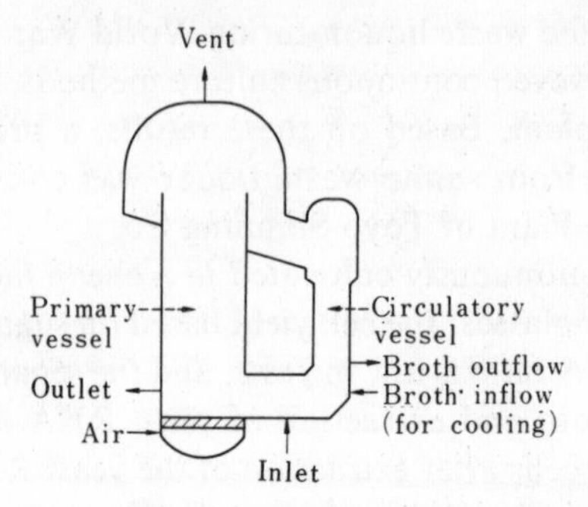

Fig. 4.2. An air-lift type fermenter.[34]

yeast from sulfite waste liquor is generally *ca.* 10% and that from molasses *ca.* 15%.

Yeast RNA can be extracted with solutions of sodium chloride, alkali, SDS, xylene sulfonate, phenol, etc., but in industrial processes sodium chloride or alkali has been used. After the filtered aqueous extracts have been concentrated, the concentrate is made acidic and then crude RNA precipitates. Crude RNA, if necessary as the sodium salt or some other salt, was then subjected to enzymatic decomposition. The content of RNA in crude RNA preparations produced as a raw material for 5′-nucleotides is about 70–90%, and the molecular weight distribution 10,000–150,000. In the case of yeast dried at high temperature, the product suspended in water can be used directly for enzymatic decomposition, and extraction and decomposition of yeast RNA can be carried out simultaneously.[6]

REFERENCES

1. J. C. Mounolou, *The Yeasts* (ed. A. H. Rose and J. S. Harrison), vol. 2, Academic, p. 316, 1971.
2. Y. Ito and T. Fujii, *Kisoseikagaku* (Japanese), p. 96, Asakura Shoten, 1971.
3. A. L. Lehninger, *Biochemistry*, p. 241, Worth, 1970.
4. A. M. Crestfield, K. C. Smith and F. W. Allen, *J. Biol. Chem.*, **216**, 185 (1955).
5. S. Osawa, *Biochim. Biophys. Acta*, **43**, 110 (1960).
6. K. Ogata, *Amino Acid and Nucleic Acid*, no. 8, 1 (1963).
7. S. S. Cohen, *J. Biol. Chem.*, **174**, 281 (1948).
8. A. N. Belozersky, *Ann. Symp. Quant. Biol.*, **12**, 1 (1947).
9. B. Malmgren and C. G. Heden, *Acta Path. Microbiol. Scand.*, **24**, 496 (1947).
10. M. L. Morse and C. E. Carter, *J. Bacteriol.*, **58**, 317 (1949).
11. H. E. Wade and D. M. Morgan, *Biochem. J.*, **65**, 21 (1957).
12. F. C. Neidhardt and B. Magasanik, *Biochim. Biophys. Acta*, **42**, 99 (1960).
13. K. Yamada, J. Takahashi and K. Kobayashi, *Agr. Biol. Chem.*, **26**, 636 (1962).
14. M. Kuno, T. Kono, M. Asai and Y. Nakao, 18th Symp. Ass. Amino Acid and Nucleic Acid (Japanese), Association of Amino Acid and Nucleic Acid, p.3, 1969.

15. B. J. Katchman and W. O. Fetty, *J. Bact.*, **69**, 607 (1955).
16. R. Chayen, S. Chayen and E. R. Roberts, *Biochim. Biophys. Acta*, **16**, 117 (1955).
17. F. J. Di Carlo and A. S. Schultz, *Arch. Biochem.*, **17**, 293 (1948).
18. Y. Kuroiwa and Y. Horie, *Bull. Agr. Chem. Soc.*, **19**, 35 (1955).
19. Y. Kuroiwa and N. Hashimoto, *ibid.*, **24**, 547 (1960).
20. K. Ogata and A. Imada, *Ann. Rept. Takeda Res. Lab.*, **21**, 31 (1962).
21. K. Yamaguchi and T. Mikami, *Ibaragi Daigaku Nogakubu Gakujutsu Hokoku* (Japanese),no. 13, 69 (1965).
22. K. Yamaguchi and T. Aida, *ibid.*, no. 13, 75 (1965).
23. T. Yamashita, T. Hidaka and K. Watanabe, *Agr. Biol. Chem.*, **38**, 727 (1974).
24. T. Takahashi, Y. Kawabata and K. Yamada, *ibid.*, **29**, 292 (1965).
25. M. Doi, Y. Arai, S. Akiyama, Y. Nakao and H. Fukuda, Ann. Mtg. Agr. Chem. Soc. Japan, 1974.
26. K. Ogata, H. Nishikawa, M. Ohsugi and T. Tochikura, *Hakkokogaku Zasshi* (Japanese), **48**, 470 (1970).
27. K. Yamada and T. Nakahara, *Sekiyu Hakko* (Japanese), Saiwai Shobo, p. 97, 1971.
28. K. Yamada, J. Takahashi and H. Okada, *Nippon Nogeikagaku Kaishi*, **27**, 704 (1953).
29. M. Miwa, *Kobo Riyo Kogyo* (Japanese), Kyoritsu Shuppan, p. 328, 1957.
30. M. Miwa, T. Kihara and T. Kagami, *Eiyo to Shokuryo* (Japanese), **13**, 47 (1960).
31. K. Watanabe, *Japanese Patent* No. 40–15960 (1965).
32. M. Nakamura, M. Kanazawa, Y. Nozaki and T. Tezaki, *Japanese Patent* No. 39–26044 (1964).
33. N. Sasa, T. Hata and A. Teranishi, *Japanese Patent* No. 39–29800 (1964).
34. M. Kanazawa, *Sekiyu to Biseibutsu* (Japanese), **12**, 33 (1974).

CHAPTER **5**

Extracellular Accumulation of DNA by Microorganisms*

DNA was discovered in 1869 by a Swiss scientist, Miescher,[1)] and from the first it was suggested to be the active principle that transmitted hereditary characteristics. However, rigorous evidence that the function of DNA is to transmit genetic information had to await the transformation experiments of Griffith[1)] and Avery, MacLeod and McCarty.[2)] These experiments, demonstrating the essential role of DNA in transmitting hereditary characteristics in bacteria, served to emphasize the importance of studies of the properties of DNA isolated from microorganisms under mild conditions, and therefore many experiments were carried out to look for the extracellular accumulation of DNA.

The content of DNA in microorganisms varies widely; bacteria normally contain one molecule of DNA per cell, but may contain more depending on the culture conditions, growth rate, etc. In the case of *Salmonella typhimurium*, up to four molecules of DNA can be present in a

* Fusao TOMITA, Kyowa Hakko Kogyo Co. Ltd.

single cell.[3] Contents of DNA in fungi or yeasts vary depending on their polyploidy. The DNA contents in cells have been reported to be 0.37–4.5% in bacteria, 0.03–0.52% in yeasts and 0.15–3.3% in fungi.

The content of DNA in bacteria is relatively high compared with those of animal and plant tissues. However, the isolation of bacterial DNA has been restricted by the following intrinsic difficulties: (1) difficulty in disrupting the cells under mild conditions, (2) contamination with other cellular macromolecules, and (3) depolymerization of DNA during the process of isolation.

As a result, extracellular accumulation of DNA has been studied as a means of obtaining large amounts of highly polymerized DNA. The presence of DNA slime layers in halophilic bacteria (*Micrococcus halodenitrificans*) was reported by Smithies and Gibbons.[6] Subsequently, similar observations were made with non-halophilic bacteria by Catlin *et al.*[7,8] DNA present in slime layers was strongly associated with the cells and it was necessary to treat them with 0.5% sodium deoxycholate in 4% NaCl or 0.41% sodium laurylate in order to extract DNA.

It was later shown that the cells of transforming bacteria (*Neisseria*,[9] *Pneumococcus*,[10] and *Bacillus subtilis*[11–14]) excreted DNA. Furthermore, the author recently found bacteria producing extracellular DNA and developed a method for their cultivation in order to produce large amounts of extracellular DNA. Mass production of highly polymerized DNA from bacteria thus became possible.

5.1 Bacteria Producing Extracellular DNA

Extracellular accumulation of DNA was first observed with a halophilic bacterium, *Micrococcus halodenitrificans*.[6] It had been suggested earlier that DNA slime layers might occur only in halophilic bacteria. However, many non-halophilic bacteria such as *Staphylococcus aureus*, *Flavobacterium* sp., *Alcaligenes faecalis*, etc. were also found to accumulate DNA.[7,8] Later it was reported that transforming bacteria including *Neisseria* sp.,[9] *Pneumococcus* sp.,[10] and *B. subtilis*[11–14] could excrete DNA into the medium. Furthermore, extracellular accumulation of DNA was observed in *Arthrobacter simplex*,[15] *Micrococcus ureae*,[15] and various *Pseudomonas*[15,16] Therefore, the phenomenon of DNA accumulation is not restricted to specific bacteria, but is observed among various bacteria. As will be shown in later sections, the mode of DNA accumulation can be classified into three categories: (1) DNA accumulation in slime layers associated with intact cells, (2) DNA accumulation through excre-

TABLE 5.1

Bacteria Producing Extracellular DNA

Bacteria	Extracellular DNA (μg/ml broth)	Mode of accumulation†	Ref.
Alcaligenes faecalis	no data	I	7, 8)
Alcaligenes viscosus	no data	I	7)
Arthrobacter simpl x	200	II	15)
Arthroba ter sp.	10	II	15)
Bacillus subtilis	0.5–100	II	11–14)
Flavobact rium sp	100	I	7)
Micrococcus citreus	no data	I	7)
Micrococcus halodenitrificans	2	I	6, 17)
Micrococcus ureae	24	II	15)
Neisseria sp.	2–5	II	9)
Pneumococcus sp.	2	II	10)
Pseudomonas aeruginosa	300–3800	III	15, 16)
Pseudomonas boreopolis	400	III	16)
Pseudomonas chloraphis	600	III	16)
Pseudomonas convexa	400	III	16)
Pseudomonas cruciviae	200	III	16)
Pseudomonas dacunhae	400	III	16)
Pseudomonas fairmontensis	200	III	16)
Pseudomonas fluorescens	500–6000	III	7, 15, 16)
Pseudomonas marginalis	300	III	16)
Pseudomonas melanogenum	600	III	16)
Pseudomonas oleovolans	200	III	16)
Pseudomonas polycolor	2100	III	16)
Pseudononas rubescens	500	III	16)
Pseudomonas syncyanea	500	III	16)
Staphylococcus aureus	no data	I	8)
Staphylococcus epidermidis	no data	I	8)
Vibrio costicolus	1	I	6)

† I, DNA is accumulated in slime layers strongly associated with cells; II, DNA is excreted from the cells; III, DNA is accumulated mainly by autolysis of cells.

tion by growing cells, and (3) DNA accumulation mainly by autolysis.

Bacteria producing extracellular DNA which appeared in the literature up to summer 1975 are summarized in Table 5.1, together with the amounts of DNA accumulated and the mode of accumulation.

5.2 Mechanism of Production of Extracellular DNA

The production of extracellular DNA depends largely on the culture conditions used for producer strains, since the conditions necessary for stimulating DNA excretion and for preventing the degradation of excreted DNA by inhibiting the action of DNases must be satisfied.

5.2.1 *Micrococcus halodenitrificans*[6,17)]

The production of slime layers depends on the salt concentration in the medium. At concentrations of more than 0.7 M NaCl, no production of slime layers can be observed and consequently DNA is not accumulated. On the other hand, at suboptimal concentrations for growth (0.55–0.65 M), the viscosity of the culture broth became marked and maximum accumulation of DNA was attained. Under these conditions, some disruption of bacterial cells was also observed due to autolysis, but most of the DNA accumulated was present in the slime layers and associated strongly with intact cells.

5.2.2 *Staphylococcus aureus*[8)]

In the presence of 1 M NaCl, extracellular DNases were inhibited and the accumulation of DNA increased. The optimum pH of this DNase is 8.6 and thus degradation of DNA can be prevented by keeping the pH of the medium at 5.2–7.2 during incubation. The presence of Ca^{2+}, an activator of DNase, reduced the amount of extracellular DNA due to degradation by DNase. Larger amounts of extracellular DNA were obtained by stationary culture than by shaking culture.

5.2.3 *Bacillus subtilis*[11–14)]

Spontaneous release of transforming DNA from growing cultures was first demonstrated by genetic studies. The release of DNA was found to be

TABLE 5.2

Effect of Carbon Sources on DNA Accumulation by *Arthrobacter simplex*[15]

Carbon source†	1st day		3rd day	
	Growth (A_{660})	DNA ($\mu g/ml$)	Growth (A_{660})	DNA ($\mu g/ml$)
Glucose	1.4	—	1.6	131
Fructose	3.0	0	2.5	71
Sorbitol	2.5	4	2.1	100
Sucrose	4.5	2	6.3	84
Mannose	0.9	2	2.1	36
n-Paraffin	1.1	1	9.9	41

† The composition of the medium was polypeptone (1 %), meat extract (1 %), NaCl (0.3 %) and one of the carbon sources (2 %). Experiments were carried out in 250 ml Erlenmeyer flasks containing 50 ml of the medium, with shaking.

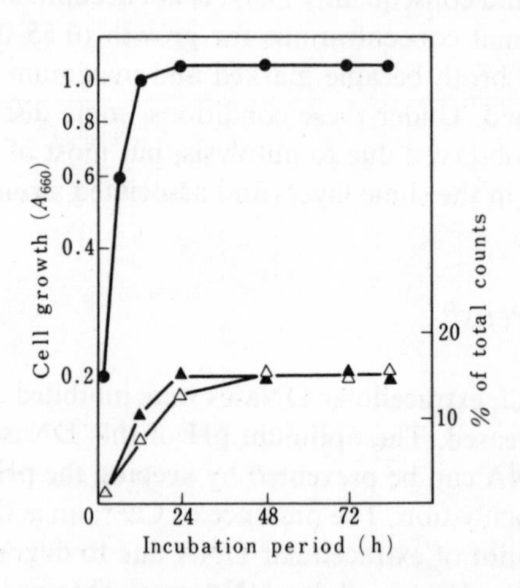

Fig. 5.1. Time course of DNA accumulation by *Arthrobacter simplex*.[15] The experiment was carried out in Monod-type L-tubes using the same medium as in Table 5.2, except that sucrose was used as a carbon source. Cells labeled with ^{3}H-TdR were transferred to medium containing ^{32}P and the time course of DNA accumulation was followed.
●—●, Cell growth; ▲—▲, ^{32}P; △—△, ^{3}H.

closely correlated to the replication of DNA, and was observed only in transformable strains. In non-transformable mutants, no release of DNA was seen. Although the mechanisms of release of DNA have not yet been determined, it is suggested that changes in cell permeability due to the diversion of essential nutrients or a shift in culture conditions may be involved in the release of DNA. Furthermore, a possible relationship between competence and the autolysis of cell walls has been considered,[18] and thus the increase of autolytic activity in competent cells may also be related to the excretion of DNA.

5.2.4 *Arthrobacter simplex*[15]

Accumulation of DNA is affected by the carbon source; glucose gave the maximum production, as shown in Table 5.2. In order to clarify the mechanism of DNA production, cells labeled with ^{3}H-TdR were incubated in the presence of ^{32}P and a kinetic study of the excretion was carried out, as shown in Fig. 5.1. Labeled DNA's (^{32}P and ^{3}H) were excreted only at the logarithmic phase and it was thus suggested that the excretion of DNA was closely related to its replication, as observed in *B. subtilis*.

5.2.5 Pseudomonads[15,16]

Many bacteria belonging to the Pseudomonadaceae accumulate extracellular DNA, as shown in Table 5.1. Among them, extensive studies have been made on *Pseudomonas fluorescens*, and *n*-paraffin was found to be the best carbon source for DNA production, as shown in Table 5.3. The presence of corn steep liquor (0.5%) and ferrous sulfate (1.0%) considerably

TABLE 5.3

Effect of the Carbon Source on DNA Accumulation by *Pseudomonas fluorescens*[15]

Carbon source	1st day		3rd day	
	Growth (A_{660})	DNA (μg/m*l*)	Growth (A_{660})	DNA (μg/m*l*)
None	2.1	0	1.7	45
Sucrose	3.7	0	2.4	78
Glucose	7.6	0	9.2	0
Sorbitol	2.7	0	1.9	98
n-Paraffin	9.6	0	9.0	210

† Experimental conditions were the same as in Table 5.2.

TABLE 5.4

Effect of pH on DNA production[16]

pH	Time (h)	High molecular weight DNA (mg/ml)	Total DNA (mg/ml)
6	0	0	0.31
	48	0.02	0.65
	60	1.0	2.2
	72	3.0	4.1
7	0	0	0.28
	48	0.09	0.64
	60	0	2.8
	72	0.02	4.6
6→8	0	0	0.32
	48	0.6	1.3
	60	4.2	5.5
	72	3.8	4.8

† Experiments were carried out in 5 liter jar fermenters using *P. fluorescens*. The medium contained 10% *n*-paraffin, 0.5% ammonium sulfate, 0.2% KH_2PO_4, 0.2% Na_2HPO_4, 0.1% K_2SO_4, 0.1% $MgSO_4 \cdot 7H_2O$, 1% $FeSO_4 \cdot 7H_2O$, 0.01% $MnSO_4 \cdot 4H_2O$, 0.01% $ZnSO_4 \cdot 7H_2O$, 0.0003% $CuCl_2 \cdot 2H_2O$, 0.5% corn steep liquor and 0.5% yeast extract. The change of pH from 6 to 8 was made after incubation for 24 h.

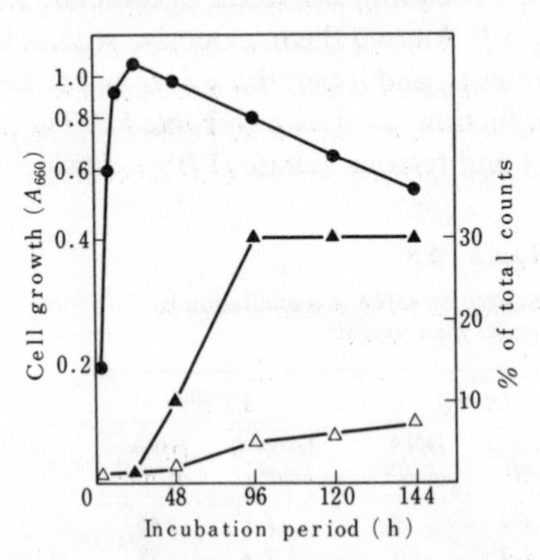

Fig. 5.2. Time course of DNA accumulation by *Pseudomonas fluorescens*.[15] The experiment was carried out as in Fig. 5.1.
●—●, Cell growth; ▲—▲, ^{32}P; △—△, ^{3}H.

stimulated the amounts of DNA in the medium. According to large-scale experiments using 5 l jar fermenters, the use of glucose in the seed culture and *n*-paraffin in the production culture gave the maximum yield of DNA. Moreover, efficient production of DNA (5.5 g/l) was obtained by controlling the pH of the broth at around 6.0 in the growth phase and then at around 8.0 during the production phase. Under these conditions, the action of DNase was inhibited. On the other hand, almost no highly polymerized DNA was recovered when the pH was controlled at around 7.0 (Table 5.4).

Accumulation of DNA by *P. fluorescens* was shown to be mainly due to autolysis, as indicated in Fig. 5.2. However, the amount of DNA accumulated exceeded that obtained from lyzed cells and it is suggested that accumulation was not through autolysis alone, but may also involve other mechanisms.

5.3 Future Prospects

Large-scale production of highly polymerized DNA has now become practical, but the recovery of DNA from the culture broth is not satisfactory, since the broth contains large amounts of by-products such as polysaccharides, RNA and proteins. Thus, it is necessary to devise methods of degrading or removing these accompanying macromolecules from the culture broth.

Although the importance of DNA in transmitting hereditary characteristics is well-known, its practical uses or applications have not yet been fully examined. Possible applications of DNA on a laboratory scale have recently been developed in such fields as affinity chromatography and radiation biology for purifing and isolating substances having a strong affinity to DNA, and in agents to reduce or prevent radiation damage.

References

1. Reviews by J. H. Taylor, in *Selected Papers in Molecular Genetics* (ed. J. H. Taylor), Academic, 1965.
2. O. T. Avery, C. M. MacLeod and M. McCarty, *J. Exptl. Med.*, **79**, 137 (1944).
3. M. Schaechter, O. Maaloe and N. O. Kjeldgaard, *J. Gen. Microbiol.*, **19**, 592 (1958).
4. *Handbook of Biochemistry* (ed. H. A. Sorber), p. H-104, The Chemical Rubber Co., 1970.
5. Y. Kurokawa and Y. Horie, *Bull. Agr. Chem. Soc.*, **19**, 35 (1955).
6. W. R. Smithies and N. E. Gibbons, *Can. J. Microbiol.*, **1**, 614 (1955).
7. B. W. Catlin, *Science*, **131**, 608 (1960).

8. B. W. Catlin and L. S. Cunningham, *J. Gen. Microbiol.*, **19**, 522 (1958).
9. B. W. Catlin, *Science*, **131**, 608 (1960).
10. E. Ottolenghi and R. D. Hotchkiss, *ibid.*, **132**, 1257 (1960).
11. I. Takahashi, *Biochem. Biophys. Res. Commun.*, **7**, 467 (1962).
12. A. L. Demain, R. W. Burg and D. Hendlin, *J. Bact.*, **89**, 640 (1965).
13. S. Borenstein and E. Ephrati-Elizur, *J. Mol. Biol.*, **45**, 137 (1969).
14. R. P. Sinha and V. N. Iyer, *Biochim. Biophys. Acta*, **232**, 61 (1971).
15. F. Tomita and T. Suzuki, *Agr. Biol. Chem.*, **36**, 133 (1972).
16. F. Tomita, T. Nakanishi and T. Suzuki, *ibid.*, **38**, 293 (1974).
17. I. Takahashi and N. E. Gibbons, *Can. J. Microbiol.*, **3**, 687 (1957).
18. F. E. Young and J. Spizizen, *J. Biol. Chem.*, **238**, 3126 (1963).

CHAPTER **6**

Hydrolysis of Nucleic Acids with Penicillium Nuclease[1)*1]

Around 1955, Kuninaka[2)] degraded RNA with a crude enzyme preparation obtained from culture filtrates of *Aspergillus oryzae*. The degradation products, 3′-nucleotides (including 3′-IMP), nucleosides, and purine bases did not have flavor activity, while IMP (i.e. 5′-IMP) extracted from animal muscle was confirmed to have marked flavor activity even in the absence of histidine.[1,3)] Thus, for the purpose of preparing IMP from RNA, screening of microorganisms capable of degrading RNA into 5′-nucleotides was systematically carried out,*2 and *Penicillium citrinum* was

*1 Akira Kuninaka, Yamasa Shoyu Co. Ltd.

*2 At that time only snake venom phosphodiesterase[5)] had been confirmed to be able to degrade nucleic acids into 5′-nucleotides. It is noteworthy that extracts of a pyrimidine oxidizing soil bacterium, strain U-1,[6)] and *Asterococcus mycoides*, the microorganism causing contagious bovine pleuropneumonia,[7)] were preliminarily reported to degrade RNA into 5′-mononucleotides in 1954 and 1957. The screening of microorganisms capable of degrading RNA into 5′-nucleotides was independently carried out by Omura *et al.* too, and a *Streptomyces* strain was selected for industrial use in 1958 (see Chapter 7).

selected in 1957.[1,4] Among 5′-nucleotides purified from the *P. citrinum* nuclease digest of RNA, GMP and XMP were found to have flavor activity, as well as IMP. In particular, the flavor activity of GMP was several times that of IMP. Furthermore, there was a marked synergistic action between monosodium glutamate (MSG) and the flavor nucleotides.[1,3] Thus an economic basis for the production of IMP and GMP from RNA was established, and their industrial production using the *Penicillium* enzyme was begun in 1961. Almost simultaneously, the industrial degradation of RNA with *Streptomyces* enzyme was also started.

Nucleolytic enzymes can be classified as shown in Table 6.1 on the

TABLE 6.1

Nucleolytic Enzymes and Their Products

Nucleolytic Enzymes		Substrates	Products
5′-P Producers	Nucleases	RNA, DNA	5′-Mononucleotides
	DNases	DNA	and/or 5′-phospho-
	RNases	RNA	oligonucleotides
3′-P Producers	Nucleases	RNA, DNA	3′-Mononucleotides
	DNases	DNA	and/or 3′-phospho-
	RNases(1)	RNA	oligonucleotides
	RNases(2)	RNA	Nucleoside 2′: 3′-cyclic phosphates and/or 2′: 3′-cyclic phospho-oligonucleotides
	RNases(3)	RNA	2′-Mononucleotides (via nucleoside 2′:3′-cyclic phosphates)

basis of their substrates and products. The main enzymes were outlined in Chapter 1. For the purpose of producing 5′-monoribonucleotides, it is necessary to select a microorganism that produces a nuclease or an RNase degrading RNA completely into 5′-mononucleotides. Nuclease P_1 purified from *P. citrinum* can degrade both RNA and DNA completely into 5′-mononucleotides. As far as is known, no other nucleolytic enzyme has so far actually been confirmed to catalyze such complete degradation of DNA.

The enzymic degradation of RNA was the first method employed for the industrial production of 5′-nucleotides. At present (1975), about 40%

of the total production of 5′-nucleotides in Japan is still obtained by this method. Other methods established so far are:

(1) Fermentative production of nucleosides, and their phosphorylation to 5′-nucleotides (from 1964; see Part III)

(2) Fermentative production of 5′-nucleotides (from 1966; see Part III)

(3) Chemical decomposition of RNA into nucleosides, and their phosphorylation to 5′-nucleotides (from 1967; see Chapter 9).

Initially, the *Penicillium* enzyme was called 5′-phosphodiesterase, because the enzyme was thought to act in the same fashion as snake venom phosphodiesterase, which had been called 5′-phosphodiesterase by Mehler.[8] Furthermore the RNA-degrading and DNA-degrading activities in the crude preparation from *Penicillium* culture were thought to be separate.[4] Recently, however, a purified preparation of the *Penicillium* enzyme was confirmed to split phosphodiester linkages in both RNA and DNA and phosphomonoester linkages in both ribo- and deoxyribonucleoside 3′-monophosphates and oligonucleotides terminated by 3′-phosphate.[9–11] Moreover, the purified enzyme was sharply distinguished from snake venom phosphodiesterase in its specificity for synthetic substrates.[13] Thus the name "5′-phosphodiesterase" has been replaced by the name "nuclease P_1".[11,16] "Nucleate 5′-nucleotidohydrolase (3′-phosphohydrolase)" seems to be the most suitable systematic name.

From the viewpoints of safety and productivity, a pigmentless mutant producing a large amount of nuclease P_1, which was induced from *P. citrinum*, has been employed for the industrial production of 5′-nucleotides since 1961 (see section 6.2). In addition, purified nuclease P_1 is now being employed for studies on the structures of nucleic acids in several laboratories (see section 6.3).

6.1 Structure, Specificity and Occurrence of Nuclease P_1

6.1.1 Structure[10,11,14)]

Nuclease P_1 was purified from an aqueous culture extract of a pigmentless mutant of *P. citrinum* on wheat bran by salting out with 90% saturated ammonium sulfate, heat treatment, salting out with 40% to 80% saturated ammonium sulfate, precipitation with 40% acetone, gel filtration on Sephadex G-100, DEAE-cellulose column chromatography, and a second gel filtration on Sephadex G-100. The purified nuclease P_1 was homogeneous on ultracentrifugation and disk gel electrophoresis.

The molecular weight of nuclease P_1 is estimated to be about 44,000, and the isoelectric point of the enzyme is at pH 4.5. The enzyme protein is rich in hydrophobic amino acid residues, especially tyrosine and tryptophan residues. The enzyme molecule contains three atoms of zinc and about 17.4% carbohydrate, consisting of mannose, galactose and glucosamine in a molar ratio of 6:2:1. The contents of α-helix, β-structure, and random coil in nuclease P_1 are estimated to be 29 to 31%, about 6%, and about 63%, respectively. The helical structure is quite stable to denaturing reagents such as urea and guanidine hydrochloride. The enzyme can be preserved in powder form or in solution, pH 5 to 8, for more than one year in a refrigerator, and it is reasonably thermostable. The high stability of nuclease P_1 may be due to its high content of hydrophobic amino acid residues, providing a compact structure, and to the presence of zinc atoms. Zinc atoms seem to be required not only for the catalytic function of nuclease P_1 but also for maintaining the active conformation. Removal of the zinc atoms from the enzyme with EDTA results in inactivation and disruption of the secondary structure; the activity is partially restored by addition of Zn^{2+}.

As inactivation of the enzyme during reaction is reduced by the presence of Zn^{2+} in the reaction mixture, it is necessary to add *ca.* 0.1 mM Zn^{2+} to a substrate solution when nucleic acids are incubated with purified nuclease P_1 at high temperatures for a long time in order to degrade nucleic acids completely. It should be noted that heat inactivation of the enzyme is stimulated by Zn^{2+} in the absence of the substrate.

6.1.2 Mode of Action and Specificity[10–13]

Nuclease P_1 can cleave substantially all 3′–5′ phosphodiester linkages of polynucleotides and 3′-phosphomonoester linkages of mono- and oligonucleotides terminated by 3′-phosphate. Nucleoside 2′-phosphates are more resistant than nucleoside 3′-phosphates. The rate of hydrolysis of 2′-AMP is less than 1/3000 that of 3′-AMP, and the rate of hydrolysis of 2′–5′ phosphodiester linkage is estimated to be about 1/100,000 that of 3′–5′ phosphodiester linkage.[16] Nuclease P_1 does not attack ribose 3-phosphate, deoxyribose 3-phosphate, nucleoside 5′-phosphate, *p*-nitrophenylphosphate, ADP, ATP, cAMP or bis(*p*-nitrophenyl)phosphate. The enzyme degrades nucleic acids endo- and exonucleolytically. Double-stranded RNA and DNA are much more resistant than single-stranded RNA and DNA, but are finally degraded completely into mononucleotides.

The optimum temperature is at around 70° C.[11] The optimum pH depends on the kind of substrate in the range of pH 4.5 to 8.0 (Tables 6.2 and 6.3). Generally, RNA, diribonucleoside monophosphates, and ribo-

TABLE 6.2
Phosphodiesterase Activity toward Various Polynucleotides at Optimal pH[12)]

	Opt. pH	Rate of hydrolysis†
Low molecular weight RNA	6.0	336.0
tRNA	5.3	262.1
rRNA	5.3	245.3
Heat-denatured DNA	5.3	218.4
Native DNA	5.3	1.1
Poly (A)	6.0	490.1
Poly (I)	4.5	541.0
Poly (C)	6.0	305.8
Poly (U)	4.0	390.0
Poly (G)	4.5	14.0
Poly (I)·poly (C)	4.8	16.8

† μmole-equivalents of phosphodiester linkages cleaved per 1 mg of the enzyme in 1 min at 37°C.

TABLE 6.3
Phosphomonoesterase Activity toward Various Nucleotides at the Optimal pH[12)]

	Opt. pH	Rate of hydrolysis†
3′-AMP	7.2	1,004.3
3′-GMP	8.5	1,440.0
3′-CMP	6.0	853.6
3′-UMP	6.0	700.3
3′-dAMP	4.5	16.7
3′-dGMP	5.0	14.0
3′-dCMP	4.5	49.6
3′-dTMP	4.5	33.6
pGp	8.5	1,296.0
dpTp	4.5	18.1
Co A	6.0	662.9
2′-AMP	6.0	0.3

† μmoles of Pi formed per 1 mg of the enzyme in 1 min at 37°C.

TABLE 6.4

Rates of Hydrolysis of Dinucleoside Monophosphates (pH 5.0)[13]

Substrate	Cleavage (μmoles/mg enzyme/min)
G-C	48
G-U	65
A-C	341
A-A	170
C-A	85
A-U	306
G-A	34
U-A	72
d(T-T)	10
d(T-G)	26
d(T-A)	22
d(A-T)	51
d(A-G)	94
d(C-C)	28
d(A-C)	75
d(C-A)	32

nucleoside 3′-phosphates are hydrolyzed faster than DNA, dideoxyribonucleoside monophosphates, and deoxyribonucleoside 3′-phosphates, respectively, and ribonucleoside 3′-phosphates are hydrolyzed faster than the corresponding homopolymers (Tables 6.2, 6.3 and 6.4).

While strict base specificity is not observed, nuclease P_1 preferentially splits phosphodiester linkages between AMP or dAMP and the adjacent nucleotides, X–A(or dA)–Y, especially A(or dA)–Y. The order of base preference of nuclease P_1 in ribonucleoside 3′-phosphates is not the same as that in deoxyribonucleoside 3′-phosphates.

The susceptibility of dinucleotide to nuclease P_1 is increased by the presence of a 5′-monophosphoryl group, but is not affected by the presence of a 2′-monophosphoryl group (Table 6.5). Dinucleotides bearing a 3′-monophosphoryl group are hydrolyzed as follows:[13]

$$\text{A–Up} \begin{cases} \nearrow \text{A–U} + \text{Pi} \searrow \\ \searrow \text{Ado} + \text{pUp} \nearrow \end{cases} \text{Ado} + \text{pU} + \text{Pi}$$

Nuclease P_1, like snake venom phosphodiesterase, produces four 5′-mononucleotides from nucleic acids. However, apart from important differences in the endo- and exonucleolytic properties and in the phospho-

TABLE 6.5

Effect of the Monophosphoryl group on the Susceptibility of Dinucleotides[13]

Substrate	Cleavage (μmoles/mg enzyme/min)
pA-A	408
A-A	170
A-Ap(2′)	165
pA-C	682
A-C	341

monoesterase activity of nuclease P_1 and snake venom phosphodiesterase, the basic substrate structural elements for these enzymes appear to be quite different: nuclease P_1 does not attack *p*-nitrophenyl TMP, which is easily split by the snake venom enzyme, but splits *p*-nitrophenyl 3′-TMP, giving rise to thymidine and *p*-nitrophenyl phosphate. Proposed structural requirements for substrates of typical phosphodiesterases that hydrolyze RNA and DNA without base specificity are shown in Fig. 6.1. The structural requirement for substrates of nuclease P_1 is similar to that for substrates of spleen phosphodiesterase. The principal difference is that nuclease P_1 cleaves the diester bond between the phosphate and 3′-carbon of the sugar, whereas the spleen enzyme cleaves the other side of the phosphate, i.e. between the phosphate and the nonspecific hydroxylic component of the diester bond. In contrast to both spleen and venom diesterases, the primary product released by nuclease P_1 hydrolysis is a derivative bearing a hydroxyl group (at the 3′ position) rather than a phosphoryl group. The 5′-phosphoryl product formed in polynucleotide hydrolysis is a secondary consequence of such cleavage. In this respect, nuclease P_1 is similar

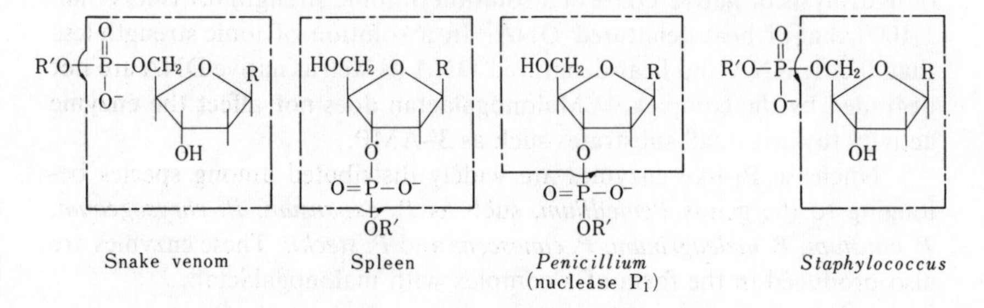

Fig. 6.1. Proposed structural requirements for the substrates of phosphodiesterases that hydrolyze RNA and DNA. The structures indicated for the venom, spleen and *Staphylococcus* enzymes are those suggested by Cuatrecasas *et al.*[18] R = Thymine; R′ = *p*-nitrophenyl.

to a 3′-P-forming nuclease from *Staphylococcus aureus*. Fig. 6.1 may be helpful in showing that nuclease P_1 splits not only 3′–5′ phosphodiester linkages but also 3′-phosphomonoester linkages and that the enzyme splits 2′-phosphomonoester linkages and 2′–5′ phosphodiester linkages at much lower rates than 3′-phosphomonoester linkages and 3′–5′ phosphodiester linkages.

Enzymes similar to nuclease P_1 are found in *Phoma cucurbitacearum*, *Monascus purpureus*, wheat seedling, mung bean, potato tubers, ginkgo nuts, tobacco, and corn (see Chapter 1, 1.1.1 B-1). These enzymes, non base-specific 5′-phosphomonoester-forming nucleases associated with 3′-nucleotidase activity, are generally highly preferential for single-stranded structure and are inactivated by EDTA. Their optimal temperatures are usually rather high. Among them, plant enzymes are not suitable for producing 5′-mononucleotides because of the low monomer contents in their RNA digests. Nuclease S_1 from *Aspergillus oryzae* (see Chapter 1, 1.1.1 B-3a) is also a single-strand specific, non base-specific 5′-phosphomonoester-forming nuclease, but the enzyme does not have 3′-phosphomonoesterase activity.

6.1.3 Occurrence (Nuclease P_1–Malonogalactan Complex)[10,15]

Nuclease P_1 is produced in the form of a complex with malonogalactan (this is a 1,5-*β*-galactofuranoside polymer partly esterified with malonic acid at position 3) when *P. citrinum* is grown on wheat bran. Nuclease P_1 protein, free from malonogalactan, can be obtained by incubating the complex with a malonogalactan-specific carboxylesterase and a galactan-hydrolyzing enzyme, both of which are present in the culture.

The activity of the complex toward native DNA is very low: the rate of hydrolysis of native DNA in a solution of ionic strength 0.1 is less than 1/1000 that of heat-denatured DNA. In a solution of ionic strength less than 0.001, RNA and heat-denatured DNA as well as native DNA are not degraded by the complex.[19] Malonogalactan does not affect the enzyme activity toward small substrates such as 3′-AMP.

Nuclease P_1-like enzymes are widely distributed among species belonging to the genus *Penicillium*, such as *P. expansum*, *P. chrysogenum*, *P. notatum*, *P. meleagrinum*, *P. canescens* and *P. steckii*. These enzymes are also produced in the form of a complex with malonogalactan.

6.2 Production of 5′-Nucleotides

6.2.1 Outline of the Production Process

A pigmentless, nuclease P_1-rich mutant of *P. citrinum* is grown on wheat bran. The culture is extracted with water. The aqueous extract contains not only thermostable nuclease P_1 but also appreciable amounts of an RNase (EC 3.1.4.23) degrading RNA into 3′-nucleotides via nucleoside 2′:3′-cyclic phosphates and a phosphomonoesterase capable of hydrolyzing 5′-nucleotides. These two enzymes can be completely inactivated by heating without any loss of nuclease P_1.

RNA solution is incubated with the heat-treated enzyme solution. During incubation, appropriate pH and temperature levels should be maintained. After incubation, the four 5′-nucleotides formed are separated by means of anion exchange resin column chromatography and purified. AMP is usually deaminated to IMP with *Aspergillus* adenyl deaminase. The pyrimidine nucleotides, CMP and UMP, are not utilized as flavor enhancers, but are used as starting materials for the production of biologically active substances.

Deoxyribonucleoside 5′-monophosphates are also produced by incubating denatured DNA with nuclease P_1 solution. The ability to degrade DNA completely into 5′-mononucleotides is an important characteristic of nuclease P_1.

Nomura *et al.*[17] reacted yeast RNA with partially purified nuclease P_1 in an impact enzyme reactor, and successfully isolated 5′-nucleotides from the reaction mixture by using a membrane ultrafilter.

6.2.2 Substrates Suitable for Nuclease P_1

If RNA is terminated by 5′-phosphate and 3′-hydroxyl groups, and consists exclusively of 3′–5′ phosphodiester linkages, 100% of the RNA can be recovered in the form of 5′-nucleotides after nuclease P_1 digestion. However, commercially available technical-grade RNA preparations are mostly terminated by 2′(3′)-phosphate and 5′-hydroxyl groups and contain several percent 2′–5′ phosphodiester linkages in addition to the 3′–5′ phosphodiester linkages, because RNA is partially depolymerized and isomerized during its extraction from yeast with hot salt water.

The presence of 2′(3′)-terminal phosphate and the absence of 5′-terminal phosphate markedly decrease the susceptibility of oligonucleotides to snake venom phosphodiesterase,* but do not greatly decrease their susceptibility to nuclease P_1 (Table 6.5). However, from nuclease P_1 digests of an oligo- or polynucleotide terminated by 2′-phosphate and 5′-hydroxyl groups, nucleoside and nucleoside 2′,5′-diphosphate are obtained together with 5′-mononucleotides. Thus the yield of 5′-nucleotides decreases with decreasing chain length of the substrate.

$$\text{pN–N–N–N} \ldots \text{N–N} \longrightarrow n\,\text{pN}$$
$$\text{N–N–N–N} \ldots \text{N–N}_p \longrightarrow \text{N} + (n-1)\text{pN} + \text{Pi}$$
$$\text{N–N–N–N} \ldots \text{N–N}^p \longrightarrow \text{N} + (n-2)\text{pN} + \text{pN}^p$$

As 2′–5′ phosphodiester linkages in an oligo- or polynucleotide chain are not attacked by usual levels of nuclease P_1, 2′→5′ dinucleotides ((2′–5′)$\text{pN}^{\alpha}\text{–N}^{\beta}$) are obtained from the nuclease P_1 digest of an isomerized oligo- or polynucleotide.[16] In addition, snake venom phosphodiesterase can split 2′–5′ phosphodiester linkages as well as 3′–5′ linkages.

As discussed above, non-depolymerized, non-isomerized RNA is the most suitable substrate for nuclease P_1 in order to obtain 5′-nucleotides in good yield. It is noteworthy therefore that Shinohara *et al.*[20] treated a bakers' yeast suspension with RNase in an impact cell mill and obtained mononucleotides corresponding to more than 80% of the total nucleic acids in the yeast.

6.3 Application of Nuclease P_1 for Structural Studies of Nucleic Acids

Recently a unique 5′-terminal structure of the type, m^7G(5′)ppp(5′)-Nm, has been found in various viral and cellular mRNA's including those of cytoplasmic polyhedrosis virus, reovirus, vaccinia, vesicular stomatitis virus, simian virus, adeno virus, mouse L cells, monkey BSC-1 cells and HeLa cells.[21,22] For example, cytoplasmic polyhedrosis virus mRNA contains blocked, methylated 5′-termini with the structure, m^7G(5′)ppp(5′)-AmpGp-,[23] reovirus mRNA contains m^7G(5′)ppp(5′)GmpCp-,[24] vaccinia virus mRNA contains m^7G(5′)pp(5′)Gmp- and m^7G(5′)pp(5′)Amp-,[25] HeLa cell mRNA contains m^7G(5′)ppp(5′) NmpNp- and m^7G(5′)ppp(5′)-NmpNmpNp-,[26] vesicular stomatitis virus mRNA contains G(5′)ppp(5′)-

* Relative amounts of snake venom phosphodiesterase required for the hydrolysis of d ($\text{pN}^{\alpha}\text{–N}^{\beta}$), d($\text{N}^{\alpha}\text{–N}^{\beta}$) and d($\text{N}^{\alpha}\text{–N}^{\beta}\text{p}$) are 1, 10 and 1000.[5]

Ap-,[27] and simian virus 40-specific RNA contains $m^7G(5')ppp(5')Nmp$-.[28] Nuclease P_1 has contributed to the identification of the above structures by splitting mRNA's to give $[m^7]G(5')pp[p](5')N[m]$ and 5′-nucleotides.

Nuclease P_1 also contributed to the identification of 2′-*O*-methylated oligonucleotides in ribosomal 18S and 28S RNA of a mouse hepatoma, MH 134, mainly by cleaving 3′-phosphomonoester linkages in the 2′-*O*-methylated tri- or tetranucleotides.[29] Nuclease P_1 was also employed for structural studies of *Bacillus subtilis* and *E. coli* tRNA's.[30,31] A phosphodiester linkage between $C_{3'}$ of a minor nucleoside and $C_{5'}$ of the adjacent nucleoside is resistant to nuclease P_1.[16] To degrade oligo- or polynucleotides containing such linkages completely, incubation should be carried out at a high temperature in the presence of a high concentration of the enzyme.[30,32]

As DNA cannot be completely decomposed chemically under mild conditions, nuclease P_1, which is capable of degrading DNA into 5′-mononucleotides, seems to be an important tool for direct determination of the base composition of DNA.

Nuclease P_1 is also used for preferential hydrolysis of single-stranded regions in nucleic acids, for the preparation of 2′-nucleotides from alkaline hydrolyzates of RNA, and for the removal of nucleic acids from cell extracts in the purification of poly(ADP-ribose) and other cellular materials.[32]

References

1. A. Kuninaka, *Shokuhin Kogyo* (Japanese), **4** (11), 9 (1961); *Hakko Kyokaishi* (Japanese), **20**, 311 (1962); *Baioteku* (Japanese), **1**, 65, 140 (1970); A. Kuninaka, M. Kibi and K. Sakaguchi, *Food Technol.*, **18**, 287 (1964); A. Kuninaka, Proc. Int. Symp. Conversion and Manufacture of Foodstuffs by Microorganisms, p. 235, Saikon, 1971.
2. A. Kuninaka, *Nippon Nogeikagaku Kaishi* (Japanese), **28**, 282 (1954); **29**, 52, 797, 801 (1955); **30**, 583 (1956); *J. Gen. Appl. Microbiol.*, **3**, 55 (1957); *Bull. Agr. Chem. Soc. Japan*, **23**, 281 (1959).
3. A. Kuninaka, *Nippon Nogeikagaku Kaishi* (Japanese), **34**, 489 (1960).
4. A. Kuninaka, S. Otsuka, Y. Kobayashi and K. Sakaguchi, *Bull. Agr. Chem. Soc. Japan,* **23**, 239 (1959); A. Kuninaka, M. Kibi, H. Yoshino and K. Sakaguchi, *Agr. Biol. Chem.*, **25**, 693 (1961).
5. M. Laskowski, Sr., *The Enzymes* IV (ed. P. D. Boyer) p. 313, Academic, 1971.
6. T. P. Wang, *J. Bact.*, **68**, 128 (1954).
7. P. Plackett, *Biochim. Biophys. Acta*, **26**, 664 (1957).
8. A. H. Mehler, *Introduction to Enzymology*, p. 251, Academic, 1957.
9. A. Kuninaka, T. Fujishima and M. Fujimoto, *Amino Acid and Nucleic Acid,* no. 16, 28, (1967).
10. M. Fujimoto, A. Kuninaka and H. Yoshino, *Agr. Biol. Chem.*, **33**, 1517 (1969); **38**, 777 (1974).

11. M. Fujimoto, A. Kuninaka and H. Yoshino, *ibid.*, **38**, 785 (1974).
12. M. Fujimoto, A. Kuninaka and H. Yoshino, *ibid.*, **38**, 1555 (1974).
13. M. Fujimoto, K. Fujiyama, A. Kuninaka and H. Yoshino, *ibid.*, **38**, 2141 (1974).
14. M. Fujimoto, A. Kuninaka and H. Yoshino, *ibid.*, **39**, 1991, 2145 (1975).
15. M. Fujimoto, A. Kuninaka, S. Yonei, T. Kohama and H. Yoshino, *ibid.*, **33**, 1666 (1969); T. Kohama, M. Fujimoto, A. Kuninaka and H. Yoshino, *ibid.*, **38**, 127 (1974); M. Ogura, T. Kohama, M. Fujimoto, A. Kuninaka, H. Yoshino and H. Sugiyama, *ibid.*, **38**, 2563 (1974).
16. A. Kuninaka, M. Fujimoto and H. Yoshino, *ibid.*, **39**, 597, 603 (1975).
17. D. Nomura, I. Hayakawa and K. Shinohara, *Hakkokogaku Zasshi* (Japanese), **52**, 35 (1974).
18. P. Cuatrecasas, M. Wilchek and C. B. Anfinsen, *Biochemistry*, **8**, 2277 (1969).
19. M. Fujimoto, A. Kuninaka and H. Yoshino, Ann. Mtg. Agr. Chem. Soc. Japan (Japanese), 1971. Abstracts, p. 250.
20. K. Shinohara, I. Hayakawa and D. Nomura, Ann. Mtg. Soc. Ferment. Technol. Japan (Japanese), 1972. Abstracts, p. 96.
21. T. Nishihara, *Tanpakushitsu Kakusan Koso* (Japanese), **20**, 608 (1975).
22. S. Muthukrishnan, G. W. Both, Y. Furuichi and A. J. Shatkin, *Nature*, **255**, 33, (1975).
23. Y. Furuichi and K. Miura, *ibid.*, **253**, 374 (1975).
24. Y. Furuichi, M. Morgan, S. Muthukrishnan and A. J. Shatkin, *Proc. Natl. Acad. Sci. U.S.A.* **72**, 362 (1975); Y. Furuichi, S. Muthukrishnan and A. J. Shatkin, *ibid.*, **72**, 742 (1975); N-L Chow and A.J. Shatkin, *J. Virol.* **15**, 1057 (1975).
25. T. Urushibara, Y. Furuichi, C. Nishimura and K. Miura, *FEBS Letters*, **49**, 385 (1975).
26. Y. Furuichi, M. Morgan, A. J. Shatkin, W. Jelinek, M. Salditt-Georgieff and J. E. Darnell, *Proc. Natl. Acad. Sci. U.S.A.*, **72**, 1904 (1975).
27. G. Abraham, D. P. Rhodes and A. K. Banerjee, *Nature*, **255**, 37, (1975).
28. S. Lavi and A. J. Shatkin, *Proc. Natl. Acad. Sci. U.S.A.*, **72**, 2012 (1975).
29. S. Hashimoto, M. Sakai and M. Muramatsu, *Biochemistry*, **14**, 1956 (1975).
30. Y. Yamada and H. Ishikura, *Biochim. Biophys. Acta*, **402**, 285 (1975).
31. Y. Yamada and H. Ishikura, *FEBS Letters*, **54**, 155 (1975).
32. A. Kuninaka, *Seikagaku Jikken Koza* (Japanese), vol. 2-II (ed. K. Imahori *et al.*), Tokyo Kagaku Dojin, 1976, in press.

CHAPTER **7**

Hydrolysis of Nucleic Acids with Streptomycete Enzymes*

Streptomycetes have become important in applied microbiology since the discovery of their ability to produce antibiotics. For this reason, however, the development and utilization of their biological activities other than the above ability lagged, compared with fungi, yeasts and bacteria. On the other hand, since Nomoto *et al.*[1] utilized *Streptomyces* for protease production, streptomycetes have been found to produce various enzymes such as protease,[2–4] aminopeptidase,[5–7] cellulase,[8] pullulanase,[9] pectate lyase,[10] α-galactosidase,[11,12] glucose isomerase,[13] arabinose isomerase,[14] fructokinase,[15] mannokinase,[15] uricase,[16] hyaluronidase,[17] phospholipase,[18] bacteriolytic enzyme[19] and nucleotide pyrophosphory-

* Yoshio NAKAO, Takeda Chemical Industries Ltd.

lase.[20,21] The application of streptomycete nuclease to fermentation processes, as will be described in this chapter, is an example of the ways in which the biochemical and enzymological features of this organism have been investigated and developed.

7.1 MICROORGANISMS

Nucleic acids are nucleotide polymers in which the adjoining 3′- and 5′-hydroxyl groups of the ribose moieties are linked via diester bonds of phosphoric acid. The enzymes shown in Fig. 7.1 are involved in the production of IMP and GMP by the hydrolysis of RNA. That is to say, the

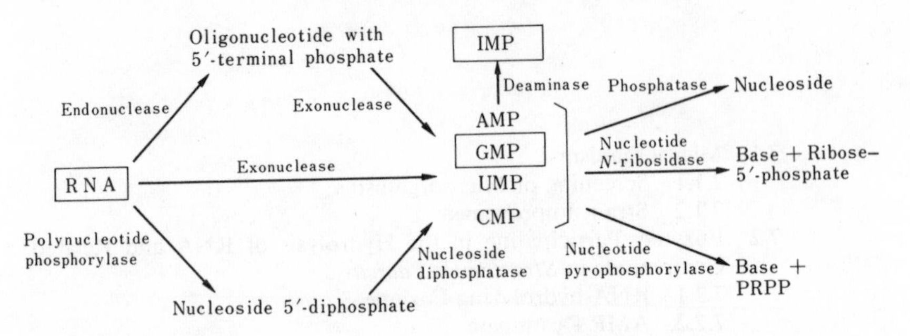

Fig. 7.1. Enzymes participating in the production of IMP and GMP by RNA hydrolysis.

following enzymes can participate in the hydrolysis of RNA to four kinds of 5′-nucleotides: (1) an endonuclease capable of hydrolyzing RNA to oligonucleotides with 5′-terminal phosphates and an exonuclease capable of hydrolyzing the oligonucleotides to 5′-nucleotides; (2) an exonuclease capable of hydrolyzing RNA directly to 5′-nucleotides; (3) a polynucleotide phosphorylase capable of degrading RNA to nucleoside diphosphates and a nucleoside diphosphatase capable of dephosphorylating the nucleoside diphosphates to 5′-nucleotides. In addition, AMP deaminase converts AMP to IMP. However, some enzymes are known to degrade 5′-nucleotides. For instance, 5′-nucleotidase or nonspecific phosphatase dephosphorylates 5′-nucleotides to nucleosides. Nucleoside *N*-ribosidase[22–27] hydrolyzes 5′-nucleotides to bases and ribose-5-phosphate. Nucleotide pyrophosphorylase[28–30] degrades 5′-nucleotides to bases and PRPP in the presence of PPi. Therefore, microorganisms which possess enzyme systems able to hydrolyze RNA to 5′-nucleotides and AMP deaminase, but which lack enzymes to degrade 5′-nucleotides, must be selected in order to produce IMP and GMP by the enzymatic hydrolysis of RNA.

7.1.1 Screening of Microorganisms

The establishment of a specific, rapid and convenient screening method is necessary to select microorganisms for a particular purpose. In order to select strains suitable for the production of IMP and GMP from RNA, the best method is to determine the 5′-nucleotides in the hydrolyzate of RNA under suitable conditions by using the culture broth of various strains as the enzyme source. A variety of methods has been established, as described in Chapter 3. Among them, the method using an enzyme having high substrate specificity is suitable for determining 5′-nucleotides rapidly and accurately in many samples. Based on the results of examination of RNA digests by *Streptomyces* culture filtrates, Ogata *et al.*[31] stated that the strains listed in Table 7.1 hydrolyzed RNA to 5′-nucleotides. 5′-Adenylic acid was found in the RNA hydrolyzates from most *Streptomyces* strains. 5′-Nucleotides were detected in RNA digests with culture broths of *Streptomyces aureus*, but little AMP was found. Therefore, it was suggested that this strain produced an enzyme capable of deaminating AMP together with the enzyme capable of hydrolyzing RNA into 5′-nucleotides. That is, the organism was considered to be a superior strain for the production of IMP and GMP by the hydrolysis of RNA. Sugimoto *et al.*[32]

TABLE 7.1
Hydrolysis of RNA by Streptomycetes

Microorganism	Degradation of RNA(%)	Formation of AMP (%)	Formation of adenosine(%)	Formation of 5′-nucleotides(%)
Streptomyces coelicolor	71.2	27.3	38.5	29.0
S. albogriseolus	90.0	20.0	66.0	40.0
S. aureus	74.2	2.0	10.7	39.3
S. gougerotii	72.0	15.0	23.0	14.0
S. griseoflavus	47.6	10.7	7.0	12.0
S. griseus	51.3	7.5	13.4	8.1
S. purpurascens	76.0	22.0	32.0	16.7
S. ruber	70.0	10.3	19.0	10.8
S. viridochromogenes	65.0	29.0	36.0	18.5

also reported that *Streptomyces* sp. No. 41 produced an enzyme hydrolyzing RNA to 5′-nucleotides.

7.1.2 Strain Improvement

As will be described in section 7.3, *S. aureus* simultaneously produces an endonuclease capable of hydrolyzing RNA to oligonucleotides having 5′-terminal phosphates, an exonuclease capable of hydrolyzing RNA or oligonucleotides formed by the above endonuclease to 5′-nucleotides, a deaminase capable of converting AMP into IMP, a 5′-nucleotidase capable of catalyzing the dephosphorylation of 5′-nucleotides, and a nonspecific alkaline phosphatase. However, neither nucleotide *N*-ribosidase nor nucleotide pyrophosphorylase was detected in the culture broth. To produce IMP and GMP from RNA in a high yield, it is necessary to prepare an enzyme solution which has high endonuclease, exonuclease and AMP deaminase activities and low 5′-nucleotidase and nonspecific alkaline phosphatase activities. First, strain improvement was carried out to enhance the production of endonuclease, exonuclease and AMP deaminase, and to reduce that of 5′-nucleotidase and alkaline phosphatase by the wild strain. The enzyme productivity of characteristic mutants is compared with that of the parent in Table 7.2. As the production of exonuclease increased, that

TABLE 7.2

Comparison of Enzyme Production by Mutants Derived from *Streptomyces aureus*

Strain	Endonuclease	Exonuclease	AMP deaminase	5′-Nucleotidase	Nonspecific alkaline phosphatase
Parent	100	100	100	100	100
K-1	310	180	110	160	20
A-5	400	210	150	180	10
S-8	100	80	100	60	10

of 5′-nucleotidase also rose. However, exonuclease production dropped in mutants having low 5′-nucleotidase productivity. The productivities of endonuclease, AMP deaminase and alkaline phosphatase changed independently. Finally, the productivities of endonuclease, exonuclease and AMP deaminase of the superior mutant, A-5, were enhanced about 4-, 2.1- and 1.5-fold, respectively, compared with those of the parent. On the other hand, alkaline phosphatase production by the mutant declined to about one-tenth, while 5′-nucleotidase production rose 1.8-fold.

7.2 Enzymes Participating in the Hydrolysis of RNA and Related Compounds in *Streptomyces aureus*

7.2.1 RNA-hydrolyzing Enzyme

Many investigations have been carried out on RNA-hydrolyzing enzymes of bacteria, fungi and yeasts, and various enzymes having interesting substrate specificities have been found. However, few studies have been carried out on RNA-hydrolyzing enzymes in *Streptomyces* species. Tanaka[33] purified the RNA-hydrolyzing enzyme of *S. erythreus* and demonstrated that the enzyme hydrolyzed RNA to oligonucleotides with terminal 3′-GMP and to 3′-GMP itself, during examination of its mode of action. Therefore, the ribonuclease of this organism is similar to the ribonuclease T_1[34–38] detected in *Aspergillus oryzae* by Egami *et al.* Yoneda[39] reported that *S. albogriseolus* also produced ribonuclease with a similar substrate specificity and mode of action to ribonuclease T_1. This is also the case with *S. aureofaciens.*[40,41] The RNA-hydrolyzing enzymes produced by *S. aureus* were investigated in detail by Yoneda[42] and Shimizu *et al.*[43] The organism produces at least two kinds of RNA-hydrolyzing enzymes; endonuclease and exonuclease. The endonuclease was purified about 130-fold from culture filtrates by CM-cellulose and DEAE-cellulose column chromatographies, gel filtration on Sephadex G-100, etc. and was finally isolated as crystals. This enzyme protein was shown to be homogeneous on ultracentrifugation and electrophoresis. The molecular weight was estimated to be about 16,000 and the isoelectric point was at pH 10.0. The optimum pH and temperature for the enzyme reaction were about 7.0 and 40° C, respectively. The enzyme was stable in the pH range 6 to 8 at room temperature, but was unstable at pH's lower than 5 and higher than 9. The enzyme activity was inhibited by EDTA and activated by Mg^{2+} or Mn^{2+}. The enzyme hydrolyzed both RNA and DNA to di- through nona-oligonucleotides bearing 5′-terminal phosphate. No base specificity was observed. With synthetic homopolymers, the enzyme hydrolyzed poly(A) rapidly, and poly(C) and poly(U) slowly. However, poly(G) was hardly hydrolyzed. The degradation product of poly(A) consisted of tri- through hexa-oligoadenylates bearing 5′-terminal phosphate; 5′-nucleotides were not produced by hydrolysis of RNA, DNA and poly(A). Therefore, the enzyme is an endonuclease and has properties similar to those of nucleases found in *Azotobacter agilis,*[44] *Bacillus brevis,*[45] *Saccharomyces fragilis*,[46,47] *Neurospora crassa*,[48] *Acrocylindrium* sp.,[49,50] judging from the activity toward RNA and DNA. However, it differs from these nucleases as regards its action toward synthetic homopolymers.

Exonuclease was partially purified from the culture filtrate by ammonium sulfate fractionation, DEAE-cellulose column chromatography and gel filtration on Sephadex G-200. The optimum pH and temperature of the enzyme reaction were about 7.5 and 60° C, respectively. The enzyme hydrolyzed both RNA and DNA, but not bis(*p*-nitrophenyl)phosphate. Assuming that the relative rate of hydrolysis of yeast RNA is 100, those of oligonucleotides with 5′-terminal phosphate (obtained by the hydrolysis of RNA using the above endonuclease) and oligonucleotides having 3′-terminal phosphate (obtained by the hydrolysis of RNA using pancreatic ribonuclease) were 400 and 100, respectively. As this enzyme hydrolyzed every substrate to 5′-nucleotides, it is considered to be an exonuclease. Sugimoto *et al.*[32] detected a similar exonuclease in *Streptomyces* sp. No. 41. Exonucleases of streptomycetes are different from those of *Penicillium*,[51–54] *Phoma*,[55] *Monascus*,[56] *Aspergillus*,[57] *Acrocylindrium*[58] and *Pellicularia*[59] in terms of optimum pH and other properties. In the production of 5′-nucleotides by the enzymatic hydrolysis of RNA, an enzyme solution containing both exonuclease and endonuclease is preferable to exonuclease alone, judging from the above results.

7.2.2 AMP Deaminase

As mentioned above, *S. aureus* produces a deaminase which converts AMP to IMP. The enzyme was partially purified from the culture filtrate by ammonium sulfate precipitation, DEAE-cellulose column chromatography and gel filtration on Sephadex G-100.

The optimum pH and temperature of the enzyme reaction were about 5.5 and 50° C, respectively. The enzyme deaminated not only AMP but also ATP, ADP and adenosine. However, it hardly reacted with 3′-AMP, and did not react with adenine. Therefore, the enzyme resembles the deaminases of *Microsporum andouini*,[60–62] *Aspergillus glaucus*[63] and *Aspergillus repens*[63] rather than the nonspecific adenosine deaminases of *Aspergillus oryzae*,[64] *Aspergillus melleus*[65] and *Aspergillus ochraceus*.[63] Sugimoto *et al.*[66] found AMP deaminase in *Streptomyces* sp. No. 41, but its substrate specificity is so far unknown.

7.2.3 Phosphatase

S. aureus produces at least two kinds of phosphatase, 5′-nucleotidase and alkaline phosphatase, in the culture broth. 5′-Nucleotidase was partially purified from the culture broth by gel filtration on Sephadex G-100 and DEAE-Sephadex A-25 column chromatography. The optimum pH and temperature of the enzyme reaction were about 8.0 and 37° C, respectively. In the presence of Ca^{2+}, the optimum temperature rose to 55° C. The enzyme was relatively unstable to heat and more than 90% of the activity was lost on heating at 45° C for 5 min (pH 8.0). However, the presence of Ca^{2+} increased the heat stability and 30% of the activity still remained after incubation for 30 min at 65° C. The enzyme hydrolyzed purine nucleotides more rapidly than pyrimidine nucleotides. It did not react with 3′-nucleotides, various sugar phosphates or synthetic substrates, such as phenylphosphate. Iwasa *et al.*[67] found a 5′-nucleotidase with similar properties in *Streptomyces* sp. No. 41. The properties of 5′-nucleotidase of *S. aureus* are different from those of other 5′-nucleotidases of bacteria such as *Escherichia coli*,[68] *Micrococcus radiodurans*[69] and of fungi such as *Neurospora crassa.*[70] In addition to this 5′-nucleotidase, *S. aureus* produces an alkaline phosphatase, which has a pH optimum of around 7.3 and an optimum temperature of about 40° C. The phosphatase attacks 3′-nucleotides, sugar phosphates and synthetic substrates, in addition to 5′-nucleotides.

7.3 Production of 5′-Nucleotides

The production of 5′-nucleotides by the hydrolysis of RNA involves three steps; i.e. RNA-hydrolyzing enzyme production, RNA hydrolysis by the enzyme, and separation and purification of 5′-nucleotides from the hydrolyzate.

7.3.1 Production of RNA-Hydrolyzing Enzyme

As described above, *S. aureus* produces endonuclease, exonuclease, AMP deaminase, 5′-nucleotidase and alkaline phosphatase. The superior mutant mentioned previously produces large amounts of endonuclease, exonuclease and AMP deaminase and a small amount of alkaline phosphatase. However, the production of 5′-nucleotidase is still rather high in the mutant. Suitable culture media and conditions were then investigated to stimulate the production of enzymes capable of hydrolyzing RNA to 5′-nucleotides and to repress the production of enzymes capable of degrading 5′-nucleotides. As in the case of general enzyme production, the

TABLE 7.3
Culture Medium for the Production of RNA-Hydrolyzing Enzymes

Enzymatic hydrolyzate of starch	3.0 %
Soybean meal	2.0 %
Corn steep liquor	1.0 %
$(NH_4)_2SO_4$	0.1 %
$MgSO_4 \cdot 7H_2O$	0.05 %
Soybean oil	0.005%
$CaCO_3$	0.5 %

kind and amount of nitrogen source affected the yield of RNA-hydrolyzing enzyme. It is well-known that Pi represses the formation of phosphatase. In the case of *S. aureus*, Pi markedly repressed the formation of the alkaline phosphatase. However, the repression of 5′-nucleotidase formation required the addition of a large amount of Pi to the medium, and under these conditions, the formation of endonuclease and exonuclease was simultaneously repressed. Various kinds of carbon and nitrogen sources are available as production media for RNA-hydrolyzing enzymes. One example of a suitable composition is shown in Table 7.3.

When *S. aureus* was cultivated in the medium shown in Table 7.3 at 28° C using a 6 m^3 fermenter, the production of each enzyme reached a maximum at 30 h, as shown in Fig. 7.2. After cultivation, the cells were separated and the cell-free culture broth was used as an enzyme solution for the hydrolysis of RNA. Sugimoto *et al.*[71,72] examined the effect of the composition of the medium and conditions of aeration and agitation on the formation of nuclease, AMP deaminase and 5′-nucleotidase using *Streptomyces* sp. No. 41, and found that the optimum oxygen transfer volume coefficient (K_d) was 2.72×10^{-5} (g mol. O_2/ml.min.atm).

7.3.2 Enzymatic Hydrolysis of RNA

Improvement of the strain, medium composition and culture conditions permits the preparation of an enzyme solution which has high endonuclease, exonuclease and AMP deaminase activities, and low 5′-nucleotide-degrading enzyme activity. Furthermore, a method of selectively inactivating 5′-nucleotidase or alkaline phosphatase in RNA-degrading enzyme solutions, prior to the enzymatic hydrolysis of RNA, has been

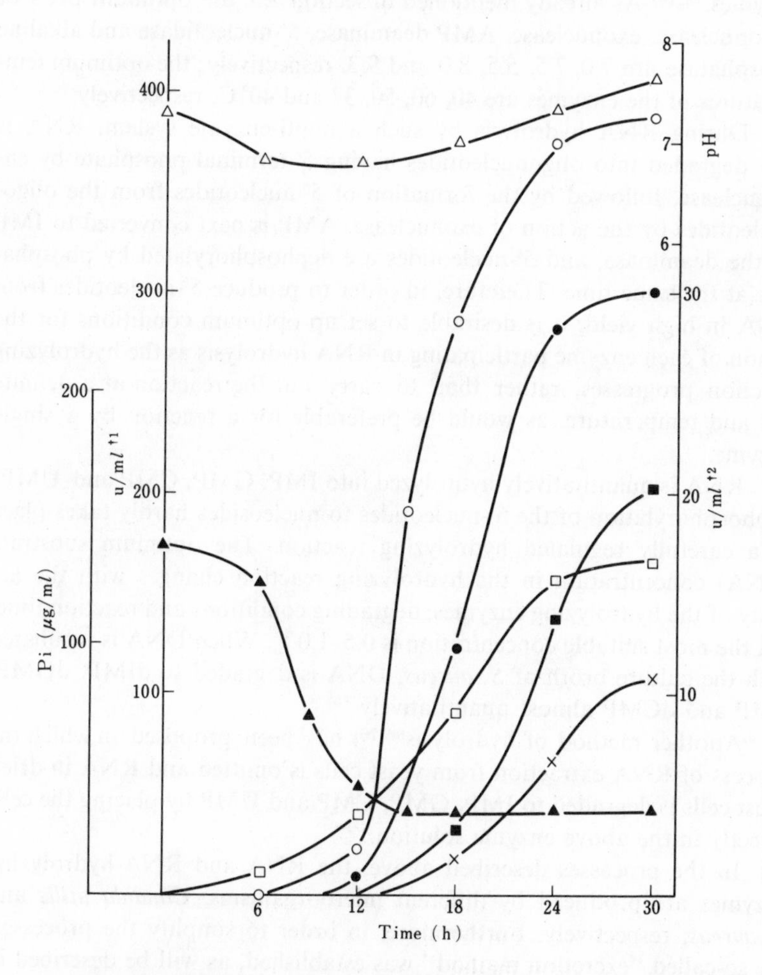

Fig. 7.2. Time course of the production of enzymes participating in the hydrolysis of nucleic acid-related substances.
○, Endonuclease (u/ml)†1; ▲, Pi; △, pH.
●, Exonuclease; □, 5′-AMP deaminase; ■, 5′-nucleotidase;
×, alkaline phosphatase (u/ml)†2.
†1The amount of enzyme capable of causing an increase in the absorbance of 0.1 at 260 nm per 30 min under optimum conditions.
†2The amount of enzyme capable of forming 1 μmole of product per min under optimum conditions.

established by making use of the differences in heat stability of these enzymes.[73,74] As already mentioned in section 7.3, the optimum pH's of endonuclease, exonuclease, AMP deaminase, 5′-nucleotidase and alkaline phosphatase are 7.0, 7.5, 5.5, 8.0 and 9.3, respectively; the optimum temperatures of the enzymes are 40, 60, 50, 37 and 40° C, respectively.

During RNA hydrolysis by such a multi-enzyme system, RNA is first degraded into oligonucleotides having 5′-terminal phosphate by endonuclease, followed by the formation of 5′-nucleotides from the oligonucleotides by the action of exonuclease. AMP is next converted to IMP by the deaminase, and 5′-nucleotides are dephosphorylated by phosphatase at the same time. Therefore, in order to produce 5′-nucleotides from RNA in high yield, it is desirable to set up optimum conditions for the action of each enzyme participating in RNA hydrolysis as the hydrolyzing reaction progresses, rather than to carry out the reaction at a definite pH and temperature, as would be preferable for a reaction by a single enzyme.

RNA is quantitatively hydrolyzed into IMP, GMP, CMP and UMP; dephosphorylation of the 5′-nucleotides to nucleosides hardly takes place in a carefully regulated hydrolyzing reaction. The optimum substrate (RNA) concentration in the hydrolyzing reaction changes with the activity of the hydrolyzing enzymes, degrading conditions and reaction time, but the most suitable concentration is 0.5–1.0%. When DNA is incubated with the culture broth of *S. aureus*, DNA is degraded to dIMP, dGMP, TMP and dCMP almost quantitatively.[75]

Another method of hydrolysis[66,76] has been proposed in which the process of RNA extraction from yeast cells is omitted and RNA in dried yeast cells is degraded to IMP, GMP, CMP and UMP by placing the cells directly in the above enzyme solution.

In the processes described above, the RNA and RNA-hydrolyzing enzymes are produced by different microorganisms, *Candida utilis* and *S. aureus*, respectively. Furthermore, in order to simplify the processes, the so-called "excretion method" was established, as will be described in Chapter 8.

In this method, RNA and its hydrolyzing enzymes are formed simultaneously in the same microbial cells and the RNA is then hydrolyzed to 5′-nucleotides by autolysis under suitable conditions.

7.3.3 Purification and Isolation of 5′-Nucleotides

In order to purify and to isolate the resulting 5′-nucleotides, IMP, GMP, UMP and CMP, from the enzymatic hydrolyzate of RNA, several methods have been developed.

As the phosphate groups of 5′-nucleotides are almost completely in the anionic form under alkaline conditions, 5′-nucleotides are adsorbed on a strongly alkaline anion-exchange resin and can be eluted fractionally with hydrochloric acid or formic acid in the order CMP, UMP, IMP and GMP, based on the differences in their charge, which correspond to the degrees of dissociation of amino groups and phosphate groups of the 5′-nucleotides under acidic conditions. Each of the 5′-nucleotides in the fractional eluates is adsorbed on charcoal for desalting, followed by elution with dilute sodium hydroxide, or is precipitated as its barium salt, followed by reaction with sodium sulfate to convert it to the sodium salt. It is then concentrated and crystallized by adding a lower alcohol to the concentrate. In general, IMP, CMP and UMP can be readily crystallized, but it is difficult to crystallize GMP. Various methods for the crystallization of GMP have been devised.[77,78]

Employment of this strongly alkaline anion-exchange resin method for industrial production is not entirely satisfactory. One of the major problems is that the phosphate groups of 5′-nucleotides are strongly adsorbed on the resin, and in addition, the differences in the affinities between each nucleotide and the resin are so small that large amounts of various solvents with different acidity or ionic strength are required in order to isolate the four 5′-nucleotides.

Another method[79,80] has been developed for the separation of 5′-nucleotides by using a strong cation-exchange resin (H^+-form). The principle of the method is to separate 5′-nucleotides from each other based on differences in the affinity of the amino groups of 5′-nucleotides for the resin. This method seems to be effective for the separation of 5′-nucleotides having amino groups from 5′-nucleotides without amino groups, though it is not satisfactory for the isolation of the present four 5′-nucleotides.[81,82] Column chromatography on ion-exchange resins can be used effectively to separate 5′-nucleotides, but is not an appropriate method for the industrial production of large amounts of 5′-nucleotides. Therefore, extensive investigations have been carried out to find a suitable method to recover both IMP and GMP, which have strong flavors, from the enzymatic hydrolyzate of RNA in a high yield.

Charcoal was introduced for use in the fractionation of 5′-nucleotides by Tanaka *et al.*[83,84] The mixture of 5′-nucleotides was adsorbed on a charcoal column and 5′-pyrimidine nucleotides were eluted first, followed by 5′-purine nucleotides, using alkaline water-miscible solvents such as a dilute ammonia or dilute methanol-ammonia solution. This is a convenient method for the separation of flavoring 5′-purine nucleotides from non-flavoring 5′-pyrimidine nucleotides.

Sanno *et al.*[85] proposed a fractional precipitation method for the separation of 5′-nucleotides. The enzymatic hydrolyzate of RNA was treated with ion-exchange resin or charcoal to remove impurities. A lower alcohol was added to the purified solution of 5′-nucleotides in an amount sufficient to precipitate 5′-purine nucleotides and 5′-pyrimidine nucleotides fractionally. The precipitate of 5′-purine nucleotides, consisting of IMP and GMP, however, was amorphous and inconvenient for industrial handling.

Ishibashi *et al.*,[86] as a result of various investigations, found that IMP and GMP could be crystallized as large mixed crystals. Mixed crystals of IMP·Na_2 and GMP·Na_2 showed the same X-ray diffraction pattern as IMP·Na_2 or GMP·Na_2 alone, and they were free from amorphous precipitate. In other words, the mixed crystals did not show a mixed X-ray diffraction pattern corresponding to both IMP·Na_2 and GMP·Na_2. Furthermore, mixed crystals obtained from a solution in which the content of IMP·Na_2 was larger than that of GMP·Na_2, showed the X-ray pattern of IMP·Na_2 alone, while those from a solution in which the content of GMP·Na_2 was larger showed the pattern of GMP·Na_2 alone. The mixed crystals were further confirmed to represent the α-crystal form of IMP·Na_2 or GMP·Na_2 by means of vibration and Weissenberg photographs of a single mixed crystal selected under a polarized microscope. It was thus shown that IMP·Na_2 and GMP·Na_2 can form mixed crystals in all proportions.[87]

Other purification methods, such as precipitation[88] as metallic (zinc, aluminum, cobalt or nickel) salts, extraction[89] with organic solvents, and purification[90] by means of decolorizing resins, have also been devised.

References

1. M. Nomoto and Y. Narahashi, *J. Biochem.* (Tokyo), **46**, 653 (1959).
2. Y. Narahashi and M. Yanagita, *ibid.*, **62**, 633 (1967).
3. Y. Narahashi, K. Shibuya and M. Yanagita, *ibid.*, **64**, 427 (1968).
4. T. Nakanishi, Y. Matsumura, N. Minamiura and T. Yamamoto, *Agr. Biol. Chem.*, **38**, 37 (1974).
5. T. Uwajima, N. Yoshikawa and O. Terada, *ibid.*, **36**, 2047 (1972).
6. T. Uwajima, N. Yoshikawa and O. Terada, *ibid.*, **37**, 2727 (1973).
7. K. D. Vosbeck, K. Chow and W. M. Awad, *J. Biol. Chem.*, **248**, 6029 (1973).
8. A. S. Perlin, *Advances in Enzymic Hydrolysis of Cellulose and Related Materials* (ed. E. T. Reese), p. 185, Pergamon, 1963.
9. S. Ueda, M. Yagisawa and Y. Sato, *J. Ferment. Technol.*, **49**, 552 (1971).
10. M. Sato and A. Kaji, *Agr. Biol. Chem.*, **39**, 819 (1975).
11. K. Oishi and K. Aida, *ibid.*, **35**, 1101 (1971).
12. K. Oishi and K. Aida, *ibid.*, **36**, 578 (1972).
13. Y. Takasaki, *ibid.*, **38**, 667 (1974).

14. K. Yamanaka and K. Izumori, *ibid.*, **37**, 521 (1973).
15. B. Sabater, J. Sebastian and C. Asensio, *Biochim. Biophys. Acta*, **284**, 406, 414 (1972).
16. Y. Watanabe, M. Yano and J. Fukumoto, *Agr. Biol. Chem.*, **33**, 1282 (1969).
17. T. Ohya and Y. Kaneko, *Biochim. Biophys. Acta*, **198**, 607 (1970).
18. T. Yamaguchi, Y. Okawa, K. Sakaguchi and N. Muto, *Agr. Biol. Chem.*, **37**, 1667 (1973).
19. K. Yokogawa, S. Kawata and Y. Yoshimura, *ibid.*, **36**, 2055 (1972).
20. S. Murao and T. Nishino, *ibid.*, **38**, 2483, 2491 (1974).
21. T. Nishino and S. Murao, *ibid.*, **39**, 1007 (1975).
22. J. Hurwitz, L. A. Heppel and B. L. Horecker, *J. Biol. Chem.*, **226**, 526 (1957).
23. A. Kuninaka, *Nippon Nogeikagaku Kaishi* (Japanese), **30**, 583 (1956).
24. A. Kuninaka, *Bull. Agr. Chem. Soc., Japan*, **23**, 281 (1959).
25. A. Imada, M. Kuno and S. Igarasi, *J. Gen. Appl. Microbiol.*, **13**, 255 (1967).
26. A. Imada, *ibid.*, **13**, 267 (1967).
27. T. Sakai, T. Watanabe and I. Chibata, *J. Ferment. Technol.*, **46**, 202 (1968).
28. A. Kornberg, I. Liebermann and E. S. Simms, *J. Biol. Chem.*, **215**, 417 (1955).
29. I. Crawford, A. Kornberg and E. S. Simms, *ibid.*, **226**, 1093 (1957).
30. S. H. Hughes, G. M. Wahl and M. R. Capecchi, *ibid.*, **250**, 120 (1975).
31. K. Ogata, Y. Nakao, S. Igarasi, E. Omura, Y. Sugino, M. Yoneda and I. Suhara, *Agr. Biol. Chem.*, **27**, 110 (1963).
32. H. Sugimoto, T. Iwasa and T. Yokotsuka, *Nippon Nogeikagaku Kaishi* (Japanese), **38**, 135 (1964).
33. K. Tanaka, *J. Biochem.* (Tokyo), **50**, 62 (1961).
34. K. Sato and F. Egami, *ibid.*, **44**, 753 (1957).
35. K. Takahashi, *ibid.*, **49**, 7 (1961).
36. K. Takahashi, *ibid.*, **51**, 95 (1962).
37. K. Takahashi, *ibid.*, **52**, 72 (1962).
38. K. Takahashi, *J. Biol. Chem.*, **240**, PC 4117 (1965).
39. M. Yoneda, *J. Biochem.* (Tokyo), **55**, 469 (1964).
40. M. Bacová, E. Zelinková and J. Zelinka, *Biochim. Biophys. Acta*, **235**, 335 (1971).
41. E. Zelinková, M. Bacová and J. Zelinka, *ibid.*, **235**, 343 (1971).
42. M. Yoneda, *J. Biochem.* (Tokyo), **55**, 475, 481 (1964).
43. M. Shimizu, S. Akiyama and B. Nakazawa, Abstracts 18th Symp. Ass. Amino Acid and Nucleic Acid, 1969.
44. A. Stevens and R. J. Hilmoe, *J. Biol. Chem.*, **235**, 3016, 3023 (1960).
45. N. Sarkar and H. Paulus, *ibid.*, **250**, 684 (1975).
46. Y. Nakao, S. Y. Lee, H. O. Halvorson and R. M. Bock, *Biochim. Biophys. Acta*, **151**, 114 (1968).
47. S. Y. Lee, Y. Nakao and R. M. Bock, *ibid.*, **151**, 126 (1968).
48. S. Linn and I. R. Lehman, *J. Biol. Chem.*, **240**, 1287, 1294 (1965).
49. I. Suhara and M. Yoneda, *J. Biochem.* (Tokyo), **73**, 647 (1973).
50. I. Suhara, *ibid.*, **73**, 1023 (1973).
51. A. Kuninaka, M. Kibi, H. Yoshino and K. Sakaguchi, *Agr. Biol. Chem.*, **25**, 693 (1961).
52. M. Fujimoto, A. Kuninaka and H. Yoshino, *ibid.*, **33**, 1517 (1969).
53. M. Fujimoto, A. Kuninaka and H. Yoshino, *ibid.*, **38**, 777, 785, 1555 (1974).
54. M. Fujimoto, K. Fujiyama, A. Kuninaka and H. Yoshino, *ibid.*, **38**, 2141 (1974).
55. H. Tone and A. Ozaki, *Enzymologia*, **34**, 101 (1968).
56. E. Soeda, A. Murata, M. Utsu and R. Saruno, *Seikagaku* (Japanese), **40**, 688 (1968).
57. J. Ando, *Biochim. Biophys. Acta*, **114**, 158 (1966).
58. I. Suhara, *J. Biochem.* (Tokyo), **75**, 1135 (1974).
59. Y. Fujimura, Y. Hasegawa, Y. Kaneko and S. Doi, *Agr. Biol. Chem.*, **31**, 92 (1967).
60. K. Aida, S. Chung, I. Suzuki and T. Yagi, *ibid.*, **29**, 508 (1965).
61. S. Chung and K. Aida, *J. Biochem.* (Tokyo), **61**, 1 (1967).
62. S. Chung, K. Aida and T. Uemura, *J. Gen. Appl. Microbiol.*, **13**, 237 (1967).

63. S. Chung, K. Aida and T. Uemura, *ibid.*, **13**, 335 (1967).
64. N. O. Kaplan, S. P. Colowick and M. M. Ciotti, *J. Biol. Chem.*, **194**, 579 (1952).
65. T. Fujishima and H. Yoshino, *Amino Acid and Nucleic Acid*, no. 16, 45 (1967).
66. H. Sugimoto, T. Iwasa, J. Ishiyama and T. Yokotsuka, *Nippon Nogeikagaku Kaishi* (Japanese), **37**, 677 (1963).
67. T. Iwasa, H. Sugimoto, M. Harada and T. Yokotsuka, *ibid.*, **41**, 386 (1967).
68. H. C. Neu, *J. Biol. Chem.*, **242**, 3896 (1967).
69. R. E. J. Mitchel, *Biochim. Biophys. Acta*, **309**, 116 (1973).
70. A. C. Olson and M. J. Fraser, *ibid.*, **334**, 156 (1974).
71. H. Sugimoto, T. Iwasa and T. Yokotsuka, *Nippon Nogeikagaku Kaishi* (Japanese), **38**, 144 (1964).
72. H. Sugimoto, S. Ishii, T. Iwasa and T. Yokotsuka, *ibid.*, **40**, 93 (1966).
73. Y. Ishida, K. Mochizuki, M. Uchida, K. Wakita and Y. Nakao, *Japanese Patent* No. 40–462113 (1965).
74. H. Sugimoto, T. Iwasa, S. Ishii and T. Yokotsuka, *Nippon Nogeikagaku Kaishi* (Japanese), **38**, 441 (1964).
75. Y. Nakao and K. Ogata, *Agr. Biol. Chem.*, **27**, 491 (1963).
76. H. Fukuda, S. Yashima, S. Akiyama, T. Fugono, B. Nakazawa and T. Tada, *British Patent* No. 933829 (1963).
77. J. Ishibashi and K. Ito, *Japanese Patent* No. 45–591077 (1970).
78. J. Ishibashi and K. Ito, *Japanese Patent Publication* No. 40–25623 (1965).
79. K. Tanaka, K. Mizuno, Y. Sanno and Y. Hamuro, *Japanese Patent Publication* No. 38–16892 (1963).
80. Y. Ueno, S. Wada, I. Imada and K: Tomoda, *Japanese Patent* No. 40–460728 (1965).
81. U. Hirose, K. Nara and M. Inoue, *Japanese Patent* No. 40–460730 (1965).
82. K. Tanaka, K. Mizuno, Y. Sanno and Y. Hamuro, *Japanese Patent* No. 41–477465 (1966).
83. K. Tanaka, E. Omura, M. Honjo, Y. Sanno and Y. Sugino, *Japanese Patent* No. 39–315762 (1964).
84. K. Tanaka, E. Omura, K. Ogata, Y. Sanno, M. Yoneda and I. Suhara, *Japanese Patent* No. 39–315920 (1964).
85. Y. Sanno, K. Nara, S. Minato and U. Hirose, *Japanese Patent* No. 39–316262 (1964).
86. J. Ishibashi, H. Kamio and M. Yoneda, *Japanese Patent* No. 40–490544 (1965).
87. H. Kamio and H. Nakamachi, *Yakugaku Zasshi* (Japanese), **87**, 1436 (1967).
88. E. Omura and M. Yoneda, *Japanese Patent Publication* No. 38–16894 (1963).
89. E. Omura, M. Yoneda, I. Suhara and Y. Hamuro, *Japanese Patent* No. 38–313212 (1963).
90. K. Miyai, N. Kameyama and R. Kamio, *Japanese Patent Publication* No. 40–3995 (1965).

CHAPTER **8**

Degradation of Cellular RNA by Endogenous Enzymes*

Various stresses upon microbial cells initiate the degradation of cellular RNA by endogenous enzymes, leading in some cases to the formation of nucleotides. The manufacture of nucleotides by such autodegradative processes has been called the autodegradation or excretion method and has been described in several reviews.[1-5]

In normally growing bacteria, for example in *Escherichia coli*, about 3% of cellular RNA exists as mRNA, about 15% as tRNA and about 80% as rRNA.[6] In the autodegradation method for producing nucleotides, the degradation of rRNA, which is the most abundant RNA species, should be the primary consideration. Ribosomes or rRNA are stable and turn over only slowly under normal growing conditions.[7] They are not readily attacked by RNA degrading enzymes, even though the substrate and the enzymes exist together in the same cell. This stability of rRNA in normally growing cells can be ascribed to the following factors: (1) rRNA in intact ribosomes is folded and is structurally resistant to enzyme attack. (2) RNA degrading enzymes are attached to ribosomes in latent or inactive forms. (3) The action of RNA degrading enzymes is suppressed in

* Akira IMADA, Takeda Chemical Industries Ltd.

cells by inhibitors such as nucleotides and proteins. (4) rRNA and the degradative enzymes are localized at cytologically separate sites (compartmentalization). (5) Ribosomes are more stable when attached to the cytoplasmic membrane as polysomes.

The degradation of RNA by endogenous enzymes is promoted by various stresses which destroy these stabilizing factors and it proceeds actively under conditions favorable for the action of the RNA degrading enzymes. Kaplan and Apirion[8] have proposed a model for the degradation of rRNA (Fig. 8.1). Ribosomes in polysomes are split into subunits via monosomes. The subunits are then denatured and finally degraded to mononucleotides by the action of the enzymes indicated. Their model, elaborated from studies on RNA catabolism in carbon-starved *E. coli*, may be generalized to include the degradation of endogenous RNA in other circumstances.

Maaløe and Kjeldgaard[9] suggested that there is a correlation between the growth rate and the content of ribosomes. The correlation is such that the ribosome content becomes zero when the growth rate is extrapolated to zero. The degradation of cellular RNA by endogenous enzymes, which is very commonly observed when microorganisms encounter unfavorable conditions for growth, may reflect a certain regulatory mechanism permitting the organisms to escape death due to unbalanced growth, i.e. overproduction of proteins by excess ribosomes. In fact the degradation of RNA

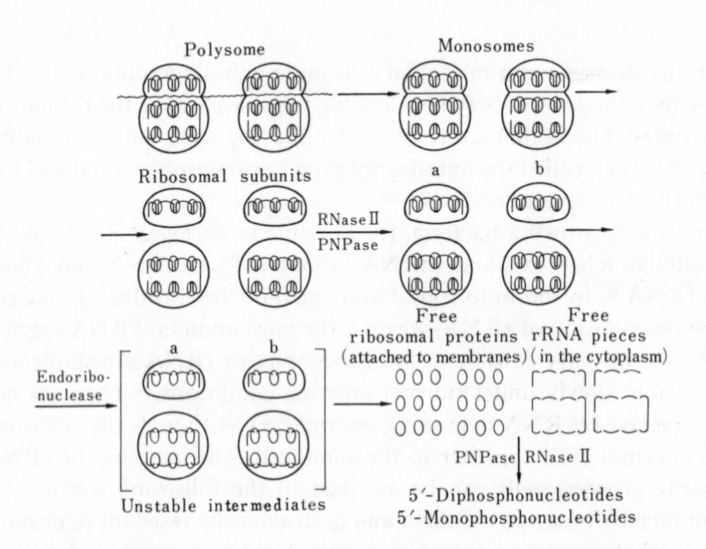

Fig 8.1. Process of degradation of rRNA. (A model elaborated from studies on metabolism in carbon-starved *E. coli*[8]). PNPase = Polynucleotide phosphorylase. (Source: ref. 8. Reproduced by kind permission of the American Society of Biological Chemists, Inc., U.S.A.)

has been shown to be necessary for *E. coli* to survive under substrate-starved conditions.[8] One can also regard the degradation of RNA as an energy generating system under conditions of endogenous respiration.[10]

8.1 Degradation of RNA as a Result of Various Stresses

The mechanism by which various stresses bring about RNA degradation can be considered to be as follows.

Temperature: The degradation of endogenous RNA by exposure to high and low temperatures has been reported to occur by damage to the permeability barrier.[11–16] In a temperature-sensitive *E. coli*, thermal damage to the permeability barrier permits the periplasmic RNase I to leak into the cytoplasm and degrade RNA.[16] The heat resistance of bacteria can be correlated with the heat stability of their ribosomes.[17] Thermal destruction of ribosomes of *Staphylococcus aureus*[18,19] and *E. coli*,[16] and thermal activation of ribosomal RNase[16,20] have been observed, and these events may lead to endogenous RNA degradation. In *Candida lipolytica*, it has been suggested that heat denaturation of an ATP-dependent proteinaceous RNase inhibitor triggers RNA degradation.[21,22]

Irradiation: *E. coli*[23–25] and yeasts[26–28] excrete UV absorbing materials on UV and X-ray irradiation. RNA degradation, however, seems to be a secondary event.[25,28]

Agents: The leakage of UV absorbing materials has been observed with bacteria, yeasts and molds treated with membrane attacking agents such as polymyxin,[29] polyene antibiotics,[30] surfactants[31–37] and organic solvents.[34,35] Endogenous RNA of *Torulopsis xylinus* was effectively converted to 3′(2′)-nucleotides by SDS treatment.[32] *Candida utilis* excretes a large amount of 5′-nucleotides on treatment with cationic surfactants.[33]

5′-Nucleotides are excreted from *E. coli* cells after streptomycin treatment.[38–43] The excreted nucleotides are formed initially by *de novo* synthesis, and later by degradation of endogenous RNA. The mechanism of streptomycin induction of nucleotide excretion is considered to involve damage to the permeability barrier. Kanamycin and neomycin showed effects similar to those of streptomycin on *E. coli*.[40] Nucleotide excretion caused by streptomycin also occurred in *Bacillus subtilis* and *Pseudomonas fluorescens*[44] but not in *Streptococcus faecalis*[44] or *S. aureus*.[45]

RNA degradation in microorganisms was observed when they were treated with inhibitors of DNA synthesis such as mitomycin[46,47] and AF-

5,[47] inhibitors of RNA synthesis such as actinomycin,[48] inducers of DNA degradation such as colicin E2,[49,50] and inhibitors of respiration such as NaF,[51] but the mechanism by which these agents bring about RNA degradation is not fully understood.

EDTA and citrate remove Mg^{2+}, which stabilizes membranes[52] or ribosomes,[53,92] and may trigger the degradation process.[54–60] Borate greatly stimulates endogenous RNA degradation, leading to 5′-nucleotide formation in *C. utilis*[33] by an unknown mechanism.

Nutrient starvation: Degradation of ribosomes or rRNA is observed under conditions of starvation of carbon,[61,62] amino acid,[63] nitrogen,[64,65] phosphate,[66–69,76] Mg^{2+},[53,69–71] K^{+},[72] uracil[73] and guanine,[74] and under complete nutrient starvation.[10,61,68,69,75–80] From the observation that divergent nutrient limitation causes the common phenomenon of ribosome degradation, Erlich *et al.*[74] presumed that a certain common signal triggering ribosome degradation is derived from the different starvation conditions. They considered that the common signal is the reduction of the intracellular pool size of nucleoside polyphosphates. The following phenomena support this hypothesis: yeast ribosomes are stabilized by ATP;[81] autodegradation of RNA in *Pseudomonas graveolens* is linked with the degradation of ATP;[82] proteinaceous RNase inhibitor in *C. lipolytica* is regulated by ATP and other purine nucleoside polyphosphates;[21,22] intracellular RNase of *B. subtilis* is inhibited by ATP and ADP;[83–86] and RNase II of *E. coli* is inhibited by ATP.[87]

Salt concentration and pH: Many bacteria actively degrade endogenous RNA when their cells are incubated in alkaline buffer, and these processes are very dependent on the salt concentration of the incubation medium.[75,82,88–90] RNA degradation in *Phytophthora infestans* is also stimulated by neutral salts.[91] The stimulatory effect of salts on RNA degradation is due to increased susceptibility of rRNA to digestive enzymes.[88,89] On raising the salt concentration, the ability of ribosomes to absorb RNase is decreased[12] and latent RNases in ribosomes are activated.[20,93,94] These facts are consistent with the view that salt-dependent denaturation makes rRNA more susceptible to the degradative enzymes.

Various yeasts undergo pH-dependent RNA degradation.[95–98] In general, 3′-nucleotide-forming enzymes are operative at acid pH and 5′-nucleotide-forming ones at alkaline pH.

Others: Degradation of endogenous RNA can be observed on lyophilization,[99,100] stirring[101,102] and other factors,[103–107] and UV-absorbing materials are sometimes excreted. Nucleotides are often formed in growth media during cultivation[82,108–116] and some of them have been shown or suggested to be accumulated as a result of degradation of endogenous RNA.[82,109,110,112,115]

8.2 Enzymes Participating in the Degradation of RNA

Microbial cells contain a variety of RNA degrading enzymes. For example, *E. coli* cells contain at least six RNA degrading enzymes, RNases I–V and polynucleotide phosphorylase (PNPase). RNase I, II and PNPase have been proved to participate in the degradation of ribosomes.[117,118] Recently RNase III, which attacks double-stranded RNA, was suggested to participate in endonucleolytic attack, as shown in Fig. 8.1.[8] RNase I of *E. coli* does not require metal ions, is stimulated by EDTA and forms 3′-nucleotides from RNA. RNase I-type enzymes have been found to participate in the degradation of RNA in *Aerobacter aerogenes*,[11] yeasts [21,22,32,54,99,119] and a mold.[51]

RNase II of *E. coli* requires Mg^{2+} for activity and forms 5′-nucleotides. Phosphodiesterases (PDase), which are similar to *E. coli* RNase II, have been shown to participitate in the degradation of RNA of *P. graveolens*[82] and *Schizosaccharomyces liquefaciens*.[102] PNPase, which is characterized by its requirement for both Mg^{2+} and phosphate ion, was first shown to participate in the autodegradation of *E. coli* ribosomes[117] and then in RNA degradation in intact or disrupted cells of a 5′-nucleotide excreter, *Bacillus megaterium*.[88,89,109] Since then, the degradation of RNA by PNPase has been observed in many bacteria[10,20,68,80,82,90] and yeasts.[100,101]

Nucleotides formed by RNA degradation are usually degraded further into nucleosides and bases. For example, in *E. coli*, which contains potent phosphomonoesterase and nucleoside phosphorylases, nucleotides are usually converted to nucleosides, then to bases. In *P. graveolens*, 5′-nucleotides formed by RNA degradation are hydrolyzed to bases and ribose-5-phosphate.[82]

8.3 Formation of 5′-Nucleotides by the Degradation of RNA

RNA degradation in microorganisms is quite widely and commonly observed. 5′-Nucleotides, however, are formed only under conditions where PDase or PNPase is operative but where nucleotide degrading enzymes are not operative.

When cells of many bacteria are incubated in Tris buffer, pH 8, glycine-NaOH buffer or carbonate buffer, pH 9–10, endogenous RNA is actively degraded, presumably by the action of PDase and/or PNPase.[82,88–90] 5′-Nucleotides are rarely excreted by enteric bacteria, which usually have potent nucleotide degrading enzymes. On the other hand, pseudomonads have weaker nucleotide degrading enzymes and can excrete 5′-nucleotides. The yield of 5′-nucleotides from RNA was usually higher in degradation at pH 9–10 than at pH 8 due to reduced 5′-nucleotide degradation at higher pH.[82,88] The process of degradation of endogenous RNA in *B. megaterium* is shown in Fig. 8.2. At pH 8.2, 5′-nucleotides, once formed, disappeared by secondary degradation, but at pH 9.8, more than 80% of the degraded RNA was recovered as 5′-nucleotides.[88] When *B. megaterium* was cultivated for 12–16 h, 3 mg/ml (as dry weight of cells) containing 10–12% RNA was obtained. On incubation of the harvested cells in glycine-NaOH buffer, pH 9.8, for 2 h, about 85% of the RNA was degraded and about 85% of the degraded RNA was recovered as nucleoside 5′-monophosphates with a small amount of nucleoside 5′-diphosphates. The yield as 5′-nucleotides was 250–300 μg/ml of culture. Incubation of disrupted cells in alkaline buffers containing phosphate yielded RNA degradation products mainly consisting of nucleoside 5′-diphosphates.[89] By a similar procedure, more than 80% of the RNA in cells of *P. graveolens*, which has an RNA content of 11%, could be converted to nucleoside 5′-monophosphates.[82]

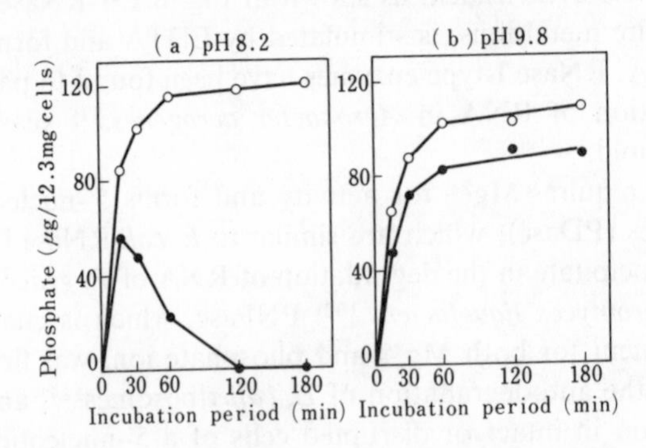

Fig. 8.2. Formation of 5′-nucleotides as a result of RNA degradation in *B. megaterium*.[88] (a) In 0.25 M Tris-HCl buffer, pH 8.2. (b) In 0.25 M glycine-NaOH buffer, pH 9.8. ○, Degradation of RNA (as the phosphate); ●, excretion of 5′-nucleotides (as phosphates).

Dried cells of *Brevibacterium sojae* degraded more than 80% of endogenous RNA when they were incubated in carbonate buffer, pH 10, and excreted nucleoside 5′-mono- and diphosphates.[90] In Tris buffer, pH 8.0, the recovery of 5′-nucleotides from degraded RNA was low. In endogenous RNA degradation at alkaline pH, PNPase operates in *B. megaterium* and *B. sojae*, and PDase in *P. graveolens*.

Yeast cells degrade endogenous RNA on incubation in buffers.[95–98] Generally 3′-nucleotides are formed at acid pH and 5′-nucleotides at alkaline pH. *Rhodotorula pallida*, as an exception, formed 5′-nucleotides at both acid and alkaline pH.[95] PDase[98] and PNPase[97] participate in 5′-nucleotide-forming RNA degradation in yeasts. Nakao *et al.*[95] obtained 5–7 mg of 5′-nucleotides from 100 ml of yeast cultures by the autodegradation method. Endogenous RNA of *C. utilis* is hardly degraded by simple incubation in buffers, but thermal shock followed by incubation in an acid medium caused extensive degradation of RNA, yielding 3′-nucleotides.[119,120] At alkaline pH, RNA in *C. utilis* could be quantitatively converted to 5′-nucleotides by incubating the cells in a buffer at pH 10 in the presence of cationic surfactant and borate ion.[33] Continuous cultivation of *C. utilis* yielded approximately 20 mg/ml/4 h of cells (as dry weight) with an RNA content of about 12%, and the cellular RNA could be converted to 5′-nucleotides. Thus 2.4 mg/ml/4 h of 5′-nucleotides is obtainable by the autodegradation method.

Excretion of 5′-nucleotides also occurs into the growth media of bacteria and yeasts. Streptomycin treatment of *E. coli* cultures yielded a small amount of 5′-nucleotides in the medium.[38–43] *B. megaterium* and *P. graveolens* formed 400–500 μg/ml of 5′-nucleotides in parallel with the loss of intracellular RNA during cultivation.[82,109] Since elongation of the bacteria was observed in 5′-nucleotide excreting cultures of *B. megaterium* and *P. graveolens*, the RNA degradation accompanying 5′-nucleotide excretion may have occurred as a result of some stress loaded on these organisms during cultivation. *Zygosaccharomyces soja* accumulated approximately 160 μg/ml of 5′-nucleotides in the medium, presumably due to the degradation of endogenous RNA.[115]

Demain *et al.* reported a unique case in which *B. subtilis*, isolated as having purine accumulating capability, excreted high molecular RNA during its slow linear growth, and the excreted RNA was degraded extracellularly by PDase or PNPase, yielding nucleoside 5′-mono- and diphosphates.[110–114] In this case, the addition of Cu^{2+} suppressed the action of 5′-nucleotidase and the yield of 5′-nucleotides was improved. The amount of accumulated RNA derivatives reached 2.5 g/l and the absorbance at 260 nm of the culture reached more than 80. They suggested that the RNA excreting *B. subtilis* is a ‘relaxed’ prototroph lacking the usual stringent control of RNA synthesis by amino acids.[114]

Brevibacterium liquefaciens[108] and several bacteria isolated from soil[116] accumulated 5′-nucleotides. Though the mechanism of their accumulation has not been examined, the presence of the four nucleotides which constitute RNA is suggestive that the nucleotides are products of degradation of endogenous RNA. Furuya *et al.*[116] reported that 5′-nucleotide excreting bacteria lack acid and alkaline phosphatase and that the most active excreter formed 587 μg of AMP per ml, 214 μg of GMP per ml, 266 μg of CMP per ml and 290 μg of UMP per ml.

8.4 CONCLUSION

The phenomenon of RNA degradation by endogenous enzymes, which is very commonly observed when microorganisms encounter stresses unfavorable for their growth, may be a reflection of a teleological regulatory mechanism permitting the organism to escape death due to unbalanced synthetic reactions or of an adaptive process for the generation of energy under endogenous respiration conditions. From an applied microbiological point of view, the autodegradation process provides an effective way to convert intracellular RNA to nucleotides without the aid of enzymes from other sources. The process, however, has not been used on an industrial scale.

The rate of RNA synthesis, which is a very important factor in the possible industrial utilization of this process, can be made quite high as a result of recent technological progress in fermentation methods. For example, *Alcaligenes marshallii* formed 10 mg/ml/6 h of intracellular RNA when grown on *n*-paraffin[121] and these cells excreted 5′-nucleotides on incubation in alkaline buffers.[122] If the endogenous RNA is quantitatively converted to 5′-nucleotides, approximately 5 mg/ml/6 h of purine 5′-nucleotides could be obtained by the autodegradation procedure. This amount may be high enough for industrial application of the autodegradation process.

The autodegradation method can be also utilized for the reduction of the nucleic acid content of single-cell protein.[119,120] Production of 5′-nucleotides could thus be carried out jointly with the reduction of the nucleic acid content in single-cell protein. Such a combined process using the autodegradation method could be of great interest.

REFERENCES

1. K. Ogata, *Amino Acid and Nucleic Acid* (Japanese), no. 8, 1 (1963).
2. K. Ogata, *Seikagaku* (Japanese), **36**, 1 (1964).
3. K. Ogata, *Kindai Kogyo Kagaku* (Japanese), vol. 23, Seibutsu Kogyo Kagaku, p. 211, Asakura Shoten, 1969.
4. A. L. Demain, *Progress in Industrial Microbiology*, **8**, 36 (1968).
5. M. Yoneda, *Hakko to Biseibutsu* (Japanese) (ed. T. Uemura and K. Aida), p. 145, Asakura Shoten, 1971.
6. T. E. Norris and A. L. Koch, *J. Mol. Biol.*, **64**, 633 (1972).
7. C. I. Davern and M. Meselson, *ibid.*, **2**, 153 (1960).
8. R. Kaplan and D. Apirion, *J. Biol. Chem.*, **250**, 1854 (1975).
9. O. Maaløe and N. O. Kjeldgaard, *Control of Macromolecular Synthesis*, Benjamin, 1966.
10. A. F. Gronlund and J. J. R. Campbell, *J. Bact.*, **86**, 58 (1963).
11. R. E. Strange and M. Shon, *J. Gen. Microbiol.*, **34**, 99 (1964).
12. R. E. Strange and J. R. Postgate, *ibid.*, **36**, 393 (1964).
13. R. J. H. Gray, L. D. Witter and Z. J. Ordal, *Appl. Microbiol.*, **26**, 78 (1973).
14. J. J. Iandolo and Z. J. Ordal, *J. Bact.*, **91**, 134 (1966).
15. M. C. Allwood and A. D. Russell, *J. Pharm. Sci.*, **59**, 180 (1970).
16. R. Nozawa, T. Horiuchi and D. Mizuno, *Arch. Biochem. Biophys.*, **118**, 402 (1967).
17. B. Pace and L. L. Campbell, *Proc. Natl. Acad. Sci. U.S.A.*, **57**, 1110 (1967).
18. P. L. Schell, *Nature*, **210**, 1157 (1966).
19. S. J. Sogin and Z. J. Ordal, *J. Bact.*, **94**, 1082 (1967).
20. R. D. Haight and Z. J. Ordal, *Can. J. Microbiol.*, **15**, 15 (1969).
21. A. Imada, A. J. Sinskey and S. R. Tannenbaum, *Biochim. Biophys. Acta*, **268**, 674 (1972).
22. A. Imada, A. J. Sinskey and S. R. Tannenbaum, Proc. 4th. Int. Fermentation Symp. *Fermentation Today*, p. 455, 1972.
23. D. Billen and E. Volkin, *J. Bact.*, **67**, 191 (1954).
24. D. Billen, *Arch. Biochem. Biophys.*, **67**, 333 (1957).
25. I. Pečevsky and Ž. Kućan, *Biochim. Biophys. Acta*, **145**, 310 (1967).
26. J. R. Loofbourow, S. Oppenheim-Errera, D. G. Loofbourow and C. A. Yeats, *Biochem. J.*, **41**, 122 (1947).
27. E. D. Thomas, F. B. Hershey, A. M. Abbate and J. R. Loofbourow, *J. Biol. Chem.*, **196**, 575 (1952).
28. N. Ghosh, *Z. Naturforsch.*, **23B**, 1273 (1968).
29. B. A. Newton, *J. Gen. Microbiol.*, **9**, 54 (1953).
30. S. C. Kinsky, *J. Bact.*, **82**, 889 (1961).
31. M. R. J. Salton, *J. Gen. Microbiol.*, **5**, 391 (1951).
32. S. Watanabe, T. Osawa and S. Yamamoto, *Hakkokogaku Zasshi* (Japanese), **46**, 538 (1968).
33. K. Kitano, S. Akiyama and H. Fukuda, *ibid.*, **48**, 14 (1970).
34. R. W. Jackson and J. A. DeMoss, *J. Bact.*, **90**, 1420 (1965).
35. M. Higuchi and T. Uemura, *Nippon Nogeikagaku Kaishi* (Japanese), **34**, 721 (1960).
36. K. Arima, T. Uozumi and M. Takahashi, *Agr. Biol. Chem.*, **29**, 1033 (1965).
37. T. Uozumi, M. Takahashi and K. Arima, *ibid.*, **29**, 1042 (1965).
38. N. Anand and B. D. Davis, *Nature*, **185**, 22 (1960).
39. H. Roth, H. Amos and B. D. Davis, *Biochim. Biophys. Acta*, **37**, 398 (1960).
40. D. S. Feingold and B. D. Davis, *ibid.*, **55**, 787 (1962).

41. D. T. Dubin and B. D. Davis, *ibid.*, **55**, 793 (1962).
42. C. L. Rosano, R. A. Peabody and C. Hurwitz, *ibid.*, **37**, 380 (1960).
43. C. Hurwitz, C. L. Rosano and R. A. Peabody, *ibid.*, **72**, 80 (1963).
44. H. Tzagoloff and W. W. Umbreit, *J. Bact.*, **85**, 49 (1963).
45. R. Hancock, *J. Gen. Microbiol.*, **23**, 179 (1960).
46. H. Kersten and W. Kersten, *Z. Physiol. Chem.*, **334**, 141 (1963).
47. N. Kato, K. Okabayashi and D. Mizuno, *J. Biochem.* (Tokyo), **67**, 175 (1970).
48. G. Acs, E. Reich and S. Valanju, *Biochim. Biophys. Acta*, **76**, 68 (1963).
49. K. Nose, D. Mizuno and H. Ozeki, *ibid.*, **119**, 636 (1966).
50. K. Nose and D. Mizuno, *J. Biochem.* (Tokyo), **64**, 1 (1968).
51. H. Sugiyama, T. Iwasa and J. Ishiyama, *Nippon Nogeikagaku Kaishi* (Japanese), **36**, 690 (1962).
52. R. E. Strange, *Nature*, **203**, 1304 (1964).
53. D. Kennell and A. Kotoulas, *J. Bact.*, **93**, 334 (1967).
54. H. C. Neu, D. F. Ashman and T. D. Price, *ibid.*, **93**, 1360 (1967).
55. M. Higuchi and T. Uemura, *Nippon Nogeikagaku Kaishi* (Japanese), **33**, 304 (1959).
56. M. Higuchi and T. Uemura, *Nature*, **184**, 1381 (1959).
57. M. Higuchi and T. Uemura, *Nippon Nogeikagaku Kaishi* (Japanese), **33**, 826 (1959).
58. M. Higuchi and T. Uemura, *Plant Cell Physiol.*, **3**, 249 (1962).
59. M. Higuchi, H. Tanaka and T. Uemura, *Nippon Nogeikagaku Kaishi* (Japanese), **36**, 971 (1962).
60. H. Tanaka, M. Higuchi and T. Uemura, *ibid.*, **36**, 978 (1962).
61. R. E. Strange, H. E. Wade and A. G. Ness, *Biochem. J.*, **86**, 197 (1963).
62. R. Kaplan and D. Apirion, *J. Biol. Chem.*, **249**, 149 (1974).
63. R. A. Lazzarini and A. E. Dahlberg, *ibid.*, **246**, 420 (1971).
64. F. Ben-Hamida and D. Schlessinger, *Biochim. Biophys. Acta*, **119**, 183 (1966).
65. R. Kaplan and D. Apirion, *J. Biol. Chem.*, **250**, 3174 (1975)
66. T. Horiuchi, *J. Biochem.* (Tokyo), **46**, 1467 (1959).
67. C. I. Hou, A. F. Gronlund and J. J. R. Campbell, *J. Bact.*, **92**, 851 (1966).
68. S. Natori, T. Horiuchi and D. Mizuno, *Biochim. Biophys. Acta*, **134**, 337 (1967).
69. H. Maruyama, M. Ono-Onitsuka and D. Mizuno, *J. Biochem.* (Tokyo), **67**, 559 (1970).
70. D. Kennell and B. Magasanik, *Biochim. Biophys. Acta*, **55**, 139 (1962).
71. S. Natori, R. Nozawa and D. Mizuno, *ibid.*, **114**, 245 (1966).
72. H. L. Ennis and M. Lubin, *ibid.*, **95**, 605 (1965).
73. R. A. Lazzarini, K. Nakata and R. M. Winslow, *J. Biol. Chem.*, **244**, 3092 (1969).
74. H. Erlich, J. Gallant and R. A. Lazzarini, *ibid.*, **250**, 3057 (1975).
75. M. Stephenson and J. M. Moyle, *Biochem. J.*, **45**, vii (1949).
76. H. Maruyama and D. Mizuno, *Biochim. Biophys. Acta*, **199**, 159 (1970).
77. H. Maruyama, M. Ono and D. Mizuno, *ibid.*, **199**, 176 (1970).
78. R. E. Strange, F. A. Dark and A. G. Ness, *J. Gen. Microbiol.*, **25**, 61 (1961).
79. E. A. Dawes and D. W. Ribbons, *Biochem. J.*, **95**, 332 (1965).
80. A. F. Gronlund and J. J. R. Campbell, *J. Bact.*, **90**, 1 (1965).
81. Y. Ohtaka and K. Uchida, *Biochim. Biophys. Acta*, **76**, 94 (1963).
82. A. Imada, Y. Nakao, I. Nogami, S. Igarasi and K. Ogata, *Amino Acid and Nucleic Acid* (Japanese), no. 8, 118 (1963).
83. M. Yamasaki and K. Arima, *Biochim. Biophys. Acta*, **139**, 202 (1967).
84. M. Yamasaki and K. Arima, *Biochem. Biophys. Res. Commun.*, **37**, 430 (1969).
85. M. Yamasaki, K. Yoshida and K. Arima, *Biochim. Biophys. Acta*, **209**, 463 (1970).
86. M. Yamasaki and K. Arima, *ibid.*, **209**, 475 (1970).
87. P. Venkov, D. Schlessinger and D. Longo, *J. Bact.*, **108**, 601 (1971).
88. K. Ogata, A. Imada and Y. Nakao, *Agr. Biol. Chem.*, **26**, 596 (1962).
89. A. Imada, Y. Nakao and K. Ogata, *ibid.*, **26**, 611 (1962).
90. T. Sugimori, Y. Tazuke and Y. Hamada, *ibid.*, **27**, 712 (1963).
91. O. T. Page and F. A. Wood, *Phytopathol.*, **53**, 946 (1963).

92. A. S. Spirin, N. A. Kiselev, R. S. Shakulov and A. A. Bogdanov, *Biokhimiya* (Engl. Transl.), **28**, 765 (1963).
93. P. Spitnik-Elson, *Biochim. Biophys. Acta*, **55**, 741 (1962).
94. M. Lindblom and H. Mogren, *Biotechnol. Bioeng.*, **16**, 1123 (1974).
95. Y. Nakao, A. Imada, T. Wada and K. Ogata, *Agr. Biol. Chem.*, **28**, 151 (1964).
96. Y. Tsukada and T. Sugimori, *ibid.*, **28**, 479 (1964).
97. Y. Tsukada and T. Sugimori, *ibid.*, **28**, 484 (1964).
98. K. Ogata, S. Song and T. Tochikura, *Proc. Agr. Chem. Soc. Japan*, p. 177 (1966).
99. J. Wagman, *J. Bact.*, **80**, 558 (1960).
100. T. J. Sinskey and G. J. Silverman, *J. Bact.*, **101**, 429 (1970).
101. H. Tanaka, H. Miyagawa and K. Ueda, *Hakkokogaku Zasshi* (Japanese), **52**, 646 (1974).
102. H. Tanaka, H. Miyagawa and K. Ueda, *ibid.*, **52**, 652 (1974).
103. J. T. Holden, *Biochim. Biophys. Acta*, **29**, 667 (1958).
104. E. D. DeLamater, K. L. Babcock and G. R. Mazzanti, *J. Bact.*, **77**, 513 (1959).
105. M. Niwa, Y. Yamadeya and Y. Kuwajima, *ibid.*, **88**, 809 (1964).
106. E. J. Herbst and B. D. Doctor, *J. Biol. Chem.*, **234**, 1497 (1959).
107. Z. A. Cohn F. E. Hahn, W. Ceglowski and F. M. Bozeman, *Science*, **127**, 282 (1958).
108. T. Okabayashi and E. Masuo, *Chem. Pharm. Bull.*, **8**, 370 (1960).
109. K. Ogata, A. Imada and Y. Nakao, *Agr. Biol. Chem.*, **26**, 586 (1962).
110. A. L. Demain, I. M. Miller and D. Hendlin, *J. Bact.*, **88**, 991 (1964).
111. A. L. Demain, R. A. Vitali, B. L. Wilker, J. W. Rothrock and T. A. Jacob, *Biotechnol. Bioeng.*, **6**, 361 (1964).
112. A. L. Demain, R. W. Burg and D. Hendlin, *J. Bact.*, **89**, 640 (1965).
113. A. L. Demain and D. Hendlin, *Appl. Microbiol.*, **14**, 297 (1966).
114. A. L. Demain, *Biochem. Biophys. Res. Commun.*, **24**, 39 (1966).
115. Y. Tsukada and T. Sugimori, *Agr. Biol. Chem.*, **28**, 471 (1964).
116. H. Furuya, K. Araki, M. Nohara, S. Abe and S. Kinoshita, *Amino Acid and Nucleic Acid*, no. 9, 24 (1964).
117. H. E. Wade, *Biochem. J.*, **78**, 457 (1961).
118. R. Kaplan and D. Apirion, *J. Biol. Chem.*, **249**, 149 (1974).
119. S. Ohta, S. Maul, A. J. Sinskey and S. R. Tannenbaum, *Appl. Microbiol.*, **22**, 415 (1971).
120. S. B. Maul, A. J. Sinskey and S. R. Tannenbaum, *Nature*, **228**, 181 (1970).
121. M. Kuno, T. Kono, M. Asai and Y. Nakao, 18th Mtg. Assoc. of Amino Acid and Nucleic Acid, 1969.
122. Y. Nakao, *personal communication.*

CHAPTER 9

Production of Nucleotides from RNA by Chemical Processes*

This chapter describes two basic processes; one is to obtain ribonucleosides from RNA by a suitable chemical process and the other is to phosphorylate these ribonucleosides to form the corresponding 5′-phosphomonoesters. Since both IMP and GMP are commercially produced as chemical seasoning agents, emphasis will be placed on the production of 5′-nucleotides.

9.1 Production of Ribonucleosides from RNA by Chemical Hydrolysis

Ribonucleosides can be obtained from RNA by several procedures, i.e. treatment of RNA with concentrated aqueous ammonia at 175–178° C in an autoclave;[1] with aqueous pyridine for a few days at room temperature;[2] with aqueous formamide;[3] or with metallic catalysts such as lead

* Tsuneo Sowa, Asahi Chemical Industry Co. Ltd.

hydroxide, calcium hydroxide or lanthanum hydroxide in aqueous solution at reflux temperature for 40–50 h.[4,5]

The chemical decomposition of RNA into ribonucleosides is in essence a hydrolytic reaction of phosphomonoesters. That is to say, the preparation of ribonucleosides by the hydrolysis of RNA requires the cleavage of the phosphate ester bonds at both the 5′- and the 3′-hydroxyl groups of ribonucleosides. It also requires a suitable choice of reaction conditions, e.g. pH, temperature and reaction time, to avoid cleavage of the nucleoside bonds.

In general, phosphodiesters such as dimethyl- and diethylphosphate are stable to alkali, as is deoxyribonucleic acid, in which 2′-deoxyribonucleoside has a phosphodiester linkage associated with the 3′- and 5′-hydroxyl groups.[6] However, RNA having a free hydroxyl group in the ribose moiety adjacent to the phosphoester linkage at the 3′-position is easily hydrolyzed by alkali, yielding a mixture of 2′- and 3′-ribonucleotides. Namely, RNA can yield a mixture of 2′- and 3′-nucleotides almost quantitatively on treatment with aqueous alkali (pH 12 to 13),[7,8] with concentrated aqueous ammonia at 45° C for eight days,[9] or with 0.5 N sodium hydroxide at 85° C for 1 h. The instability of RNA to alkali is probably due to the 2′-hydroxyl group adjacent to the 3′-hydroxyl moiety of ribose. The 2′-hydroxyl group participates in the hydrolytic cleavage of the phosphoester of RNA to give the 2′:3′-cyclic phosphomonoester of ribonucleosides as an intermediate by ester exchange of the 5′-phosphomonoester bond with the 2′-hydroxyl group. The intermediate is easily hydrolyzed by alkali to give a mixture of 2′- and 3′-nucleotides.

On the other hand, it is well-known that phosphomonoesters are stable to alkali, and that their hydrolytic reactivity increases in acidic aqueous solution, reaching a maximal value in the neighborhood of pH 4.[12,13] Takami *et al.*[14] obtained ribonucleosides quantitatively from the corresponding nucleoside 5′-phosphomonoesters in 1 M formic acid buffer solution. At reflux temperature, the liberation of Pi reached a maximum at pH's between 4.0 and 5.5.

Ribonucleosides are relatively stable in aqueous solution at an initial pH of 8.0. In weakly acidic solution at an initial pH of 5.0 to 6.0, however, hydrolysis of the phosphomonoester is completed in 2–4 h, yielding ribonucleosides almost quantitatively.[15] The *N*-glycosidic linkage of purine nucleosides is extremely sensitive to acid and heterocycles are readily produced.[16] If the initial pH of the reaction is as low as 5.0, therefore, the purine nucleosides thus formed gradually decompose and the yields fall. Further, CMP is deaminated in neutral or weak acidic solutions[14] and forms about 30% uridine in the 2–4 h during which the hydrolysis of the phosphomonoester bond of cytidine is completed.

Ribonucleosides can also be obtained by heating RNA at 130° C in the presence of calcium hydroxide in water. For example, when 20 g of RNA is heated with 1.8 g of calcium hydroxide at 130° C, with agitation in an autoclave, the pH of the reaction media gradually falls to about 5.5 and ribonucleosides are obtained in 3–4 h. Insoluble complexes of ribonucleotides with calcium can be produced under mild conditions. This suggests that the hydrolysis of RNA with calcium hydroxide also proceeds through two steps via the phosphomonoesters of nucleosides as intermediates. It is believed that the formation of calcium salts of nucleotides acts to keep the pH of the reaction media acidic and to make the hydrolysis of ribonucleotides easier.

In order to isolate individual ribonucleosides from the resulting reaction mixture, guanosine can be isolated first as an insoluble precipitate, then adenosine, cytosine and uridine can be isolated by separation on ion-exchange resins,[18] followed by concentration. In the method using calcium hydroxide, it is easy to separate the individual ribonucleosides because insoluble phosphate is formed with calcium.

9.2 Production of 5′-Nucleotides by the Direct Phosphorylation of Nucleosides

The phosphorylation of nucleosides can be compared with the synthesis of esters of carboxylic acids with alcohols. There are two differences from ordinary phosphorylation: (1) phosphoric acid is a tribasic acid and may thus react with three moles of alcohol, and (2) nucleosides possess two or more hydroxyl groups which can form esters with acid. Thus, in order to obtain 5′-nucleotides by chemical phosphorylation of the 5′-hydroxyl group of nucleosides, the introduction of a suitable protecting group for the functional hydroxyl groups not participating directly in the phosphorylation is necessary. This protecting group should be capable of being eliminated without cleavage of the nucleosidic linkage or the phosphomonoester.

5′-Nucleotides are generally prepared by the phosphorylation of individual nucleosides with an appropriate phosphorylating agent, usually an activated derivative of phosphoric acid or of pyrophosphoric acid. Suitable activated phosphoric acid derivatives include various kinds of phosphorodi- or monochloridates in which one hydroxyl group is substituted by a phenyl or *p*-nitrophenyl group,[19] morpholine,[20] and phosphoryl chloride.[21] Pyrophosphoric acid derivatives include bis(*p*-

nitrophenyl)pyrophosphate[23] and tetrachloropyrophosphate.[22] Of these phosphorylating agents, phosphoryl chloride is the only reagent available commercially in large amounts and at relatively low cost.

There are several methods for the protection of hydroxyl groups not participating in the phosphorylation, using protecting groups such as acetyl,[24] benzyl,[25] isopropylidene,[26] or benzylidene.[27] The introduction and elimination of the protecting groups and the phosphorylation of protected nucleosides are described in Chapter 12, and will be omitted here.[28]

Several attempts at direct and selective phosphorylation of the 5′-hydroxyl group (the primary hydroxyl group) of nucleosides unprotected at both the 2′- and 3′-hydroxyl groups (secondary hydroxyl groups) have been made. For example, AMP[29] and GMP[30] were prepared from adenosine and guanosine, respectively, by direct phosphorylation with phosphoryl chloride in pyridine. Tenner[31] studied the reaction of thymidine, 2′-deoxycytidine or 2′-deoxyadenosine with 2′-cyanoethylphosphate in the presence of *N,N′*-dicyclohexylcarbodiimide (DCC), yielding the corresponding 5′-nucleotides, and indicated that, in each case, preferential phosphorylation of the 5′-hydroxyl group occurred.

Several attempts at the selective phosphorylation of the 5′-hydroxyl group of nucleosides without the introduction of any protecting groups have been made using phosphoryl chloride as the phosphorylating agent. These are industrially practicable. Namely, 5′-nucleotides were prepared from the corresponding nucleosides by direct phosphorylation with phosphoryl chloride in trialkyl phosphate in the presence of adequate amounts of water. High yields were obtained, as shown in Table 9.1.[32] This table shows that IMP was obtained in 68% yield by reacting inosine with phosphoryl chloride in triethyl phosphate. However, by the addition of suitable amounts of water to the reaction media, the yield of 5′-phosphomonoester was increased to 91%. In the case of guanosine too, the addition of small amounts of water was found to increase the conversion of GMP from 85% to 90%. By the addition of adequate amounts of water, competitive phosphorylation of the 2′-(or 3′-) hydroxyl group was strongly inhibited and, as a result, the yields of 5′-phosphomonoesters increased.

In the cases of adenosine, cytidine, uridine, and xanthosine, phosphorylation is so slow that the corresponding 5′-phosphomonoesters can be obtained in yields between 80 and 90% without the addition of water. In the phosphorylation of nucleosides with phosphoryl chloride in trialkyl phosphate, it is not clear what role the water plays in the reaction mechanism. Trialkylphosphates are considered to increase the solubility of the nucleosides in the reaction medium and to activate phosphoryl chloride by complex formation, as shown in the following equation:[33]

TABLE 9.1

Direct Phosphorylation of Unprotected Nucleosides in Trialkyl Phosphate†

Nucleoside	$POCl_3$					$P_2O_3Cl_4$			
	R of $(RO)_3PO$	H_2O (mmol)	Time (h)	Yield of nucleoside phosphate (mole %) 5′-mono	higher	R of $(RO)_3PO$	Time (h)	Yield of nucleoside phosphate (mole %) 5′-mono	higher
Inosine	CH_3CH_2	0	2	68	18	CH_3	1	54	46
Inosine	CH_3CH_2	2.0	2	91	8				
Guanosine	CH_3	0	6	85	9	CH_3	4	59	41
Guanosine	CH_3	2.0	6	90	5				
Adenosine	CH_3CH_2	0	6	84	11	CH_3CH_2	4	57	33
Xanthosine	CH_3	0	9	80	5	CH_3	12	64	22
Uridine	CH_3	0	12	89	8	CH_3	6	73	25
Cytidine	CH_3CH_2	0	1	88	8	CH_3	0.5	86	13
Deoxyinosine	CH_3	0	5	73	6				
Deoxycytidine	CH_3	0.5	6	64	19				
AICARP	CH_3	0	4	91	5				

† Conditions: nucleoside, 2 mmole; $POCl_3$ or $P_2O_3Cl_4$, 4 mmole (6 mmole in the case of inosine and guanosine); trialkyl phosphate, 5 ml; at 0°C.

$$(RO)_3PO + POCl_3 \longrightarrow [(RO)_3\text{-}\overset{\displaystyle O}{\overset{\|}{P}}\text{-}O\text{-}PCl_2]^+Cl^- \tag{9.1}$$

m-Cresol was useful as a solvent in place of acetonitrile, and 5′-nucleotides were again obtained in high yields by reacting unprotected nucleosides with phosphoryl chloride in the presence of small amounts of water.[34)]

As a direct phosphorylation method, the use of phosphoryl chloride in acetonitrile in the presence of pyridine and water or alcohols such as *tert*-butanol has proved useful in producing 5′-phosphomonoesters.[35)]

In the above reaction, where 5′-nucleotides are obtained by reacting nucleosides with phosphoryl chloride, maximal formation of 5′-nucleotides is obtained by mixing phosphoryl chloride, water and pyridine in a molar ratio of 2:1:2 at a low temperature, or by mixing phosphoryl chloride, alcohols and pyridine in a molar ratio of 2:1:1, then adding the nucleosides. Yields of more than 85% were obtained, as shown in Table 9.2. Phosphoryl chloride reacts with water in the presence of pyridine in acetonitrile to give tetrachloropyrophosphate via dichlorophosphoric acid as an intermediate.[36,37)]

A stoichiometry of these reactants for phosphomonoester formation

TABLE 9.2

Direct Phosphorylation by $POCl_3$ in the Presence of Water and Pyridine†

Nucleoside	H_2O (mmole)	Time (h)	Yield of 5′-phosphate (mole %)	Selectivity for 5′-phosphate (%)
Guanosine	25	6	89.5	93.0
Adenosine	25	4	91.3	92.9
Inosine	28	4	92.4	93.3
Xanthosine	24	6	90.0	99.4
Uridine	25	6	86.3	89.1
Cytidine	20	2	98.6	100
Arabinouracil	25	6	84.8	90.0
Arabinocytosine	20	2	97.5	99.1
Deoxyadenine	25	4	—	—
Deoxycytidine	20	2	95.1	98.4
Deoxyuridine	25	4	85.1	90.6
2′-*S*-Cyclo-adenosine	25	4	98.8	100
Cyclouridine	25	5	81.3	94.3
8-Bromoadenosine	25	4	84.3	91.7

† Conditions: CH_3CN, 10 ml; $POCl_3$, 44 mmole; C_5H_5N, 48 mmole; H_2O, as listed; nucleoside, 10 mmole; reaction temp., 2°C.

of 2:1:2 (phosphoryl chloride, water and pyridine) can be accounted for as follows:

$$\begin{aligned} POCl_3 + H_2O + C_5H_5N &\longrightarrow POCl_2(OH) + C_5H_5N \cdot HCl \\ POCl_2(OH) + POCl_3 + C_5H_5N &\longrightarrow P_2O_3Cl_4 + C_5H_5N \cdot HCl \end{aligned} \tag{9.2}$$

On the other hand, the reaction of *tert*-butanol is known to give dichlorophosphoric acid.[33)] Since *tert*-butanol acts as an acceptor of the chlorine atom, the stoichiometry of these three reactants should be 2:1:1 (phosphoryl chloride, *tert*-butanol and pyridine), as follows:

$$\begin{aligned} POCl_3 + t\text{-}C_4H_9OH &\longrightarrow POCl_2(OH) + t\text{-}C_4H_9Cl \\ POCl_2(OH) + POCl_3 + C_5H_5N &\longrightarrow P_2O_3Cl_4 + C_5H_5N \cdot HCl \end{aligned} \tag{9.3}$$

Thus, in the phosphorylation methods described above, tetrachloropyrophosphate is the real phosphorylating species; this is further supported by the fact that *tert*-butanol could be replaced by methanol or ethanol without affecting the reaction yields and selectivities of phosphorylation, whereas the isomeric *n*-butanol gives poor results.

Some attempts at the direct phosphorylation of nucleosides with tetrachloropyrophosphate alone in trialkylphosphate,[33] *m*-cresol[38] or acetonitrile[35,39] have been made. However, when guanosine was reacted with tetrachloropyrophosphate in acetonitrile, the reaction went essentially to completion after half an hour and the corresponding phosphomonoester was obtained in only 60% yield. Considerable amounts of the higher phosphorylated product, guanosine 2′(or 3′),5′-diphosphate were formed (Fig. 9.1). Therefore, in the phosphorylation of nucleosides with phosphoryl chloride in the presence of water and pyridine or of *tert*-butanol and pyridine, pyridinium chloride seems to play an important role together with tetrachloropyrophosphate in the efficient selective phosphorylation of the 5′-hydroxyl group of nucleosides. When guanosine was reacted with tetrachloropyrophosphate with various amounts of pyridinium chloride in acetonitrile, as shown in Fig. 9.1, the formation of the 5′-phosphomonoester clearly depended on the amount of pyridinium chloride in the reaction medium, and the addition of 1 to 2 moles per mole of pyridinium chloride gave the maximal yield of the 5′-phosphomonoester. As shown in this figure, the addition of pyridinium chloride inhibits the formation of higher phosphorylated products, guanosine 2′(or 3′),5′-diphosphate, and, as a result, influences the selectivity of phosphorylation at the 5′-hydroxyl group of guanosine, as does the conductivity of the reaction medium.

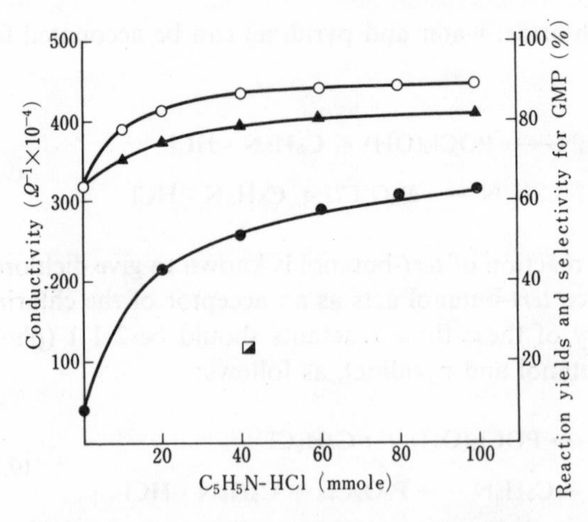

Fig. 9.1. Effect of pyridinium chloride on the phorphorylation of guanosine with tetrachloropyrophosphate. Experimental conditions: CH_3CN, 10 ml; $P_2O_3Cl_4$, 22 mmole; nucleoside, 10 mmole; at 2°C. ●, Conductivity in the absence of guanosine; ○, selectivity for GMP; ▲ reaction yield of GMP; ◪, conductivity in the absence of $P_2O_3Cl_4$ and guanosine.

Pyridinium chloride is readily dissociated into charged species in solution, and H^+ derived as indicated in the following equation acts to increase the conductivity of the mixture and also to increase the selectivity for the 5′-hydroxyl group by decreasing the reactivity of the hydroxyl groups in the sugar moiety of nucleosides in the adduct.

$$C_5H_5N \cdot HCl \rightleftharpoons C_5H_5NH^+ + Cl^- \quad (9.4)$$

$$C_5H_5NH^+ \rightleftharpoons C_5H_5N + H^+ \quad (9.5)$$

$$Cl_2P(O){-}O{-}P(O)Cl_2 + C_5H_5N \longrightarrow \left[Cl_2P(O){-}O{-}P(O)(Cl){-}{}^+NC_5H_5\ Cl^- \right] \quad (9.6)$$

It is also believed that pyridine derived as in Eq. 9.5 reacts with tetrachloropyrophosphate to form trichloropyrophosphopyridinium chloride, which acts as the real phosphorylating species in the phosphorylation of unprotected nucleosides. Thus, pyridinium chloride has an important role in increasing the yields of 5′-nucleotides as well as in the selectivity.

Similar results can be seen when phosphoryl chloride is used in place of tetrachloropyrophosphate.[35,39] Namely, when guanosine is reacted with phosphoryl chloride in acetonitrile, the addition of pyridinium chloride also increases 5′-phosphomonoester production, and the yield of the 5′-phosphomonoester reaches over 90% on the addition of 2 moles of pyridinium chloride per mole of phosphoryl chloride. In this case, pyridine derived from pyridinium chloride (Eq. 9.5) seems to activate phosphoryl chloride by producing dichlorophosphopyridinium chloride.

The advantages of direct phosphorylation of nucleosides in the unprotected form are clear in the preparation of 5′-phosphomonoesters of 2′-deoxynucleosides,[32,33,35,39] arabinonucleosides, 8,2′-*O*- or 8,3′-*O*-anhydroadenine nucleosides,[39,40] and 2,2′-*O*-anhydropyrimidine nucleosides.[39] These phosphorylations act on the primary hydroxyl group selectively, and show great promise.

For the preparation of the 5′-phosphomonoesters of nucleosides without protecting the 2′- and 3′-hydroxyl groups, two methods have been reported, in which tris(8-quinoyl)phosphate[41] or di(2-*tert*-butylphenyl)-phosphorochloridate[42] is used as the phosphorylating agent. There are two methods for the preparation of 2′-(or 3′-) phosphomonoesters of nucleosides directly by treatment with phosphoryl chloride in the presence of calcium hydroxide[43] in aqueous solution or with trichloromethane-phosphonate[44] in the presence of pyridine. The reaction mechanism of the former is interesting in comparison with the selective preparations of 5′-

phosphomonoesters of nucleosides with phosphoryl chloride in organic solvents. However, at the present stage, the mechanism of the former reaction is not definitely established.

REFERENCES

1. P. A. Levene and W. A. Jacobs, *Chem. Ber.*, **43**, 3150 (1910).
2. H. Bredereck, A. Martini and F. Richter, *ibid.*, **74**, 694 (1941).
3. D. H. Heyes, *J. Chem. Soc.*, **1960**, 1184.
4. K. Dimroth, L. Jaenicke and D. Heinzel, *Ann.*, **566**, 206 (1950).
5. K. E. Pfitzner and J. G. Moffatt, *J. Am. Chem. Soc.*, **85**, 3027 (1963); F. Egami, H. Ishihara and M. Shimomura, *Z. Phys. Chem.*, **295**, 349 (1953); E. Bamann, F. Fisher, *Biochem. Z.*, **326**, 89 (1954); *ibid.*, **326**, 87, (1954); E. Bamann and W.D. Mutterlein, *Chem. Ber.*, **91**, 471 (1958).
6. C. A. Bunton, M. M. Mhala, K. G. Oldham and C. A. Vernon, *J. Chem. Soc.*, **1960,** 3293.
7. E. Chargaff, B. Magasanik, E. Visher, C. Green, R. Doniger and D. Elson, *J. Biol. Chem.*, **186**, 51 (1950).
8. P. Boulanger and J. Montevil, *Bull. Soc. Chim. Biol.*, **33,** 784 (1951).
9. P. Boulanger and J. Montevil, *ibid.*, **33,** 791 (1951).
10. J. Kumamoto, J. R. Cox, F. H. Wetheimer, *J. Am. Chem. Soc.*, **77,** 2420 (1955).
11. J. Lecog, *Compt. Rend.*, **242,** 1902 (1956)
12. A. Desjobert, *Bull. Soc. Chim.*, **14,** 809 (1947).
13. M. C. Bailly, *ibid.*, **9,** 427 (1942).
14. A. Takami, M. Imazawa, M. Irie and T. Ukita, *Yakugaku Zasshi* (Japanese), **85,** 658 (1965).
15. S. Ouchi and T. Sowa, in press.
16. J. D. Smith and R. Markham, *Biochem. J.*, **46,** 509 (1950).
17. T. Sowa, M. Ojiro, S. Fukuhara and S. Ouchi, *Japanese Patent Publication* No. 47–23296 (1972).
18. T. Numata, M. Kishi, T. Sato, N. Yano, S. Senoo and H. Hiro, *Japanese Patent Publication* No. 44–25587 (1969).
19. P. Brigl and H. Hüller, *Ber.*, **72,** 2121 (1939); E. Baer, *Biochem. Prep.*, **1,** 50 (1959); J. M. Gulland and G. L. Hobday, *J. Chem. Soc.*, **1940**, 746; M. Ikehara, E. Ohtsuka and F. Ishikawa, *Chem. Pharm. Bull.*, **9,** 173 (1961); J. M. Gulland and M. Smith, *J. Chem. Soc.*, **1947**, 338; A. F. Tenner and H. G. Khorana, *J. Am. Chem. Soc.*, **81,** 4651, (1959).
20. H. A. G. Montogomery and J. H. Turnbell, *Proc. Chem. Soc.*, **1957**, 178; H. A. C. Montogomery, *ibid.*, **1958**, 1963; M. Ikehara and E. Ohtsuka, *Chem. Pharm. Bull.* **10**, 536 (1962).
21. P. A. Leven and R. S. Tipson, *J. Biol. Chem.*, **106**, 113 (1934); *ibid.*, **111**, 313 (1935); *ibid.*, **121**, 131, (1937).
22. H. Grunze, *Z. Anorg. Chem.*, **296**, 63, (1958).
23. R. W. Chamber, J. G. Moffatt and H. G. Khorana, *J. Am. Chem. Soc.*, **79**, 3747 (1957); *ibid.*, **77**, 3416 (1955).
24. M. L. Wolfrom, H. G. Garg and D. Horton, *Chem. Ind.*, **1964**, 930.
25. B. E. Griffin, C. B. Reese, G. F. Stephenson and D. R. Trentan, *Tetr. Lett.*, **1966**, 4349.
26. R. L. C. Brimacombe and C. B. Leese, *J. Chem. Soc.*, **1966**, 588; M. Yoshikawa and T. Kato, *Chem. Pharm. Bull.*, **40**, 2849 (1967).
27. M. Smith, D. H. Rammler, I. H. Goldberg and H. G. Khorana, *J. Am. Chem. Soc.*, **84**, 430 (1962); A. M. Schrecker, J. A. R. Meed and M. J. Urohel, *Biochem. Pharmacol.*, **15**, 1443 (1966).

28. Y. Mizuno and T. Nomura, *Kakusan Kagaku* (Japanese), vol, 2, Asakura Shoten,1970; Y. Mizuno, *Kakusan no Kagakugosei*, *Seibutsukagaku Jikken Koza*, vol. X, Nucleic Acid, Protein, Enzyme (Japanese), Kyoritsu Shuppan, 1968; A. M. Michelson, *The Chemistry of Nucleosides and Nucleotides*, p. 110, Academic, 1963.
29. C. R. Baker and G. E. Fall, *J. Chem. Soc.*, **1957**, 3798.
30. A. F. Tenner and H. G. Khorana, *J. Am. Chem. Soc.*, **81**, 4651 (1959).
31. G. M. Tenner, *ibid.*, **83**, 159 (1961).
32. M. Yoshikawa, T. Kato, and T. Takenishi, *Tetr. Lett.*, **1967**, 5065.
33. M. Yoshikawa, T. Kato and T. Takenishi, *Bull. Soc. Chem. Japan*, **42**, 3505 (1969).
34. Y. Sanno, Y. Kanai, A. Nohara, H. Honda and Y. Miyata, *Ann. Takeda Res. Lab.* (Japanese), **30**, 217 (1971).
35. T. Sowa, *Koenyoshishu*, *Yukigoseikagaku Kyokaishi*, *Yamaguchi Shibu* (Japanese), p. 11, 1972.
36. M. Yoshikawa and T. Kato, *Chem. Pharm. Bull.*, **40**, 2849 (1967); M. Viscontini and E. Ehrkardt, *Angew. Chem.*, **66**, 717 (1954).
37. R. W. Chamber, J. G. Moffatt and H. G. Khorana, *J. Am. Chem. Soc.*, **79**, 3747 (1957).
38. K. Imai, S. Fujii, K. Takanohashi, Y. Furukawa, T. Masuda and M. Honjo, *J. Org. Chem.*, **34**, 1547 (1969).
39. T. Sowa and S. Ouchi, *Bull. Chem. Soc. Japan*, **48**, 2084 (1975).
40. A. Hampton and T. Sasaki, *Biochemistry*, **12**, 2188 (1973).
41. H. Takaku and Y. Shimada, *Tetr. Lett.*, **1974**, 1279.
42. J. Hes and M. P. Mertes, *J. Org. Chem.*, **39**, 3768, (1974).
43. Y. Sanno, Y. Kanai, A. Nohara, H. Honda and Y. Miyata, *Ann. Takeda Res. Lab.* (Japanese), **30**, 217 (1971).
44. A. Holy, *Tetr. Lett.*, **1972**, 157.

PART III

PRODUCTION OF NUCLEIC ACID-RELATED SUBSTANCES BY FERMENTATION

CHAPTER **10**

Production of IMP*

Following the successful production of amino acids by fermentation, which had been developed and established largely in Japan,[1] extensive research was carried out in order to establish fermentative procedures for the production of ribonucleotides with flavor enhancing activity, such as IMP and GMP. However, nucleotide fermentation differs in several respects from amino acid fermentation. Namely, nucleotides contain three constituents, purine base, ribose and phosphate, and enzymes having the ability to degrade nucleotides are widely distributed in microorganisms as well as in higher organisms. Moreover, as shown in Fig. 10.1, salvage pathways are also involved in the biosynthesis of nucleotides, in addition

* Akira FURUYA, Kyowa Hakko Kogyo Co. Ltd.

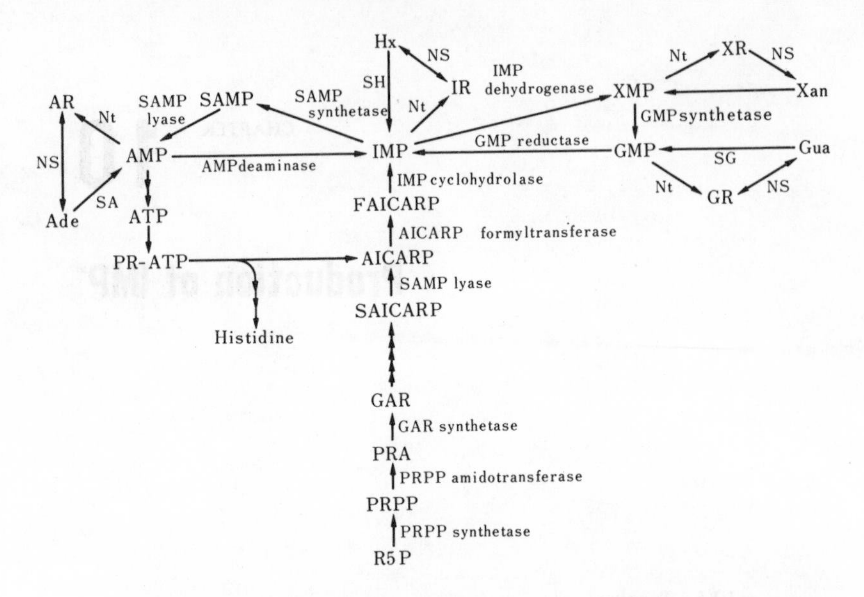

Fig. 10.1. Biosynthetic pathway of purine nucleotides. Hx, Hypoxanthine; AR, adenosine; GR, guanosine; XR, xanthosine; IR, inosine; Ade, adenine; Xan, xanthine; Gua, guanine; R5P, ribose-5-phosphate; PRPP, 5-phosphoribosylpyrophosphate; PRA, 5-phosphoribosylamine; GAR, glycineamide ribotide; SAICARP, 5-amino-4-imidazole-*N*-succino carboxamide ribotide; AICARP, 5-amino-4-imidazole carboxamide ribotide; FAICARP, 5-formamido-4-imidazole carboxamide ribotide; SAMP, adenylosuccinic acid; PR-ATP, 1-(5′-phosphoribosyl)-ATP; Nt, nucleotidase, phosphatase; NS, nucleosidase, nucleoside phosphorylase; SH, IMP pyrophosphorylase; SG, GMP pyrophosphorylase; SA, AMP pyrophosphorylase.

to the main *de novo* pathway. Since nucleotides are phosphoesters of sugar which do not permeate readily through cell membranes, it may be difficult to accumulate nucleotides efficiently in culture media across the permeability barrier. To overcome these complex problems, several different approaches have been made, as follows:

(1) Direct accumulation of IMP from a carbon source by the *de novo* biosynthetic pathway.

(2) Fermentative production of inosine from a carbon source and conversion of the inosine to IMP by a chemical or enzymatic procedure.

(3) IMP formation from hypoxanthine by the salvage pathway.

(4) Adenosine or AMP accumulation by a fermentative procedure, then conversion to inosine or IMP by a chemical or enzymatic process.

Among these procedures, (1) and (2) are apparently superior to the others and, in fact, have been used in industrial production. Therefore, these two procedures will be considered in this chapter. Since adenosine and AMP have pharmacological activities and are also important as starting materials for the chemical synthesis of pharmacological agents, the final section will deal with these two compounds. The salvage synthesis of nucleosides and nucleotides is described in Chapter 16.

10.1 INOSINE

10.1.1 Inosine Producing Microorganisms

In 1957, Gots and Gollub[2] and Partridge and Gile[3] reported the accumulation of hypoxanthine by adenine requiring mutants of *Escherichia coli* and *Neurospora crassa*, respectively. Since 1960, investigations aiming at the fermentative production of inosine were started in Japan and developed quickly, resulting in the establishment of an industrial process. The reason why fermentative production of inosine instead of IMP was attempted is that IMP does not permeate readily through cell membranes, and IMP decomposing enzymes are widely distributed in microorganisms, making it difficult to accumulate IMP effectively in culture media. In fact, inosine and/or hypoxanthine accumulating strains were obtained from various mutants of microorganisms; however, few IMP accumulating ones were found.[4–6] Therefore, an approach to the industrial production of IMP was made by chemical or enzymatic phosphorylation of inosine, which could be fermentatively accumulated without much difficulty.

As inosine fermentation was developed, the properties and applications of inosine were closely examined and it is now used in the medical field as a drug for hepatic disease (see Chapter 21).

As shown in Table 10.1, most inosine producers are adenine requiring mutants and further genetic improvement was carried out in order to increase inosine accumulation. Most of the inosine producing mutants were derived from the genus *Bacillus*, though mutants of *Brevibacterium ammoniagenes* accumulated reasonable amounts of inosine.[28,32] As a special case, it should be noted that a mutant of *Corynebacterium petrophilum* accumulated inosine from *n*-paraffin.[29–31]

The accumulation of hypoxanthine derivatives by an adenine requiring mutant of *Bacillus subtilis* was first reported by Uchida *et al.*,[9] and this was followed by a series of reports by Aoki *et al.*[10–15] Aoki *et al.* employed *B. subtilis* No. 2 (adenine requiring) as a parent strain, which accumulated a relatively small amount of inosine, and induced amino acid requiring mutants by X-ray irradiation. These they examined for inosine production. As shown in Table 10.2, C-30 strain accumulated 6.3 g

TABLE 10.1

Accumulation of Inosine and Hypoxanthine by Microorganisms

Accumulated substance[†1]	Microorganism	Genetic character[†2]	Ref.
Hypoxanthine	*Escherichia coli*	ade	2)
Hypoxanthine	*Neurospora crassa*	ade	3)
Hypoxanthine, AIR	Yeast	(biotin deficient)	7)
Hypoxanthine, AIR	Yeast	(biotin deficient)	8)
Hypoxanthine, IR, IMP	*Bacillus subtilis*	ade	9)
Inosine, Hypoxanthine	*B. subtilis*	ade+his+tyr	10~15)
Inosine	*B. subtilis*	ade+try+GMPred+dea +8AG^r	16, 17)
Inosine, Hypoxanthine	*B. subtilis*	ade+lys	4)
Inosine, Hypoxanthine	*B. subtilis*	ade	6, 18)
Inosine, Hypoxanthine	*B. subtilis*	ade	19)
Inosine, Hypoxanthine	*B. subtilis*	ade	20)
Inosine, Hypoxanthine	*B. subtilis*	ade+leu+his+GMPred+dea +8AX^r	21)
Inosine	*Bacillus pumilus*	ade	22)
Inosine, Hypoxanthine	*Bacillus* sp.	ade+his+thr+dea	23~25)
Inosine, Guanosine	*Bacillus* sp.	ade+his+GMPred+8AG^r	26, 27)
Inosine	*Brevibacterium ammoniagenes*	ade^L+6MG^r	28, 32)
Inosine, Hypoxanthine	*Corynebacterium petrophilum*	ade	29~31)
Inosine, Hypoxanthine	*Streptococcus coelicolor*	ade	4)
Inosine, Hypoxanthine	*Saccharomyces cerevisiae*	ade	4)
Inosine, Hypoxanthine	*Coprinus lagopus*	ade	4)

†1 AIR, Aminoimidazole riboside; IR, inosine.
†2 8AG^r, 8-Azaguanine resistant; 8AX^r, 8-azaxanthine resistant; 6MG^r, 6-mercaptoguanine resistant; ade^L, adenine leaky; dea, deaminase; GMPred, GMP reductase.

TABLE 10.2

Production of Inosine by Various Mutants derived from *Bacillus subtilis* Strain No. 2

Mutant[†1]	Requirement	Amounts of product (g/l)	
		Inosine	Hypoxanthine
B-4	ade+his	4.46	<0.1
S-26821	ade+his	0.70	0.2
B-1	ade+his+asp	2.45	<0.1
C-30	ade+his+tyr[†2]	6.30	<0.1
A-43	ade+met	2.10	0.9
S-3291	ade+amino	1.56	<0.1
A-28	ade+amino acids	1.11	0.26
A-9	ade+X[†3]	1.59	0.51
No. 2 (original strain)	ade	0.21	1.03

†1 The microorganisms were cultivated at 30°C for 72 h in flasks with shaking.
†2 Tyr marker was added to B-4 strain by further X-ray treatment.
†3 Unknown requirement (normal nutrients tested were not effective).

of inosine per liter. Recently, Shiio and Ishii[16,17] induced mutants resistant to low concentrations of 8-azaguanine (8AG) from an adenine requiring and adenine deaminase lacking strain of *B. subtilis* and obtained accumulation of 16–18 g of inosine per liter, which corresponded to a 50 to 70% increase of inosine production over the parent strain.

Nogami *et al.*[26,27] reported inosine plus guanosine producing mutants from *Bacillus* sp. No. 102 and observed that adenine requirement and 8AG resistance promoted the accumulation of inosine, resulting in the production of 13 g of inosine per liter. Suzuki *et al.*[22] induced an adenine requiring mutant from *Bacillus pumilus* and obtained 16 g of inosine accumulation per liter after optimization of the culture conditions.

Komatsu *et al.*[24,25] induced inosine plus guanosine producing mutants employing *Bacillus* sp. No. 1043 as a parent strain and obtained Ad-1 strain (ade + his + thr + dea) which accumulated 10.8 g of inosine per liter.

Furuya *et al.*[28,32] induced mutants resistant to low concentrations of 6-mercaptoguanine (6MG) from an IMP producing strain, KY 13102 of *B. ammoniagenes*, and observed the appearance of inosine accumulating mutants at high frequency. They obtained a maximal inosine accumulation of 13.6 g per liter.

10.1.2 Conditions for Inosine Production

Culture media: Table 10.3 shows the composition of fermentation media which were employed for inosine production, yielding relatively high

TABLE 10.3

Composition of Media Employed for Inosine Accumulation

Microorganisms / Compounds	*B. subtilis*[15] ade+his +tyr	*B. subtilis*[17] ade+trp +GMPred +dea +8AG^r	*B. pumilus*[22] ade	*Bacillus* sp.[26] ade+his +GMPred +8AG^r	*Bacillus* sp.[24] ade+his +thr +dea	*B. ammoniagenes*[32] ade^L+gua +6MG^r
Glucose	60~70 g/l	70 g/l	120 g/l		100 g/l	130 g/l
Maltose				100 g/l		
NH_4Cl	20 g/l	15 g/l			20 g/l	
$(NH_4)_2SO_4$			22.5 g/l	20 g/l		
$(NH_2)_2CO$			2.5 g/l			4 g/l
KH_2PO_4	1 g/l	1 g/l			2 g/l	10 g/l
K_2HPO_4						10 g/l
$CaHPO_4$			5 g/l	5 g/l		
$Ca_3(PO_4)_2$			5 g/l	5 g/l		
$MgSO_4 \cdot 7H_2O$	0.4 g/l	0.4 g/l	0.5 g/l	2 g/l	0.5 g/l	10 g/l
$FeSO_4 \cdot 7H_2O$	2ppm(Fe^{2+})	2ppm(Fe^{2+})				10 mg/l
$ZnSO_4 \cdot 7H_2O$						10 mg/l
$MnSO_4$	2ppm (Mn^{2+})	2ppm (Mn^{2+})				
$MnCl_2 \cdot 4H_2O$					0.01 g/l	10 mg/l
Cysteine						20 mg/l
Tryptophan		300 mg/l				
Thiamine						5 mg/l
Pantothenate						10 mg/l
Biotin			30 μg/l	0.2 mg/l		30 μg/l
Nicotinic acid						5 mg/l
Mieki	4 g/l	2 g/l				
Casamino acids					1 g/l	
Yeast extract					10 g/l	
Meat extract						10 g/l
Dry yeast	14 g/l		12 g/l	10 g/l		
Adenine		100 mg/l				100 mg/l
Guanine						100 mg/l
$CaCO_3$	2%	2.5%	2%	2%	5%	
pH	6.0	7.0		7.6		8.3
Max. inosine accumulated	10.5 g/l	18 g/l	16 g/l	14.1 g/l	10.8 g/l	13.6 g/l

levels of accumulation. As shown in the table, glucose was used as a carbon source in most cases, but cheaper materials such as starch hydrolyzate have been applied for the industrial production of inosine. Nogami *et al.*[26] employed maltose as a carbon source, but this material does not appear practical for industrial production. An adenine requiring mutant has been induced from a hydrocarbon assimilating bacteria, *C.*

petrophilum, and accumulated 1.6 g of inosine per liter from *n*-paraffin, C_{12} to C_{16}.[30]

As a nitrogen source, NH_4Cl, $(NH_4)_2SO_4$ and urea have generally been used. Since the nitrogen content in inosine is fairly high (20.9%), the nitrogen source in the fermentation media is important. In the industrial production of inosine, ammonia gas is used as a pH control agent as well as a nitrogen source.

Among inorganic salts, the effects of phosphate salts were reported by Suzuki *et al.*,[22] who examined inosine accumulation using an adenine requiring mutant of *B. pumilus*. When a soluble phosphate salt such as potassium phosphate was used, inosine accumulation was severely depressed, but it was stimulated by the addition of an insoluble salt such as calcium phosphate. In contrast, a mutant of *B. ammoniagenes* accumulated a considerable amount of inosine even in the presence of 2% potassium phosphate.[28]

Since most inosine producing strains require adenine for growth, a supplement of adenine to the media was essential. Yeast extract has often been used as an adenine source; in the case of inosine production by mutants belonging to *Bacillus*, dry yeast has been added as a cheaper adenine source, since these mutants can readily decompose dry yeast. Adenine is a precursor of adenine nucleotides, which strongly regulate IMP biosynthesis *in vivo*, so the concentration of adenine markedly affects cell growth as well as inosine accumulation. As shown in Fig. 10.2,[16]

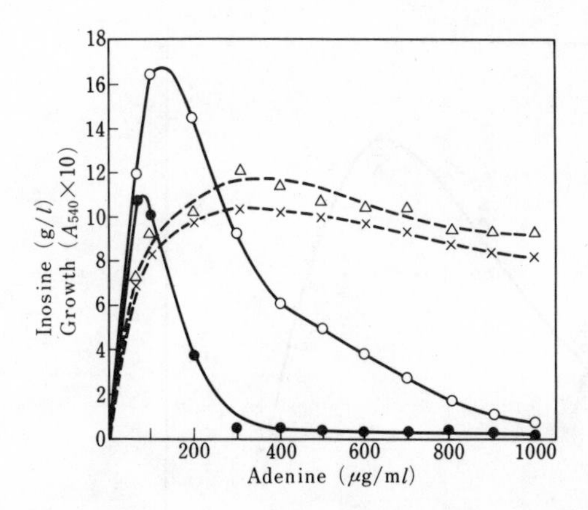

Fig. 10.2. Effect of adenine concentration on the growth and inosine accumulation of *B. subtilis*. Bacteria were grown in 20 ml of the media supplemented with various amounts of adenine in a 500 ml flask for 4 days at 30°C. ×, Growth of strain RDA-3; △, growth of strain No. 231; ●, inosine accumulation of strain RDA-3; ○, inosine accumulation of strain No. 231.

optimal concentrations of adenine for inosine accumulation were lower than those giving maximal cell growth. Yamanoi *et al.*[14,15] reported that an amino acid mixture promoted inosine accumulation and reduced the adenine requirement. It was found that histidine was essential and that seven other amino acids (isoleuciue, leucine, methionine, glycine, valine, phenylalanine and lysine) were stimulative. The seven amino acids could be replaced by relatively high concentrations of phenylalanine. The amino acids appeared not to be directly involved in purine nucleotide biosynthesis but to act through the stimulation of cell growth.

Other culture conditions: In addition to the composition of the media, pH control, temperature, aeration and agitation are also important factors for inosine accumulation. In inosine accumulation in flasks by an adenine requiring mutant, C-30 strain of *B. subtilis*, it was found that the optimal pH range was fairly wide, 5.0 to 7.0, while in production using large fermenters, the optimal pH range was narrow, 6.0 to 6.2, when the pH was controlled by ammonia gas[15] (Fig. 10.3).

As for cultivation temperature, 30° C was optimal for inosine accumulation in the case of C-30 strain of *B. subtilis*[15] and 32° C was the optimum for an adenine requiring mutant of *B. pumilus*.[22]

Aoki *et al.*[15] examined conditions of aeration for inosine accumulation by C-30 strain using 500 ml flasks and various volumes of media from 20 to 40 ml, and obtained a maximal accumulation at an oxygen transfer rate of 7.0–5.8 $\times$ 10^{-6} (mole/atom.ml.min). Suzuki *et al.*[22] also examined the effects of aeration on inosine accumulation by an adenine requiring

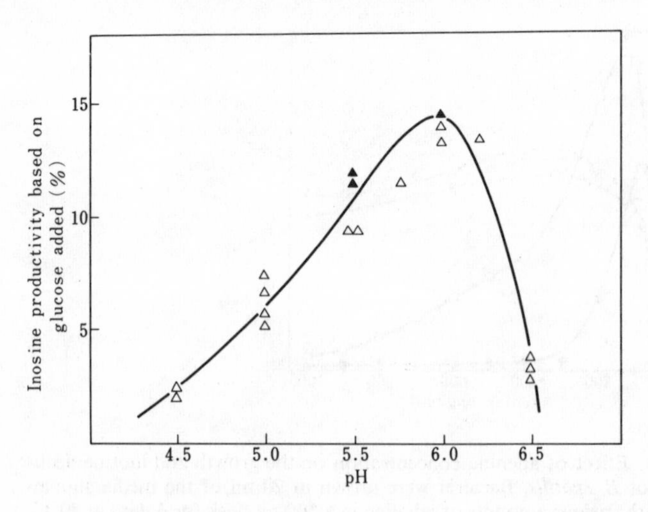

Fig. 10.3. Effect of pH on inosine productivity. △, 10 liter jar fermenter; ▲, 1 kiloliter tank fermenter.

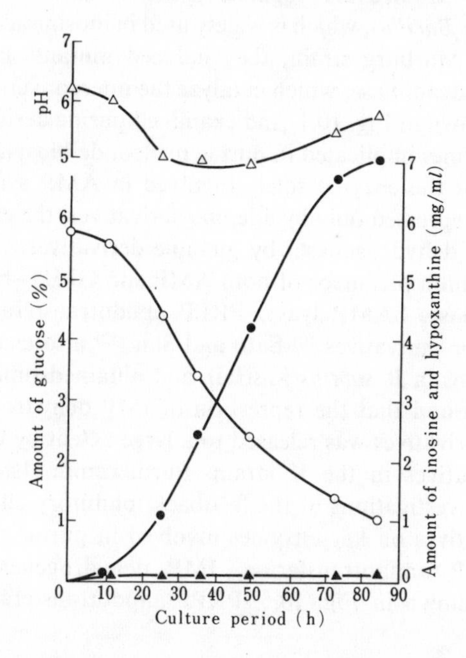

Fig. 10.4. Time course of inosine fermentation with C-30 strain. ●, Inosine; ▲, hypoxanthine; ○, glucose; △, pH.

mutant of *B. pumilus* and showed that maintenance of an oxygen transfer level of more than 4×10^{-7} (g.mole O_2/min.ml) was necessary for maximal production. Furthermore, Hirose *et al.*[33-36] showed, after extensive investigations, that aeration was not only important in inosine accumulation from the viewpoint of oxygen supply but also played a significant role in decreasing the partial carbon dioxide pressure in the media, increase of which severely depressed inosine accumulation. These points will be described in detail in Chapter 19. Fig. 10.4 shows the time course of inosine accumulation by *B. subtilis* C-30 strain.[12]

10.1.3 Mechanisms of Inosine Accumulation

It is generally accepted that inosine is accumulated in media by the dephosphorylation of IMP synthesized through the *de novo* pathway shown in Fig. 10.1. General features of the biosynthesis of purine nucleotides and related regulatory mechanisms are described in Chapter 11. The mechanisms directly involved in inosine accumulation will be described here.

Momose *et al.*[37–41] studied the regulatory mechanism of purine nucleotide biosynthesis in *Bacillus*, which is widely used in inosine accumulation. From *B. subtilis* Marburg strain, they induced mutants lacking GMP reductase or AMP deaminase, which catalyze the interconversion of purine nucleotides, as shown in Fig. 10.1, and examined purine derivatives for repression of the enzymes implicated in purine nucleotide biosynthesis. It was demonstrated that the enzyme solely involved in AMP synthesis (SAMP synthetase) was repressed only by adenine derivatives, the enzyme in GMP synthesis (IMP dehydrogenase) by guanine derivatives and the enzymes in IMP—a common precursor of both AMP and GMP—biosynthesis (IMP transformylase, SAMP lyase, PRPP amidotransferase) by both adenine and guanine derivatives.[40] Sato and Shiio[42] also examined the regulatory mechanisms in *B. subtilis* K strain and obtained similar results. However, it was found that the repression of IMP dehydrogenase formation by guanine derivatives was released to a large extent by the addition of adenine derivatives in the K strain. Furthermore, Ishii and Shiio[43–46] carried out investigations of the feedback inhibitory effects of purine nucleotide derivatives on key enzymes involved in purine nucleotide biosynthesis—PRPP amidotransferase, IMP dehydrogenase and SAMP synthetase. As shown in Fig. 10.5, PRPP amidotransferase was

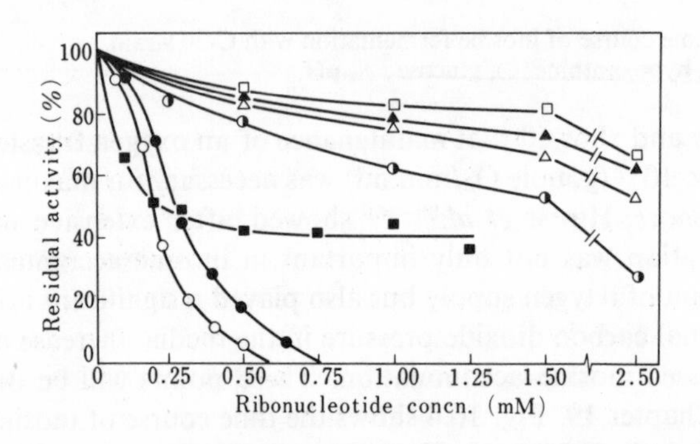

Fig. 10.5. Inhibition of PRPP amidotransferase activity by purine nucleotides at various concentrations.
■, GTP; ▲, XMP; ●, ADP; ◑, GMP; □, IMP; △, CMP; ○, AMP.

inhibited specifically and completely by AMP and ADP, while the inhibition by GMP, XMP or GTP was fairly mild. The inhibition by GMP was only 60% at maximum, showing a different pattern from that observed in *Aerobacter aerogenes*.[47] IMP dehydrogenase was most strongly inhibited by GMP (58%) and also by ATP (50%), XMP (39%) and GTP (26%).[43] Ishii and Shiio[46] induced inosine, xanthosine or adenosine accumulating

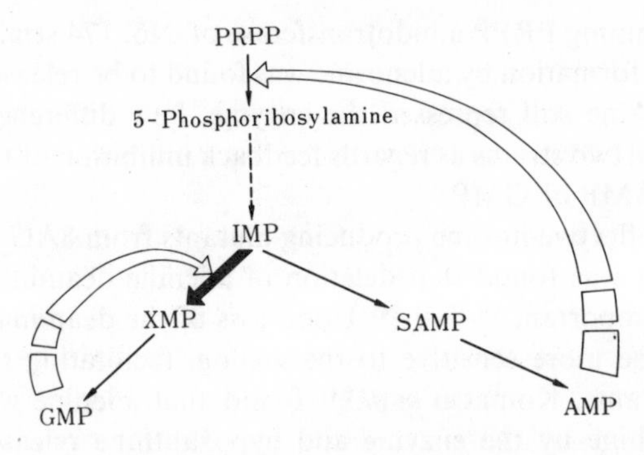

Fig. 10.6. Regulatory mechanisms for purine nucleotide biosynthesis in *Bacillus subtilis*. ⇨, Main regulation points; ➡, preferential synthesis.

mutants from *B. subtilis* K strain and examined the effects of adenosine or guanosine on the accumulation of these ribonucleosides by the mutants. Based on the results obtained, they presented a pattern of metabolic regulations operating in purine nucleotide biosynthesis in *B. subtilis*, as shown in Fig. 10.6.[48] The characteristic features are that the biosynthesis of AMP is regulated by feedback inhibition of PRPP amidotransferase by AMP and by repression of the enzyme formation by AMP involved in IMP and AMP biosynthesis, and that the biosynthesis of GMP is regulated by feedback inhibition of IMP dehydrogenase by GMP and by repression of the enzyme formation by GMP involved in IMP and GMP biosynthesis. Moreover, it was assumed that GMP was preferentially synthesized from IMP, which is a common precursor of both AMP and GMP.

From these regulatory mechanisms, it is apparent that the adenine concentration in the media should be kept at a low level, provided that this does not suppress cell growth too much, in order to maintain intercellular adenine nucleotides at low levels, since they act as the main regulatory factor on purine nucleotide biosynthesis. The relationship between adenine concentration in the media and inosine accumulation, shown in Fig. 10.2, supports the regulatory mechanisms described above. Therefore, it appears reasonable to assume that mutants lacking this regulatory mechanism would accumulate inosine more efficiently. In general, regulatory mutants have been selected from among analog resistant mutants.[49,50] Mutants resistant to 8AG[16,17,26,51] or 8-azaxanthine (8AX)[21,24,52] were employed in order to increase inosine productivity or to select guanosine producers from inosine producing strains. Shiio and Ishii[16,17] induced mutants resistant to low concentrations of 8AG from *B. subtilis* RDA-16 (ade + trp + GMP red + dea) and obtained No. 174 strain, which showed a better than 60–80% increase in inosine productivity compared to the parent. Upon examining PRPP amidotransferase of No. 174 strain, repression of the enzyme formation by adenosine was found to be released in the mutant but guanosine still repressed the enzyme. Few differences were observed between the two strains as regards feedback inhibition of the enzyme by AMP, 8-aza-AMP or GMP.

For the selection of effective inosine producing mutants from 8AG or 8AX resistant mutants, it was found that deletion of adenine deaminase in the parent strain was important.[16,17,21,24] Upon loss of the deaminase activity, the strain became more sensitive to the analog, facilitating the isolation of resistant mutants. Komatsu *et al.*[24] found that adenine was deaminated to hypoxanthine by the enzyme and hypoxanthine released the inhibitory effects of 8AX on cell growth.

Finally, the effect of inosine degrading activity on inosine accumulation will be described. Aoki *et al.*[10] induced various mutants from *B. subtilis* No. 2 (ade) and selected the C-20 strain as an effective inosine producer, as shown in Table 10.1. The C-20 strain was found to have much less inosine degrading activity than the parent strain. Nogami *et al.*[27] induced nucleoside phosphorylase deletion mutants from Nt 1011 B strain (ade + his + GMP red + 8AG^r + ade & AR^r + streptomycin^r) derived from *Bacillus*, resulting in the selection of effective inosine accumulating strains such as T-180 and T-780, as shown in Table 10.4. Komatsu *et al.*[25] also isolated mutants devoid of nucleoside degrading activity during the breeding of inosine plus guanosine producing strains from *Bacillus* sp. No. 1043 and observed that the nucleoside productivity of the mutants increased in parallel with the disappearance of hypoxanthine and guanine accumulation.

TABLE 10.4

Accumulation of Purine Nucleosides by Mutants Devoid of Nucleoside Phosphorylase (NP)

Microoganism†	Accumulated substance (mg/ml)		
	Hypoxanthine	Inosine	Guanosine
Nt 1011 B (Sm^{r}NP$^+$)	2.0	4.6	7.9
T-180 (Sm^{r}NP-dea$^-$)	0	12.6	5.4
T-780 (Sm^{r}NP$^-$)	0	10.8	5.8

† Sm^r = Streptomycin resistant.

10.2 IMP

It is quite reasonable to assume that the direct fermentative production of IMP from a carbon source is likely to be the most advantageous procedure for industrial production, if a reasonable accumulation of IMP can be obtained. However, as described earlier, effective accumulation of IMP is difficult chiefly due to its low permeability through the cell membrane and the wide distribution of IMP degrading enzymes. In order to overcome these difficulties, two types of investigations were carried out. (1) Genetic deletion of the nucleotide degrading activity of inosine producing mutants derived from *Bacillus* sp., leading to IMP producers. (2) Genetic and environmental improvement of the IMP productivities of microorganisms which accumulate small amounts of IMP and hypoxanthine. At present, the industrial production of IMP by direct fermentation is being carried out with mutants induced from *B. ammoniagenes*, based on the second approach.

10.2.1 IMP Producing Microorganisms

Table 10.5 lists IMP producing microorganisms. The induction of IMP producing mutants from inosine accumulators of *Bacillus* sp. was

TABLE 10.5

Accumulation of IMP by Microorganisms

Accumulated Substances[†1]	Microorganism	Genetic character[†2]	Ref.
IMP, IR, Hx	*Bacillus subtilis*	ade+Nt^W	9, 53, 54)
IMP, Hx	*B. subtilis*	ade+Nt^{VW}	55, 56)
IMP, Hx	*B. subtilis*	ade+Nt^W	57)
IMP, IR, Hx	*B. subtilis*	ade+Nt^W	58)
IMP, Hx	*Bacillus* sp.	ade	59)
IMP, Hx	*Corynebacterium glutamicum*	ade+$6MG^r$	60~62)
IMP, Hx	*C. glutamicum*	ade+xan	63~65)
IMP, Hx	*Brevibacterium ammoniagenes*	ade+gua	66, 78)
IMP, Hx	*B. ammoniagenes*	ade^L	68, 76)
IMP, Hx	*B. ammoniagenes*	ade^L+Mn^I	73, 78)
IMP, Hx	*Corynebacterium equi*	ade	79)
IMP	*Streptomyces* sp.	thiamine	80, 82)
IMP	*Streptomyces* sp.	(Na-barbiturate added)	81)

†1 IR = Inosine; Hx = hypoxanthine.
†2 Nt^W, Nucleotide degrading activity weak; Nt^{VW}, nucleotide degrading activity very weak; Mn^I, manganese ion insensitive.

reported by Fujimoto *et al.*,[53,54] Momose *et al.*[57] and Fujiwara *et al.*[58] In these studies, IMP producers were selected from mutants which showed colony formation on adenine-containing agar but not on AMP agar after mutagenic treatment of inosine producing strains or by a similar procedure. In spite of the fact that the mutants selected contained considerably decreased nucleotide degrading activities, good IMP producers were not obtained, probably due to the marked effects of the remaining small degradative activities on IMP accumulation. Fujimoto *et al.*[54] subsequently induced a mutant, A-1-25, having only trace activity to degrade 5′-nucleotides from an adenine requiring mutant, A-1 strain, which had considerably reduced degrading activity and obtained 1.73 mM accumulation of IMP by the A-1-25 mutant, corresponding to about twice that obtained with the A-1 strain. Akiya *et al.*[56] examined the culture conditions for IMP accumulation by A-1-25Z strain, derived from the A-1-25 strain, and obtained good accumulation of IMP (9.82 mM). This is the highest amount of IMP accumulated by mutants induced from inosine producers of *Bacillus* species.

Imada *et al.*[59] induced an adenine requiring mutant from *Bacillus* sp. No. 17–5 and observed the accumulation of small amounts of IMP. It was found that No. 17–5 strain had the ability to excrete 5′-nucleotides corresponding to components of RNA, so No. 17–5 strain appears to belong to microorganisms of type (2), even though it is classified as *Bacillus*. Other microorganisms having the ability to accumulate IMP instead of inosine upon acquisition of an adenine requirement by mutation include *Brevibacterium*, *Corynebacterium* and *Streptomyces*. These microorganisms appeared to have much smaller nucleotide degrading activities in comparison with *Bacillus* species.[83] Nakayama *et al.*[60–62] induced an adenine requiring mutant, No. 534–348, from *Corynebacterium glutamicum*, a glutamic acid producer, and observed the accumulation of IMP and hypoxanthine. Furthermore, they induced mutants resistant to 6MG from No. 534–348 strain and selected an IMP accumulator producing 2 mg of IMP per ml in a medium containing 10% glucose. Demain *et al.*[63] also obtained an adenine requiring mutant, MB 1807, from *Micrococcus glutamicus* (*C. glutamicum*) and obtained an accumulation of 0.4–0.6 mg of IMP per ml. From MB 1807 strain, a mutant with deletion of IMP dehydrogenase was selected by xanthine requirement for cell growth and was found to accumulate 0.8–0.9 mg of IMP per ml.[64] This is the first report to show that a genetic block in IMP dehydrogenase in addition to SAMP synthetase stimulates IMP accumulation.

Nara *et al.*[66] observed IMP accumulation by an adenine requiring mutant, KY 7208, derived from *B. ammoniagenes* ATCC 6872 and obtained more than 5 mg of IMP accumulation per ml after studies of the culture conditions. Furuya *et al.*[68] isolated a leaky mutant (cell growth was stimulated by adenine) from *B. ammoniagenes* ATCC 6872 and reported a maximal accumulation of 12.8 mg of IMP per ml. The accumulation of nucleotides by ATCC 6872 strain or its mutants was found to be severely affected by very low concentrations of Mn^{2+} irrespective of biosynthetic mechanism—*de novo* or salvage pathway.[67,68,72,76] Furuya *et al.* succeeded in isolating Mn^{2+} insensitive mutants, KY 13105[73] and KY 13171,[78] in which IMP accumulation was not affected by Mn^{2+}, an important step in the establishment of an industrial procedure for IMP production. Nara *et al.*[70] confirmed that the addition of a guanine requirement (deletion of IMP dehydrogenase) to an adenine requiring mutant of *B. ammoniagenes* stimulated IMP productivity, as observed in *C. glutamicum*.[64]

In addition, Schwartz and Margalith[80–82] reported 5′-nucleotide accumulation by *Streptomyces* sp. 772 strain and its mutants. It is noteworthy that the parent and its mutants accumulated a few milligrams of IMP per ml in a culture medium supplemented with sodium barbiturate.

10.2.2 Conditions for IMP Production

In connection with the accumulation of IMP by A-1-25 strain, which was isolated from an adenine requiring mutant of *B. subtilis* after genetic treatments and was largely devoid of 5′-nucleotide degrading activity, Fujimoto *et al.*[53] investigated the effect of adenine concentration in the media and obtained a maximal accumulation of 1.73 mM of IMP. Akiya *et al.*[56] further examined the conditions of IMP accumulation using A-1-25Z strain, isolated from A-1-25 strain by single colony selection. On addition of adequate amounts of yeast extracts (10–12 g/l), IMP accumulation increased to 3.0 mM. The effect of cultivation temperature was examined and, as shown in Fig. 10.7, upon shifting the temperature from 30° C to 40° C, accumulation of IMP sharply increased to 9.82 mM, accompanied by a drastic decrease in hypoxanthine accumulation. Since the effective conversion of hypoxanthine added to the media into IMP was observed at 40° C, it was assumed that salvage biosynthetic activity from hypoxanthine to IMP was stimulated at high temperature,[56] though it is also possible that the remaining weak activity to degrade IMP was inactivated completely at high temperature. In any event, the effect of temperature is important in nucleotide accumulation by *Bacillus* species.

Concerning IMP accumulation by an adenine requiring mutant, KY 7208, of *Brevibacterium ammoniagenes*, Nara *et al.* examined suitable culture conditions and obtained the following important results:[66,67,72,74]

(1) Fairly large amounts of KH_2PO_4 and K_2HPO_4 (1% each) in addi-

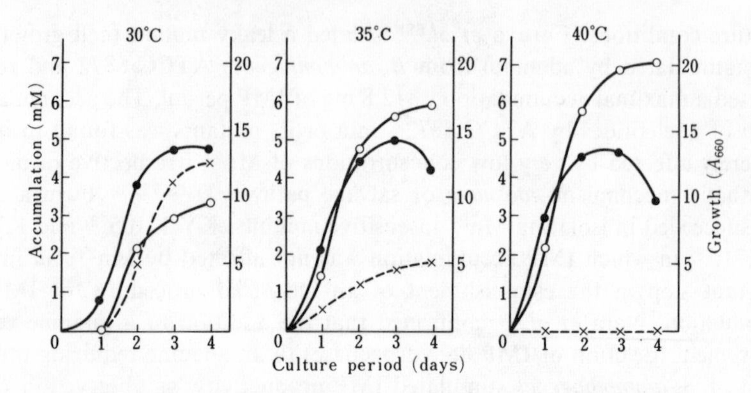

Fig. 10.7. Effect of culture temperature on the growth of strain A-1-25Z and IMP accumulation. Cultivation was carried out at the indicated temperature using the basal culture medium containing 10 g/l of yeast extract. ●—●, Growth; ○—○, IMP; ×---×, hypoxanthine.

tion to $MgSO_4 \cdot 7H_2O$ (about 1%) were necessary for effective IMP accumulation (Table 10.6).

(2) On addition of phosphate and Mg^{2+} at high concentrations, Mn^{2+}, thiamine and pantothenate were necessary for cell growth as well as IMP accumulation (Table 10.7).

(3) In medium containing high levels of phosphate and Mg^{2+}, casein hydrolyzate stimulated both cell growth and IMP accumulation. Casein hydrolyzate could be replaced by a mixture containing 400 mg each of histidine, lysine, homoserine, glycine and alanine per liter.

(4) For IMP accumulation, Mn^{2+}, Zn^{2+}, Fe^{2+} and Ca^{2+} are necessary; the effect of Mn^{2+} is particularly marked, and strict control of Mn^{2+} at optimal levels was essential. Excessive levels of Mn^{2+} caused a drastic decrease in IMP accumulation.

TABLE 10.6

Effect of Phosphate and Magnesium Levels on IMP and Hypoxanthine Production†

K_2HPO_4 KH_2PO_4 (each %)	$MgSO_4 \cdot 7H_2O$ (%)	Dry cell weight (mg/ml)	IMP produced (mg/ml)	Hypoxanthine produced (mg/ml)
0.2	0.1	11.9	1.37	1.31
0.4	0.1	10.7	1.78	0.97
0.4	0.4	9.7	2.65	1.46
0.6	0.1	11.6	1.24	0.87
0.6	0.6	10.2	3.52	0.80
1.0	1.0	16.4	4.55	0.10

† *B. ammoniagenes* KY 7208 (ade) was cultivated at 30°C for 7 days.

TABLE 10.7

Effect of Pantothenate and Thiamine on Purine Nucleotide Production

Supplement to "F" medium†		ATCC 6872						KY 7208
		Adenine added			Guanine added			
P. A.	V. B_1	AMP	ADP (mg/ml)	ATP	GMP	GDP (mg/ml)	GTP	IMP (mg/ml)
− + −	− − +	tr.	tr.	tr.	tr.	tr.	tr.	tr.
+	+	1.03	1.87	2.02	0.36	0.71	3.11	4.69

† "F" medium: fermentation medium containing Mn^{2+} 7.5 μg/l was supplemented with Ca pantothenate (P.A.) 10 μg/ml, or thiamine·HCl 5 μg/ml, as indicated. In the case of KY 7208 strain, adenine 200 μg/ml added to both media. Analyses were made for 4 to 5 day culture broth. Adenine or guanine (2.5 mg/ml) was added 3 days after inoculation.

Furuya *et al.*[68,76] investigated the effects of Mn^{2+} and adenine on IMP accumulation by KY 13102 strain (adenine leaky mutant) of *B. ammoniagenes.* As shown in Figs. 10.8 and 10.9, very low concentrations of Mn^{2+} had marked effects on IMP accumulation as well as cell morphology. On addition of excess Mn^{2+}, IMP accumulation changed to hypoxanthine accumulation. Adenine was found to act on IMP accumulation and cell growth as a regulatory factor for purine nucleotide biosynthesis, independently of Mn^{2+}.Excessive levels of adenine stimulated cell growth but repressed IMP accumulation. The time course of IMP ac-

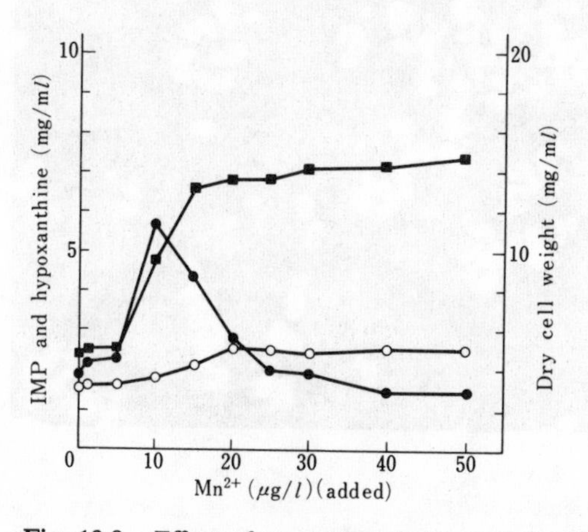

Fig. 10.8. Effect of manganese ions on the accumulation of IMP. The organism was cultivated at 30°C for 7 days. ■, Dry cell weight; ●, IMP; ○, hypoxanthine. (Source: ref. 68. Reproduced by kind permission of the American Society for Microbiology, U.S.A.)

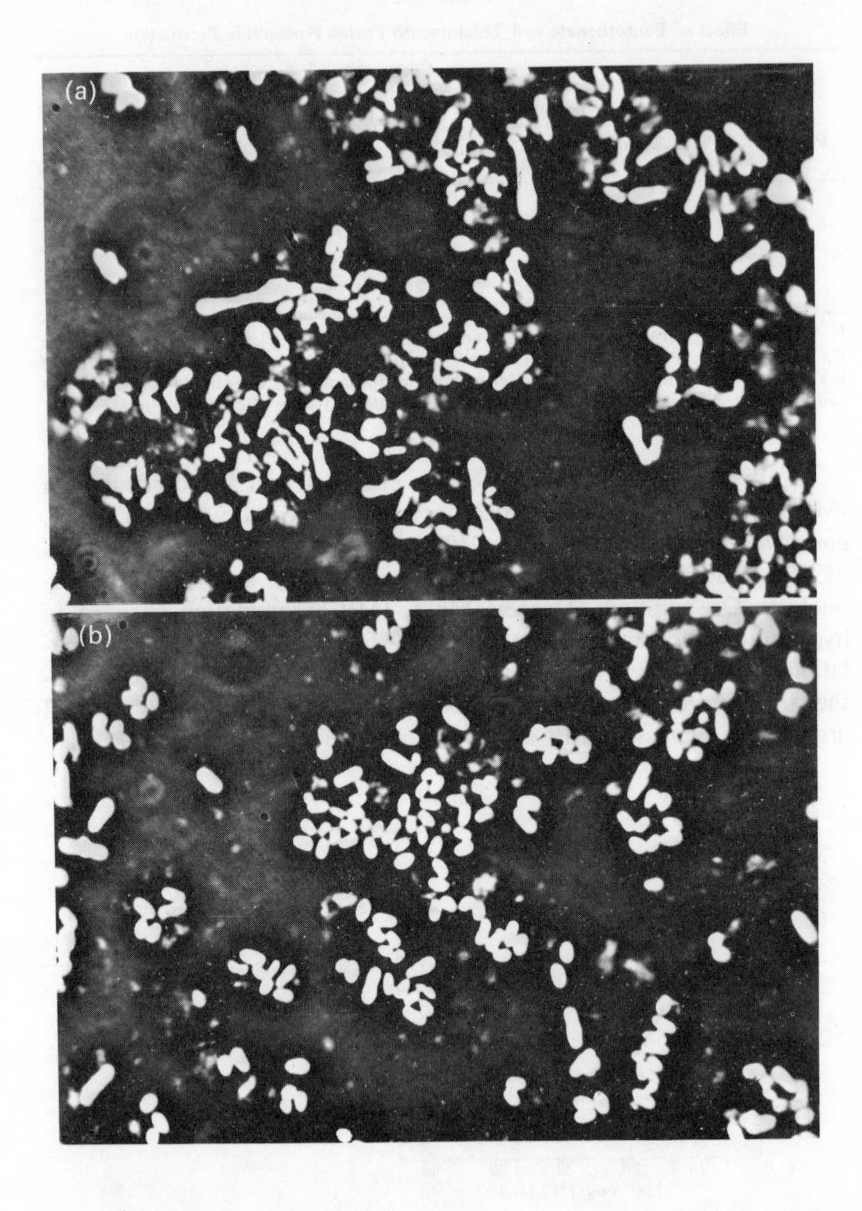

Fig. 10.9. Effect of manganese ions on cell morphology. (a) Cells cultivated in the basal accumulation medium supplemented with 10μg/*l* of Mn^{2+} at 30°C for 7 days. (b) Cells cultivated in the basal accumulation medium supplemented with 500 μg/l of Mn^{2+} at 30°C for 7 days.

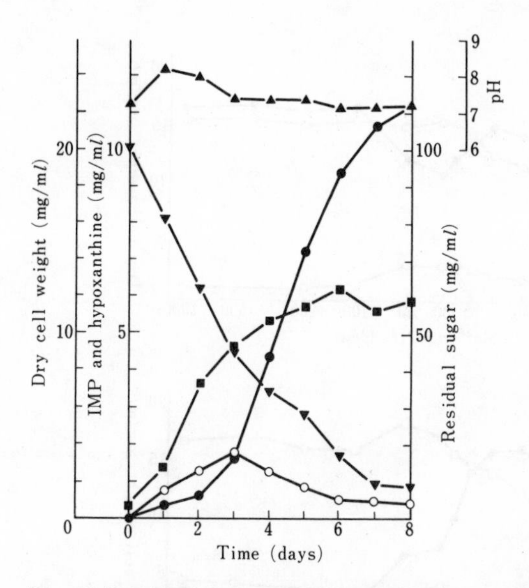

Fig. 10.10. Time course of IMP accumulation. ■, Dry cell weight; ●, IMP; ○, hypoxanthine; ▲, pH; ▼, residual sugar. (Source: ref. 68. Reproduced by kind permission of the American Society for Microbiology, U.S.A.)

cumulation by KY 13102 strain under optimal conditions is shown in Fig. 10.10[68]. The characteristic features of this fermentation are that, in an early stage of cultivation (2–3 days), accumulation of hypoxanthine is observed and after three days abrupt accumulation of IMP occurred accompanied by a decrease of hypoxanthine. Abnormal cells, as shown in Fig. 10.9, increased in number after two days, suggesting a strong correlation with IMP accumulation.

As described above, strict control of the Mn^{2+} level in the fermentation media was necessary for IMP production by mutants of *B. ammoniagenes*. However, it is rather difficult to control the Mn^{2+} level at such low concentrations as 10–20 μg per liter in large-scale fermenters during industrial production, and in addition some of the raw materials are known to contain significant amounts of Mn^{2+}, making it difficult to use them for fermentation. In order to solve this problem, two procedures, genetic and environmental, were applied. The application of antibiotics or surfactants, which had proved to be effective for glutamic acid accumulation in the presence of excess biotin, was tried. Nara *et al.*[75] examined the time of addition and concentrations of these agents in the presence of excess Mn^{2+}. Streptomycin, cycloserine, mitomycin C and penicillin were effective for IMP accumulation and polyoxyethylene stearylamine and hydroxyethylglyoxamine were selected among the surfactants. Meanwhile,

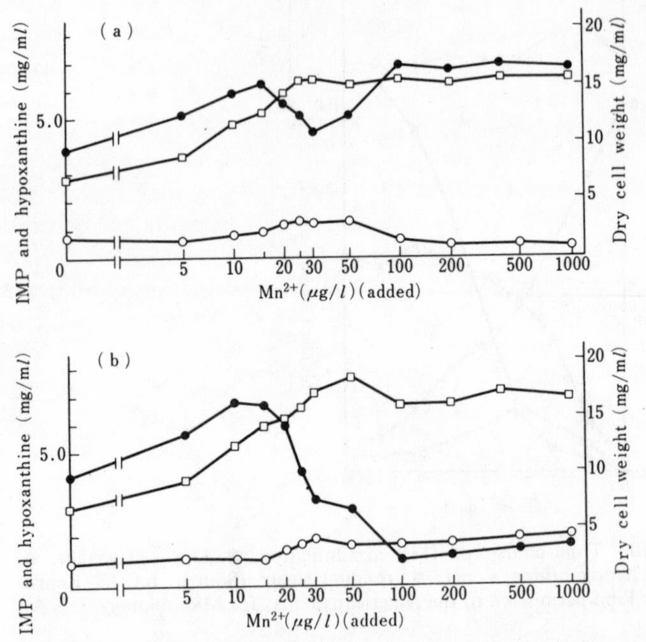

Fig. 10.11. Effect of manganese ions on IMP accumulation by manganese insensitive mutants. (a) KY 13105 (Mn^{2+} insensitive) was cultivated at 30°C for 6 days. (b) KY 13102 (Mn^{2+} sensitive) was cultivated at 30°C for 6 days. □, Dry cell weight; ●, IMP; ○, hypoxanthine.

Furuya *et al.*[73,78] tried to isolate mutants (Mn^{2+} insensitive) in which IMP accumulation was not affected by excess Mn^{2+} and obtained KY 13105 strain. Fig. 10.11 shows the effects of Mn^{2+}on the accumulation of IMP by KY 13105 strain and the parent. As can be seen, IMP accumulation by KY 13105 strain responded to low levels of Mn^{2+} in the same way as the parent, but accumulation by the mutant was not affected at all in the presence of excess Mn^{2+}. The cell forms of the mutant in the presence of excess Mn^{2+} were abnormal, like those shown in Fig. 10.9. It was found that KY 13105 strain showed stable IMP accumulation over fairly wide ranges of concentration of Zn^{2+}, Fe^{2+} and Ca^{2+} (essential factors) and the properties of the mutant were suitable for the industrial production of IMP.

10.2.3 Mechanisms of IMP Accumulation

In order to isolate IMP producing mutants from inosine producers of *B. subtilis*, attempts were made to block the 5′-nucleotide degrading activity of inosine producing strains genetically.[53–58] Enzymes participating in the degradation of 5′-nucleotides include 5′-nucleotidase, acid phosphatase, alkaline phosphatase, etc. Therefore, it was important to elucidate which enzyme was actually involved in the degradation of IMP formed by the *de novo* pathway and where it was located in the cells. Momose *et al.*[57] examined the intracellular distribution of the enzymes of *B. subtilis* C-30 strain. When the cells were converted to protoplasts, all the enzyme activity was released from the cells and recovered in the supernatant, indicating localization of the enzyme at the cell surface. The optimal pH range of the enzyme was almost the same as that of alkaline phosphatase. Fujiwara *et al.*[58] induced a mutant having reduced 5′-nucleotide degrading activity from *B. subtilis* No. 93 strain (adenine requiring and inosine producing) and obtained a small accumulation of IMP. The nucleotide degrading activity of the mutant was less than one-third of that of the parent. On the basis of the pH response of the enzyme activity and repression of the enzyme formation by phosphate, most of the activity was attributed to alkaline phosphatase. On the other hand, Demain and Hendlin[84] analyzed the phosphohydrolase activities of an inosine producing strain, MB-1839, of *B. subtilis* and found that four kinds of phosphohydrolase could be detected in the strain. The enzyme involved in IMP degradation was found to be located at the cell surface and was non-repressible by phosphate. It was suggested that the enzyme was of 5′-nucleotidase type.

The accumulation of IMP by mutants of *B. ammoniagenes* was markedly affected by the adenine concentration in the media.[68,70,76] Nara *et al.*[71] investigated the regulatory mechanism on PRPP amidotransferase in *B. ammoniagenes*, which is a key enzyme in the IMP biosynthetic pathway. They observed that the enzyme formation was repressed by adenine and its activity was inhibited by 70 to 100% by ATP, ADP, AMP and GMP. The results shown in Fig. 10.10 suggested that salvage synthesis of IMP from hypoxanthine played a significant role in IMP accumulation. The regulatory mechanisms controlling the salvage activities were also analyzed by Nara *et al.*[71] From these results, the overall regulatory mechanism of purine nucleotide biosynthesis in *B. ammoniagenes* was summarized as shown in Fig. 10.12.[71]

It was found that accumulation of 5′-nucleotides by *B. ammoniagenes* was severely affected by very low concentrations of Mn^{2+}, irrespective of the biosynthetic route—*de novo* or salvage pathway.[67,68,72,76] The mechanisms of action of Mn^{2+} were extensively studied by Nara *et al.*,[67,69] Furuya *et al.*[76] and Oka *et al.*[85] When the microorganism was cultivated in media containing the optimal amount of Mn^{2+} for nucleotide accumulation, abnormally shaped cells always appeared. Nara *et al.*[69] observed that phosphoribose pyrophosphokinase, hypoxanthine pyrophosphotrans-

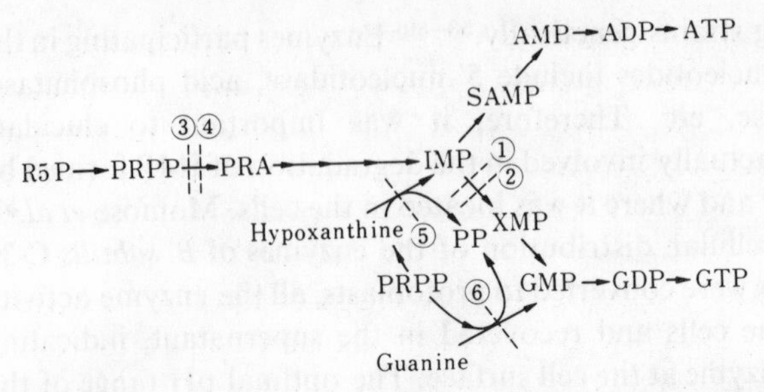

Fig. 10.12. Regulation of purine nucleotide synthesis by *Brevibacterium ammoniagenes*. The dashed lines show steps that are subject to enzyme repression and steps that are susceptible to feedback inhibition.
① repressed by guanine ② inhibited by GMP ③ repressed by adenine
④ inhibited by ATP, ADP, AMP and GMP ⑤ inhibited by GTP and ATP
⑥ inhibited by GTP

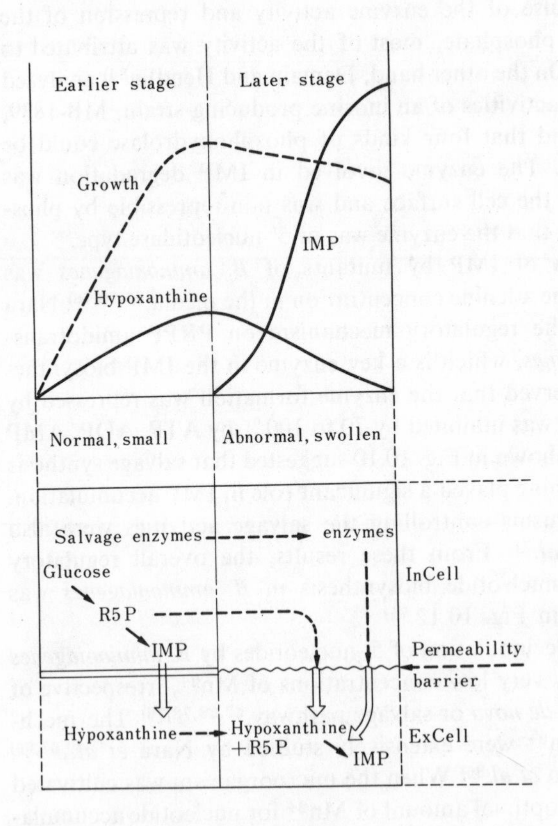

Fig. 10.13. Hypothetical diagram for direct IMP fermentation with an adenineless mutant of *B. ammoniagenes*.

ferase and ribose-5-phosphate, which are implicated in IMP formation by the salvage pathway, were excreted into the media in parallel with the appearance of the abnormal cells, though such excretion was not observed in the presence of excess Mn^{2+}. This accounts well for the fact that hypoxanthine accumulated in the early stage was converted to IMP on appearance of the abnormal cells, as indicated in Fig. 10.10. Furthermore, excretion of the enzymes and ribose-5-phosphate was also observed with cells cultivated in the presence of excess Mn^{2+} when a surfactant was added to the culture broth. These results strongly suggest that Mn^{2+} affects the cell membrane permeability. It seems that the formation of cell membranes (or walls) was disturbed by the limitation of Mn^{2+}, resulting in the appearance of abnormally formed cells with increased membrane permeability due to random membrane damage, and permitting the excretion of enzymes and metabolites into the culture media. In fact, it was found that the nucleotide pool of abnormal cells was much smaller than that of cells cultivated in the presence of excess Mn^{2+}.[76] On the basis of these results, Nara *et al.*[69] proposed a model for direct IMP accumulation by *B. ammoniagenes* as shown diagrammatically in Fig. 10.13. The characteristic features of this model are that, by limitation of Mn^{2+}, abnormal cells were induced in a later stage of cultivation and IMP accumulation occurred extracellularly by salvage biosynthesis, mainly due to the breakdown of the permeability barrier. Later, Furuya *et al.*[76,78] proposed in addition that IMP formed by the *de novo* pathway in the later stage of cultivation was directly excreted into the culture media, based on data for IMP accumulation by various mutants of *B. ammoniagenes.*

The appearance of abnormally formed cells on limitation of Mn^{2+} suggested that Mn^{2+} was directly involved in the formation of the cell membrane or wall. In glutamic acid fermentation, it has been shown that biotin regulates the formation of phospholipid in the cell membrane, resulting in control of the permeability of cell membranes to glutamic acid.[86] Since *B. ammoniagenes* requires biotin for growth, the effects of biotin and adenine on IMP accumulation by KY 13105 strain (Mn^{2+} insensitive) were studied.[73] As shown in Fig. 10.14, limitation of biotin affected only the cell growth, but a considerable accumulation of IMP was obtained in the presence of excess biotin and optimal amounts of adenine. In glutamic acid producing microorganisms, it was reported that a decrease in the fatty acid content in the cells was observed on limitation of biotin,[86] so the effect of Mn^{2+} on the fatty acid content of mutants of *B. ammoniagenes* was examined.[76] However, the amount of fatty acid increased somewhat in Mn^{2+} limited cells. Therefore, it was assumed that the mechanism of regulation of the permeability barrier for IMP by Mn^{2+} was different from that by biotin in glutamic acid fermentation. Misawa *et al.*[87] reported the

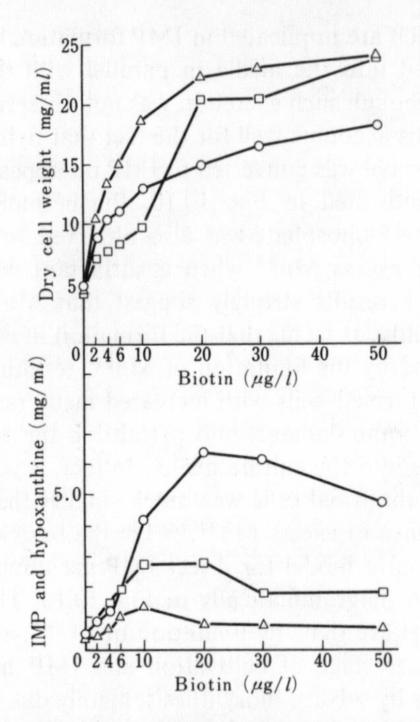

Fig. 10.14. Effect of adenine and biotin on IMP accumulation. KY 13105 was cultivated at 30°C for 6 days, in basal accumulation media (Mn^{2+}, 500 μg/l) containing various concentrations of adenine and biotin. □, Adenine 15 mg/l; ○, adenine 25 mg/l; △, adenine 50 mg/l.

accumulation of XMP by a guanine requiring mutant, KY 7450, induced from *B. ammoniagenes* ATCC 6872. Accumulation of XMP by the mutant was stimulated by the addition of large amounts of Mn^{2+}, and the cells formed were normal short rods. The only difference between XMP and IMP is the presence of a hydroxy group at the 2-position of the purine base. Nevertheless, IMP could not permeate through membranes of cells cultivated in the presence of excess Mn^{2+} but XMP could do so freely. It is not clear whether the ability of the nucleotides to permeate depends on their chemical structures or on some specificity of the membrane or both. However, this is a useful system for investigation of the permeation of nucleotides from the inside to the outside of cells.

10.3 Adenosine and AMP

In the production of IMP by enzymatic hydrolysis of RNA, AMP is formed as an intermediate, and is converted to IMP by deamination. Thus, it appears that the conversion of adenosine or AMP to inosine or IMP is a practical procedure for the production of IMP if adenosine or AMP can be efficiently accumulated by fermentation. In fact, in an early stage of investigations for IMP production, Koaze and Hara[88–93] published a series of papers dealing with the microbial conversion of chemically synthesized adenine to IMP through adenosine and AMP. Since the fermentative production of inosine or IMP has developed considerably and the importance of adenosine or AMP as a drug or as a raw material for medical agents has been recognized, the production of adenosine or AMP by fermentation has been investigated independently of IMP production. Considerable advances have been made in the fermentative production of adenosine, which will mainly be described here. In addition, several reports on the fermentative production of AMP, ADP and ATP by salvage synthesis (Part IV), and on 3′:5′-cyclic AMP production (Chapter 12) are noteworthy. A few papers have reported the accumulation of AMP by the *de novo* pathway, but the amounts were fairly low and further development is necessary.

10.3.1 Adenosine or AMP Producing Microorganisms

In Table 10.8, microorganisms accumulating adenine derivatives by *de novo* biosynthesis are listed. The first report appeared in 1953,[94] describing the accumulation of adenine by a guanine requiring mutant of *Ophiostoma multiannulatum,* but the accumulated amounts were very small (10 to 20 μg per ml). Later, Onoda *et al.*[97] isolated an adenine accumulating mutant among mutants resistant to sulfaisoxazole of *Brevibacterium flavum*, a glutamic acid producer. They identified the substance accumulated as adenine and by investigating suitable culture conditions for accumulation they obtained a maximal yield of 3.0 mg of adenine per ml. Recently, Shimojo *et al.*[98] obtained adenine accumulating mutants among 2-fluoroadenine (2FA) resistant mutants of *Corynebacterium, Brevibacterium* and *Microbacterium.* Using strain KY 10478 induced from *C. glutamicum,* a maximal accumulation of 2.0 mg of adenine per ml was obtained. Araki *et al.*[99] found that 2FA resistant mutants of a histidine producing strain, KY 10260, derived from *C. glutamicum* accumulated 0.5–2.0 mg of adenine per ml in addition to histidine. The accumulation of adenine was not compatible with histidine accumulation and high histidine accumulators showed relatively low adenine accumulation.

Concerning adenosine fermentation, Konishi *et al.*[100] reported the accumulation of adenosine by an isoleucine requiring mutant induced from *B. subtilis* K strain. On examination of suitable culture conditions, a maxi-

TABLE 10.8
Accumulation of Adenine Derivatives by Microorganisms

Accumulated substance†	Microorganism	Genetic character	Ref.
Adenine	*Ophiostoma multiannulatum*	*gua*	94)
Adenine	*Salmonella typhimurium*	2, 6-diamino purine^r	95)
Adenine, Hypoxanthine	*Aerobacter aerogenes*		96)
Adenine	*Bacillus megaterium*		96)
Adenine	*Bacillus cereus*		96)
Adenine	*Brevibacterium flavum*	sulfoisoxazole^r	97)
Adenine	*Corynebacterium glutamicum*	2-fluoroadenine^r	98)
Adenine	*Corynebacterium murisepticum*	2-fluoroadenine^r	98)
Adenine	*Brevibacterium flavum*	2-fluoroadenine^r	98)
Adenine	*B. ammoniagenes*	2-fluoroadenine^r	98)
Adenine	*Microbacterium* sp.	2-fluoroadenine^r	98)
Adenine Histidine	*Corynebacterium glutamicum*	Triazolealanine^r +2-fluoroadenine^r	99)
Adenosine	*Bacillus subtilis*	ile	100)
Adenosine	*Bacillus* sp.	his+thr+xan+dea+GMP red +8AX^r	101, 102)
Adenine nucleotides	*Streptomyces* sp.		103)
Adenine nucleotides	*Streptomyces phaeochromogenes*		104)

† Ade, Adenine; Hx, hypoxanthine; His, histidine; AR, adenosine.

mal accumulation of 2.5 mg of adenosine per ml was obtained in a glycerol-peptone medium. Even though the relationship between isoleucine requirement and adenosine accumulation is not yet clarified, it was confirmed that adenosine accumulation was not due to the degradation of cellular RNA. Haneda *et al.*[101,102] isolated adenosine accumulating mutants—A + 2-P13, A + 1-PI, P53 and P48—from an inosine producing strain, Ad-1, induced from *Bacillus* sp. No. 1043 according to the procedure shown in Fig. 10.15. A maximal accumulation of 16.27 mg of adenosine per ml was obtained. It was found that the important requirement for adenosine accumulation was a xanthine requirement caused by genetic deletion of IMP dehydrogenase.

Furuya *et al.*[103] observed the accumulation of AMP in addition to small amounts of 3′-AMP and ADP by *Streptomyces* sp. No. 327 strain isolated from soil. The amount of AMP accumulated was 0.85 mg per ml. Igarashi *et al.*[104] also observed the accumulation of AMP, adenosine-3′,5′-

1043 (wild)
↓
1043-226 (ad⁻)
↓
Ad-1 (ad⁻, his-thr, -, adenine dea⁻)
↓
AX-3 (ad⁻, his⁻, thr⁻, adenine dea⁻, 8AXr)
↓
A+2, A+1 (ad+, his⁻, thr⁻, adenine dea, 8AXr)
↓ ↘ ↘
A+2-P13, A+1-Pl (his⁻, thr⁻, adenine dea⁻, 8AXr, xn⁻)
↓
A+2-Pl (his⁻, thr⁻, adenine dea⁻, 8AXr, hx⁻)
↓
Rt-4 (his⁻, thr⁻, adenine dea⁻, 8AXr, hx⁻, Rtase⁻)
↓
Rt-4-H+3 (his⁻, thr⁻, adenine dea⁻, 8AXr, hx+, Rtase⁻)
↓
P53; P48 (his⁻, thr⁻, adenine dea⁻, 8AXr, Rtase⁻, xn⁻)

Fig. 10.15. Derivation of adenosine-producing mutants. xn⁻, Xanthine requirement.

diphosphate, adenosine-5′-phosphosulfate and adenosine-3′-phosphate-5′-phosphosulfate by *S. phaeochromogenes*.

10.3.2 Conditions and Mechanisms of Adenosine Accumulation

Konishi *et al.*[100] examined adenosine accumulation by an isoleucine requiring mutant, No. 717, of *B. subtilis*. As shown in Fig. 10.16, cell growth was stimulated as the concentration of isoleucine increased but the accumulation of adenosine reached a maximal level on addition of 0.2–0.3 mg of isoleucine per ml. The relationship between isoleucine requirement and adenosine accumulation is not clear. In place of 0.03% isoleu-

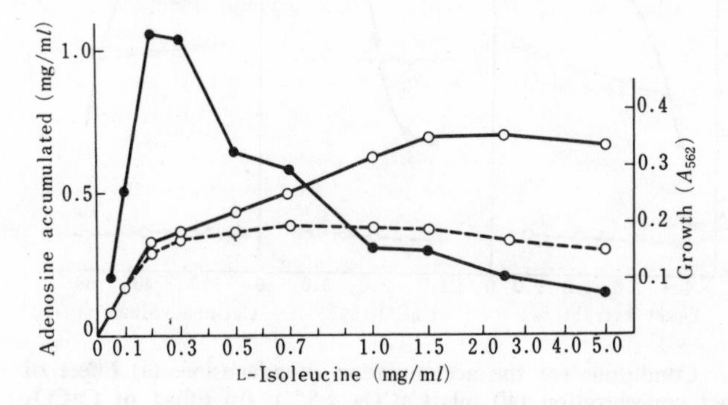

Fig. 10.16. Effect of L-isoleucine concentration on the accumulation of adenosine. Various amounts of L-isoleucine were added to the basal medium. ●—●, Adenosine; ○—○, A_{562} at 64 h; ○ --- ○, A_{562} at 24 h.

cine, addition of 0.2% peptone increased adenosine accumulation from 1.65 mg per ml to 2.08 mg. Furthermore, a maximal accumulation of 2.45 mg of adenosine per ml was obtained in a medium containing 0.5% peptone and 10% glycerol as a carbon source.

Haneda *et al.*[101,102] also examined adenosine accumulation by P53 strain induced from *Bacillus* sp. 1043 strain following the procedure shown in Fig. 10.15. The concentrations of yeast extract and $CaCO_3$ and aeration were important factors affecting the accumulation: their effects are shown in Fig. 10.17. The time course of adenosine accumulation by P53–18 strain is shown in Fig. 10.18.

Later, it was found that adenosine accumulation by the mutants shown in Fig. 10.15 was very unstable.[102] Drastic decreases in adenosine accumulation were observed after successive transfers of the mutants on preservation medium. This phenomenon was analyzed and was ascribed to reversion of the xanthine requirement to prototrophy, resulting in a change to an adenosine non-accumulator. At the same time it was confirmed that the amino acid requiring character (histidine and threonine) of the adenosine producing mutant was not essential for adenosine accumulation. In order to stabilize the adenosine productivity of the mutants, two approaches were tried. One involved improvement of the preservation medium. On supplementing the preservation medium with 0.1% yeast extract and 50 mg of guanosine per liter (300 g of potato per liter of water), the appearance of revertants was markedly suppressed. The other approach was genetic improvement of the mutants. A stable strain was selected

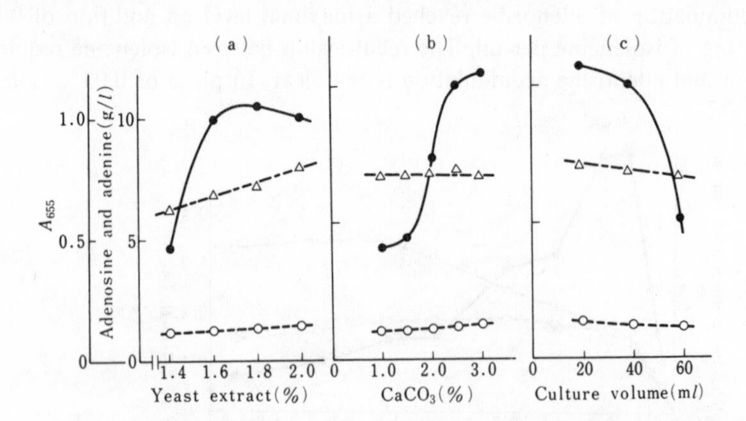

Fig. 10.17. Conditions for the accumulation of adenosine. (a) Effect of yeast extract concentration (40 ml, $CaCO_3$ 2.5%). (b) Effect of $CaCO_3$ concentration (yeast extract 1.6%, 40 ml). (c) Effect of culture volume (yeast extract 1.6%, $CaCO_3$ 2.5%). ●—●, Adenosine; ○---○, adenine; △—△, A_{655}

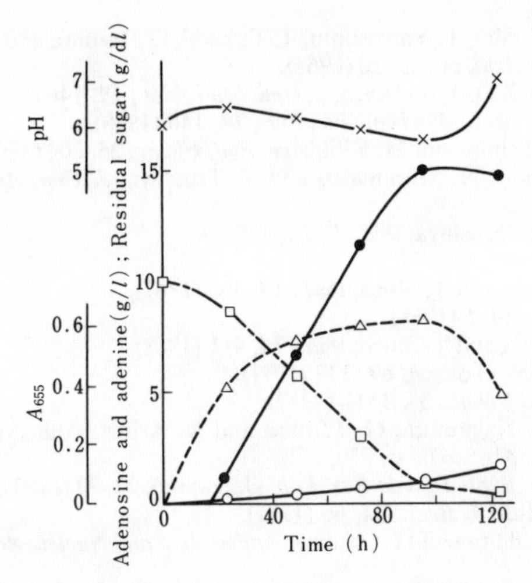

Fig. 10.18. Time course of adenosine fermentation (strain: *Bacillus* sp. P53–18). ●—●, Adenosine; □—□, residual sugar; ○—○, adenine; △---△, A_{655} ; ×—×, pH.

after successive transfers on non-supplemented potato medium and NB-1, -3, and -17 strains were obtained; these did not segregate xanthine non-requiring revertants even under conditions stimulating reversion. On examination of the most stable strain, NB-1, it was found that the strain was also genetically devoid of GMP synthetase in addition to IMP dehydrogenase.[102] To confirm this, genetic deletion of GMP synthetase was induced by mutagenic treatment using an unstable adenosine producing strain, and GX-2 strain was obtained, which showed stable adenosine productivity. It was therefore concluded that genetic deletion of IMP dehydrogenase was essential for adenosine accumulation and that genetic deletion of GMP synthetase was very effective for the stabilization of adenosine productivity.

References

1. *The Microbial Production of Amino Acids* (ed. K. Yamada *et al.*), Kodansha, Wiley, 1972.
2. J. S. Gots and E. G. Gollub, *Proc. Natl. Acad. Sci. U.S.A.*, **43**, 826 (1957).
3. C. W. H. Partridge and N. H. Gile, *Arch. Biochem. Biophys.*, **67**, 237 (1957).
4. T. Nara, M. Misawa, K. Nakayama and S. Kinoshita, *Amino Acid and Nucleic Acid*, no. 8, 94 (1963).
5. T. Fukami, H. Imanaka, M. Yokota, M. Fujiwara, G. Tamura and K. Arima, *Nippon Nogeikagaku Kaishi* (Japanese), **37**, 505 (1963).

6. M. Fujiwara, H. Nakamura, K. Shu, T. Yamamoto, T. Fukami, G. Tamura and K. Arima, *Amino Acid and Nucleic Acid*, no. 8, 110 (1963).
7. D. P. Lanes, C. Rainknow and J. D. Woodward, *J. Gen. Microbiol.*, **19**, 146 (1958).
8. H. Friedman and A. G. Moat, *Arch. Biochem. Biophys.*, **78**, 146 (1958).
9. K. Uchida, A. Kuninaka, H. Yoshino and M. Kibi, *Agr. Biol. Chem.*, **25**, 804 (1961).
10. R. Aoki, H. Momose, Y. Kondo, N. Muramatsu and Y. Tsuchiya, *J. Gen. Appl. Microbiol.*, **9**, 387 (1963).
11. R. Aoki, N. Muramatsu and Y. Tsuchiya, *ibid.*, **9**, 397 (1963).
12. R. Aoki, *ibid.*, **9**, 403 (1963).
13. A. Yamanoi, Y. Hirose, M. Aoki and T. Shiro, *ibid.*, **13**, 365 (1967).
14. A. Yamanoi and T. Shiro, *ibid.*, **14**, 1 (1968).
15. A. Yamanoi, Y. Hirose, M. Aoki and T. Shiro, *ibid.*, **14**, 411 (1968).
16. I. Shiio and K. Ishii, *J. Biochem.* (Tokyo), **69**, 339 (1971).
17. K. Ishii and I. Shiio, *Agr. Biol. Chem.*, **36**, 1511 (1972).
18. M. Fujiwara, V. Bykovsky, H. Nakamura, G. Tamura and K. Arima, Ann. Mtg. Agr. Chem. Soc. Japan , 1966. Abstracts, p. 250,
19. I. Nogami and S. Igarashi, *Ann. Rept. Takeda Res. Lab.* (Japanese), **22**, 99 (1963).
20. A. L. Demain and D. H. Hendlin, *J. Bact.*, **94**, 66 (1967).
21. K. Komatsu, A. Nishijo, R. Kodaira and H. Ohsawa, *Amino Acid and Nucleic Acid*, no. 22, 54 (1970).
22. T. Suzuki, I. Nogami, Y. Kitahara, M. Ishikawa and M. Yoneda, *Ann. Rept. Takeda Res. Lab.*, **26**, 126 (1967).
23. A. Hirano, T. Akimoto and T. Ohsawa, *Nippon Nogeikagaku Kaishi* (Japanese), **42**, 60 (1968).
24. K. Komatsu, H. Haneda, A. Hirano, R. Kodaira and H. Ohsawa, *J. Gen. Appl. Microbiol.*, **18**, 19 (1972).
25. K. Komatsu and R. Kodaira, *ibid.*, **19**, 263 (1973).
26. I. Nogami, M. Kida, T. Iijima and M. Yoneda, *Agr. Biol. Chem.*, **32**, 144 (1968).
27. I. Nogami and M. Yoneda, *Kagaku to Seibutsu* (Japanese), **7**, 371 (1969).
28. A. Furuya, S. Abe and S. Kinoshita, *Appl. Microbiol.*, **20**, 263 (1970).
29. T. Iguchi, T. Watanabe and I. Takeda, *Agr. Biol. Chem.*, **31**, 569 (1967).
30. T. Iguchi, T. Watanabe and I. Takeda, *ibid.*, **31**, 574 (1967).
31. T. Iguchi, R. Kodaira and I. Takeda, *ibid.*, **31**, 885 (1967).
32. A. Furuya, F. Kato and K. Nakayama, *ibid.*, **39**, 767 (1975).
33. H. Shibai, A. Ishizaki, H. Mizuno and Y. Hirose, *ibid.*, **37**, 91 (1973).
34. A. Ishizaki, H. Shibai, Y. Hirose and T. Shiro, *ibid.*, **37**, 99 (1973).
35. A. Ishizaki, H. Shibai, Y. Hirose and T. Shiro, *ibid.*, **37**, 107 (1973).
36. H. Shibai, A. Ishizaki and Y. Hirose, *ibid.*, **37**, 2083 (1973).
37. H. Momose, H. Nishikawa and N. Katsuya, *J. Gen. Appl. Microbiol.*, **10**, 343 (1963).
38. H. Momose, H. Nishikawa and N. Katsuya, *ibid.*, **11**, 211 (1965).
39. H. Momose, H. Nishikawa and I. Shiio, *J. Biochem.* (Tokyo), **59**, 325 (1966).
40. H. Nishikawa, H. Momose and I. Shiio, *ibid.*, **62**, 92 (1967).
41. H. Momose, *J. Gen. Appl. Microbiol.*, **13**, 39 (1967).
42. H. Sato and I. Shiio, *J. Biochem.* (Tokyo), **68**, 763 (1970).
43. K. Ishii and I. Shiio, *ibid.*, **63**, 661 (1968).
44. I. Shiio and K. Ishii, *ibid.*, **66**, 175 (1969).
45. K. Ishii and I. Shiio, *ibid.*, **68**, 171 (1970).
46. K. Ishii and I. Shiio, *Agr. Biol. Chem.*, **37**, 287 (1973).
47. D. P. Nierlich, B. Magasanik, *J. Biol. Chem.*, **240**, 358 (1965).
48. I. Shiio, *Seikagaku* (Japanese), **44**, 7 (1972).
49. A. L. Demain, *Advan. Appl. Microbiol.*, **8**, 1 (1966).
50. K. Nakayama, *Hakko to Biseibutsu* (Japanese) (ed. T. Uemura and K. Aida), vol. 1, p. 133, Asakura Shoten, 1971.
51. S. Konishi and T. Shiro, *Agr. Biol. Chem.*, **32**, 396 (1968).
52. H. Momose and I. Shiio, *J Gen Appl. Microbiol.*, **15**, 399 (1969).

53. M. Fujimoto and K. Uchida, *Agr. Biol. Chem.*, **29**, 249 (1965).
54. M. Fujimoto, M. Morozumi, Y. Midorikawa, S. Miyakawa and K. Uchida, *ibid.*, **29**, 918 (1965).
55. Y. Midorikawa, T. Akiya, A. Kuninaka and H. Yoshino, *ibid.*, **37**, 1595 (1973).
56. T. Akiya, Y. Midorikawa, A. Kuninaka, H. Yoshino and Y. Ikeda, *ibid.*, **36**, 227 (1972).
57. H. Momose, H. Nishikawa and N. Katsuya, *J. Gen. Appl. Microbiol.*, **10**, 343 (1964).
58. M. Fujiwara, V. Bykovsky, H. Nakamura, G. Tamura and K. Arima, *ibid.*, **13**, 1 (1967).
59. S. Imada, I. Nogami, Y. Nakao and S. Igarashi, *Ann. Rept. Takeda Res. Lab.* (Japanese), **23**, 54 (1964).
60. K. Nakayama, T. Suzuki, Z. Sato and S. Kinoshita, *J. Gen. Appl. Microbiol.*, **10**, 133 (1964).
61. K. Nakayama, T. Nara, H. Tanaka, Z. Sato, M. Misawa and S. Kinoshita, *Agr. Biol. Chem.*, **29**, 234 (1965).
62. Z. Sato, K. Nakayama, H. Tanaka and S. Kinoshita, *ibid.*, **29**, 412 (1965).
63. A. L. Demain, M. Jackson, R. A. Vitali, D. Hendlin and T. A. Jacob, *Appl. Microbiol.*, **13**, 757 (1965).
64. A. L. Demain, M. Jackson, R. A. Vitali, D. Hendlin and T. A. Jacob, *ibid.*, **14**, 821 (1966).
65. A. L. Demain, *Biotechnol. Bioeng.*, **10**, 291 (1968).
66. T. Nara, M. Misawa and S. Kinoshita, *Agr. Biol. Chem.*, **31**, 1351 (1967).
67. T. Nara, M. Misawa and S. Kinoshita, *ibid.*, **32**, 1153 (1968).
68. A. Furuya, S. Abe and S. Kinoshita, *Appl. Microbiol.*, **16**, 981 (1968).
69. T. Nara, M. Misawa, T. Komuro and S. Kinoshita, *Agr. Biol. Chem.*, **33**, 358 (1969).
70. M. Misawa, T. Nara and S. Kinoshita, *ibid.*, **33**, 514 (1969).
71. T. Nara, T. Komuro, M. Misawa and S. Kinoshita, *ibid.*, **33**, 739 (1969).
72. T. Komuro, T. Nara, M. Misawa and S. Kinoshita, *ibid.*, **33**, 1018 (1969).
73. A. Furuya, S. Abe and S. Kinoshita, *Appl. Microbiol.*, **18**, 977 (1969).
74. T. Nara, T. Komuro, M. Misawa and S. Kinoshita, *Agr. Biol. Chem.*, **33**, 1030 (1969).
75. T. Nara, M. Misawa, T. Komuro and S. Kinoshita, *ibid.*, **33**, 1198 (1969).
76. A. Furuya, S. Abe and S. Kinoshita, *ibid.*, **34**, 210 (1970).
77. K. Takayama, A. Furuya and S. Abe, *Amino Acid and Nucleic Acid*, no. 22, 15 (1970).
78. F. Kato, A. Furuya and S. Abe, *Agr. Biol. Chem.*, **35**, 1061 (1971).
79. H. Sasaki, E. Nakazawa and S. Okumura, Ann. Mtg. Agr. Chem. Soc. Japan, 1969. Abstracts, p. 177.
80. J. Schwartz and P. Margalith, *J. Appl. Bact.*, **34**, 348 (1971).
81. J. Schwartz and P. Margalith, *ibid.*, **35**, 83 (1972).
82. J. Schwartz and P. Margalith, *ibid.*, **35**, 271 (1972).
83. M. Misawa, T. Nara and K. Nakayama, *Nippon Nogeikagaku Kaishi* (Japanese), **38**, 167 (1964).
84. A. L. Demain and D. Hendlin, *J. Bact.*, **94**, 66 (1967).
85. T. Oka, K. Udagawa and S. Kinoshita, *ibid.*, **96**, 1760 (1968).
86. S. Fukui and M. Ishida, *The Microbial Production of Amino Acids* (ed. K. Yamada *et al.*), p. 123, Kodansha, Wiley, 1972.
87. M. Misawa, T. Nara, K. Udagawa, S. Abe and S. Kinoshita, *Agr. Biol. Chem.*, **33**, 370 (1969).
88. T. Hara, Y. Koaze, Y. Yamada and M. Kojima, *ibid.*, **26**, 61 (1962).
89. M. Kojima, Y. Koaze and T. Hara, *ibid.*, **26**, 656 (1962).
90. Y. Koaze, Y. Yamada, M. Kojima and T. Hara, *ibid.*, **26**, 740 (1962).
91. T. Hara, Y. Koaze, Y. Yamada, M. Kojima, K. Sato and Y. Aoyama, *ibid.*, **26**, 747 (1962).

92. Y. Koaze, Y. Yamada, M. Kojima, H. Goi and T. Hara, *ibid.*, **26**, 754 (1962).
93. M. Kojima, Y. Koaze and T. Hara, *ibid.*, **26**, 758 (1962).
94. N. Fries, *J. Biol. Chem.*, **200**, 325 (1953).
95. G. P. Kalle and J. S. Gots, *Proc. Soc. Exptl. Biol. Med.*, **109**, 281 (1962).
96. T. Fukami, H. Imanaka, M. Yokota, M. Fujiwara, G. Tamura and K. Arima, *Nippon Nogeikagaku Kaishi* (Japanese), **37**, 505 (1963).
97. T. Onoda, H. Kamijo and K. Kubota, Ann. Mtg. Agr. Chem. Soc. Japan, 1973. Abstracts, p. 117.
98. S. Shimojo, K. Araki and K. Nakayama, *Nippon Nogeikagaku Kaishi* (Japanese), **48**, 63 (1974).
99. K. Araki, S. Shimojo and K. Nakayama, *Agr. Biol. Chem.*, **38**, 837 (1974).
100. S. Konishi, K. Kubota, R. Aoki and T. Shiro, *Amino Acid and Nucleic Acid*, no. **18**, 15 (1968).
101. K. Haneda, A. Hirano, R. Kodaira and S. Ohuchi, *Agr. Biol. Chem.*, **35**, 1906 (1971).
102. K. Haneda, K. Komatsu, R. Kodaira and H. Ohsawa, *ibid.*, **36**, 1453 (1972).
103. A. Furuya, S. Abe and S. Kinoshita, *Amino Acid and Nucleic Acid*, no. 8, 100 (1963).
104. S. Igarashi, Y. Takeuchi, S. Imada and I. Nogami, *Ann. Rept. Takeda Res. Lab.* (Japanese), **23**, 64 (1964).

CHAPTER 11

Production of GMP*

Several enzymes involved in the pathway of purine nucleotide biosynthesis have been shown to be repressed by the end products. Synthesis of these enzymes, especially PRPP amidotransferase, IMP dehydrogenase and GMP synthetase, is repressed by guanine derivatives, and the reactions of these enzymes are markedly inhibited by GMP. Furthermore, GMP is easily decomposed to guanosine or guanine by 5′-nucleotidase or nucleosidase. For these reasons, it is difficult to establish a procedure for the industrial production of GMP. The production of nucleosides such as

* Teruo SHIRO, Ajinomoto Co. Ltd.

AICAR, xanthosine and guanosine, on the other hand, is relatively easy to establish: the synthesis of AICAR or xanthosine is only slightly inhibited by the end products themselves or their derivatives, and furthermore guanosine tends to precipitate, due to its low solubility, which is clearly favorable for guanosine accumulation.

Possible processes of GMP production are described below:
(1) GMP can be synthesized by chemical methods from AICAR produced by a fermentative process.
(2) GMP can be synthesized by chemical phosphorylation from guanosine produced by a fermentative process.
(3) GMP can be synthesized by enzymatic conversion from XMP or xanthosine produced by a fermentative process.
(4) GMP can be produced directly by a fermentative process.

11.1 Production of GMP from AICAR

In 1942, Fox[1] made the important discovery that *Escherichia coli* grown in the presence of bacteriostatic concentrations of a sulfonamide accumulated a diazotizable amine. This was subsequently isolated[2] and identified[3] as AICA. AICA and related compounds were subsequently found to be accumulated in culture broth of various bacteria. Such accumulation can be obtained (1) by the use of antimetabolites such as sulfonamides and aminopterine, which cause breakdown of purine formation, (2) by using purine requiring mutants having genetic blocks in purine synthesis, (3) by using a suitable wild type strain of *E. coli*. These methods are summarized in Table 11.1. A series of studies on the accumulation of AICA and its derivatives by microorganisms did not contribute greatly to the development of *de novo* synthesis of purine but did contribute to the achievement of industrial production of AICAR by purine requiring mutants.

Kinoshita *et al.*[21] isolated a purine requiring mutant, *Bacillus megaterium* No. 366 (ATCC 15117) that was an excellent accumulator of AICAR. At the same time, Yamazaki *et al.*[21,22] developed a novel and convenient method for the chemical synthesis of GMP from AICAR. A process for manufacturing GMP that combined AICAR production by fermentation and conversion of AICAR to GMP by chemical synthesis was therefore devised.

Extensive studies on the fermentative production of AICAR were carried out due to an increase in the importance of AICAR as a starting

TABLE 11.1

Accumulation of AICAR and AICAR-related Substances

Type of accumulation	Accumulated substance	Microorganism	Ref.
By drug action	AICA	*E. coli*	1~5, 15)
	AICAR	*E. coli, Aerobacter aerogenes, Bacillus subtilis, Pseudomonas aeruginosa*, etc.	7, 8, 19)
By purine requiring mutant	AICA	*E. coli*	6, 11, 13)
	AICAR	*E. coli*	10, 12, 14, 16, 18)
By wild type strain	AICA	*E. coli*	8, 9, 13)
	AICAR	*E. coli*	19, 20)
	AICARP	*E. coli*	17)

compound for synthesis of the flavoring reagent, GMP. For the commercial production of AICAR by fermentation, important features such as the derivation of AICAR producing microorganisms, control of sporulation, suppression of reverse mutation, establishment of medium constituents, and determination of oxygen supply conditions had to be elucidated.

An outline of the production process with *Bacillus megaterium* No. 336 will be described below. Investigations from the standpoint of bioengineering are discussed in Chapter 13.

11.1.1 Selection and Derivation of AICAR Producing Microorganisms

The culture fluids of purine requiring mutants of *E. coli* were found to contain insufficient AICAR (about 30 μM) for industrial production.[18] Shiro *et al.*[23] isolated a purineless mutant, D-421, by X-ray irradiation of the auxotrophic mutants from *Bacillus subtilis* No. 2093. The amount of AICAR accumulated by the selected mutant, D-421, reached an acceptable industrial level. Furthermore, Kinoshita *et al.* obtained many auxotrophic mutants derived from *B. subtilis* IAM 1523, *B. megaterium* IAM 1245, and *Brevibacterium flavum* No. 2247 by X-ray or ultraviolet irradiation. AICAR accumulators were found only among purine requiring mutants. The particular strain named *B. megaterium* No. 336 of purineless mutants from *B. megaterium* IAM 1245 was the best producer of AICAR. These results are shown in Table 11.2.

As for the regulatory mechanism of purine biosynthesis, the main re-

Table 11.2

Derivation of Auxotrophic Mutants

Parent	Requirement (Number of strains)		Product (Number of strains)	
Bacillus subtilis IAM 1523	Adenine	(3)	Hypoxanthine	(1)
			SAICAR[†2]	
	Guanine	(4)	Xanthosine	(4)
	Purine	(2)	**AICAR**	(1)
	Uracil or cytosine	(2)	Orotic acid	(1)
	Amino acids	(13)		
	Vitamins	(4)		
Bacillus megaterium IAM 1245	Adenine	(5)	Hypoxanthine	(2)
			SAICAR[†2] +SAICARP[†3]	(2)
	Guanine	(3)	Xanthosine	(2)
	Guanine or xanthine	(1)		
	Purine	(8)	AICAR	(3)
	Uracil or cytosine	(4)	Orotic acid	(2)
	Amino acids	(45)	UV absorption compounds	(19)
			Ninhydrin positive compounds	(8)
	Vitamins	(49)	UV absorption compounds	(4)
			Ninhydrin positive compounds	(339)
	Amino acids+ vitamins	(907)	UV absorption compounds	(132)
			Ninhydrin positive compounds	(339)
	Unknown	(34)		
Brevibacterium flavum No. 2247	Adenine	(5)	Hypoxanthine+IMP	(3)
			SAICA[†1]	(1)
	Purine	(14)	**AICAR**	(2)
	Uracil or cytosine	(6)		
	Amino acids	(32)		
	Amino acids+ bases	(2)		

†1 SAICA = *N*-Succino-5(4)-amino-4(5)-imidazole carboxamide.
†2 SAICAR = SAICA riboside.
†3 SAICARP = SAICA ribotide.

quirements for obtaining AICAR producing mutants are as follows:
(1) Purine synthesizing capacity in the parent microorganisms should be high.
(2) The mutants should lack AICARP formyltransferase, which is responsible for the conversion of AICARP to FAICARP.
(3) The mutants should lack control mechanisms consisting of feedback inhibition and repression of biosynthetic enzymes, especially of PRPP amidotransferase by intracellular purine nucleotides.
(4) The mutants should be deficient in AICAR hydrolyzing activity.

The excellent producer of AICAR obtained by Kinoshita *et al.* satisfies all four conditions.

Shirafuji *et al.*[24] chose AICAR producing mutants of purineless mutants from an inosine forming adenine auxotroph of *Bacillus pumilus*. The amount of AICAR accumulated by the purineless mutants (pur$^-$) was less than that of inosine accumulated by their adenine auxotrophs (ade$^-$). They isolated a number of further mutants (ade$^-$, pur$^-$), of which more than half accumulated a large amount of AICAR as compared with their parent purine auxotrophs.

11.1.2 Control of Sporulation

B. megaterium No.336 showed spore forming characteristics similar to those of its parent. AICAR accumulation depended on the rate of sporulation during fermentation.

On the occasion of a high frequency of sporulation, the rate of AICAR production decreased, glucose consumption decreased, and AICAR accumulation was low. When sporulation occurred at a low rate during fermentation, the fermentation process followed one of two patterns:
(1) The rate of AICAR production directly paralleled the rate of glucose consumption, and consequently AICAR was produced satisfactorily.
(2) AICAR production occurred when glucose could no longer be detected as a reducing sugar, but the level of production was low.

Fig. 11.1 shows the relationship between these fermentation patterns and the rate of sporulation. Thus, because of variation of the sporulation rate, stable AICAR fermentation could not be obtained. Measures to control sporulation at the optimal rate must be taken in order to produce AICAR satisfactorily.[25] Although asporogenous mutants from spore forming *B. megaterium* No. 336 could be obtained by X-ray irradiation, these mutants could not be adopted for commercial production because of their relatively poor AICAR accumulation. Antisporulating reagents which could be used included saturated fatty acids, surfactants, and anti-

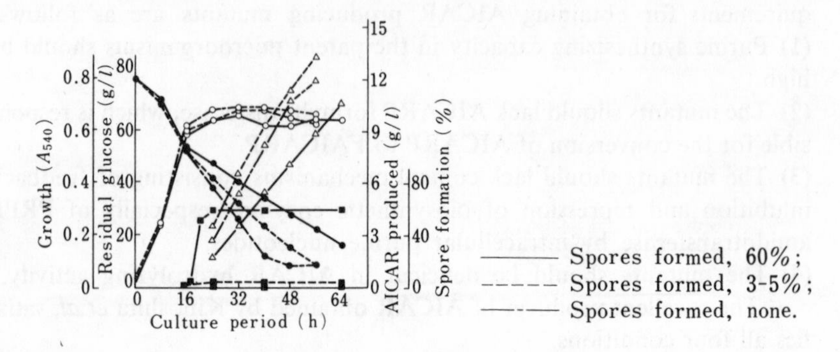

Fig. 11.1. Time course of AICAR fermentation. ○, Growth; ●, residual glucose; △, AICAR produced; ■, Spore formation

biotics. Among saturated fatty acids having two to eighteen carbon atoms, those with three to nine carbon atoms (propionate to pelargonate) had antisporulating activity. As for surfactants, nonionic surfactants such as polyoxyethylene nonyl phenol ether had activity.

Antibiotics such as chloramphenicol and tetracycline also showed activity. In general, antibiotics are known to have such effects as inhibition of cell wall, cell membrane, protein and nucleic acid synthesis. The antisporulating activity of these antibiotics cannot, however, be explained by these known modes of action. Among the antisporulating reagents, butyric acid was expected to be suitable for commercial use due to its low price and the ease of treatment. Thus, the antisporulating mechanism of butyric acid was investigated.[26] AICAR fermentation was divided into three phases; the former and latter halves of the logarithmic growth phase, and the AICAR production phase. Butyric acid was effective in controlling sporulation only in the latter half of the logarithmic growth phase. Thus, it can be considered that the latter half of the logarithmic growth phase determines in some way whether sporulation will occur or not. Butyric acid was neither decomposed nor assimilated during fermentation. It inhibited the endogenous oxygen uptake of washed cells harvested in the latter half of the logarithmic growth phase.

The effect of oxygen supply on sporulation was also investigated. These results are discussed in Chapter 13.

11.1.3 Suppression of Reverse Mutation

Revertant cells appeared and increased in number to some extent (10^3–10^5/ml), even when the fermentation proceeded normally with marked

AICAR accumulation. In some cases, a very high frequency of revertants was observed, resulting in reduced production of AICAR.[21,27]

Fig. 11.2 shows the effect of revertants on AICAR fermentation. In order to isolate the nonreverting strains, the original mutants were irradiated with X-rays or ultraviolet rays. Nonreverting strains could be obtained, but they were not suitable for commercial use because of a considerable decrease in AICAR accumulation. Attempts to give resistance to antibiotics and purine analogs also failed to yield nonreverting strains from the

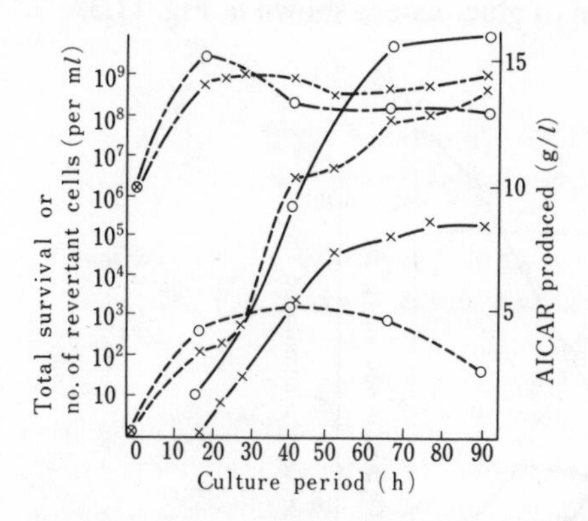

Fig. 11.2. Time course of AICAR fermentation: comparison between cultures with greater and lesser numbers of revertants. ×, More revertants; ○, less revertants. -•-•-, Total survival (per ml); ---, no. of revertant cells (per ml); —, AICAR formed (g/l).

AICAR producer. Therefore, practical methods for minimizing the appearance of revertants were developed as follows:

(1) Choice of an organism with as low a reversion rate as possible by monocell or monocolony isolation at regular intervals.

(2) Counting revertants and contaminants with minimal medium plates periodically during fermentation (distinction between revertants and contaminants was made in terms of their shapes observed under a microscope, the features of colonies grown on a minimal medium plate, and the action of phage specific to *B. megaterium* No. 336).

(3) Addition of erythromycin to inhibit the growth of revertants specifically in the period from the end of the logarithmic growth phase to the early stationary phase.

11.1.4 Gluconate as an Intermediate

In this fermentation, AICAR sometimes continued to be produced when glucose could no longer be detected as a reducing sugar. Over 50% of the total AICAR accumulation occurred after the disappearance of glucose in an extreme case. It was anticipated that some intermediate was formed from glucose in the production of AICAR, and it was shown that gluconate was accumulated in the culture fluids by isolation and identification.[28] It was found that the rate of AICAR production from gluconate was the same as that from glucose with washed cells harvested in the stationary growth phase. AICAR was produced by consuming gluconate after the disappearance of glucose. A typical fermentation pattern with accumulation and consumption of gluconate is shown in Fig. 11.3.

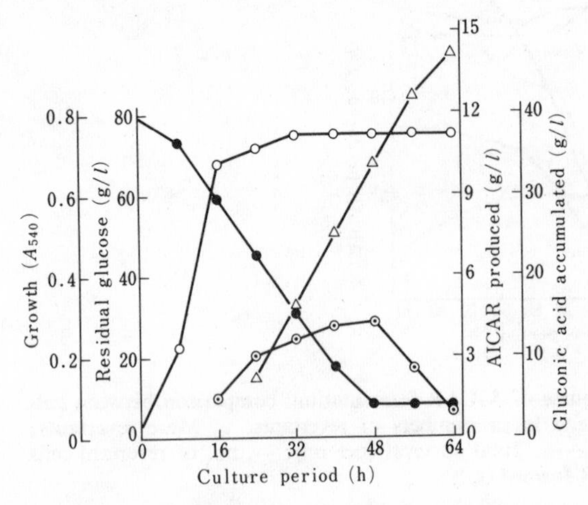

Fig. 11.3. Time course of accumulation of gluconic acid in AICAR fermentation. ○, Growth; ▲, glucose; △, AICAR; ⊙, gluconic acid.

11.1.5 Effect of Medium Components on AICAR Accumulation

In order to establish the most economic medium, various kinds of components such as purine, phosphate, potassium, sulfate, and amino acids were evaluated.

A. Source of purines and purine concentration

The maximal production of AICAR was obtained in a medium containing a suboptimal concentration of any purine (dry yeast, RNA, purine bases such as adenine, guanine, hypoxanthine, and xanthine, and the corresponding nucleosides or nucleotides). The optimal molar concentrations of bases, nucleosides and the corresponding nucleotides were identical. RNA, adenine, and inosine were suitable as commercial sources of purines as regards cost and ease of treatment.

B. Concentration of phosphate

In case of RNA medium, where RNA was used as a purine source, the addition of KH_2PO_4 resulted in inhibition of AICAR production. Fig. 11.4 shows the relation between KH_2PO_4 and RNA concentrations. When adenine was used as a purine source, the optimal concentration of KH_2-PO_4 for AICAR production was 35 to 40 mg/dl, and the addition of excess KH_2PO_4 caused a considerable decrease in AICAR productivity in the same way as in the RNA medium. RNA contains about 25% of phosphate ions, so variation of RNA concentration results in a parallel variation of purine and phosphate concentrations. With adenine or inosine as a purine source, optimal concentrations of purine and phosphate could be obtained independently for AICAR production.

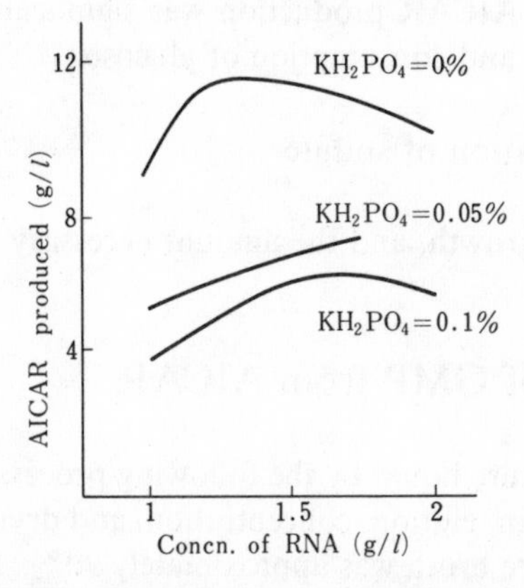

Fig. 11.4 Relation between KH_2PO_4 and RNA concentration. Basal medium: D-glucose, 8%; NH_4Cl, 1.5%; urea, 0.3%; NH_4NO_3, 0.2%; AJI-EKI (soy bean protein hydrolyzate), 6 ml/l; $MgSO_4 \cdot 7H_2O$, 0.04%; Fe^{2+}, Mn^{2+}, 2 ppm; $CaCO_3$, 5%. pH adjusted to 7.0 with KOH. (Source: ref. 21. Reproduced by kind permission of J. Wiley and Sons, Inc., U.S.A.)

C. Concentration of potassium

The initial pH of the media was adjusted to about 7.0 with potassium hydroxide solution. The amount of potassium ions supplied to media by neutralization was approximately 1.4 g/dl. In the commercial process,

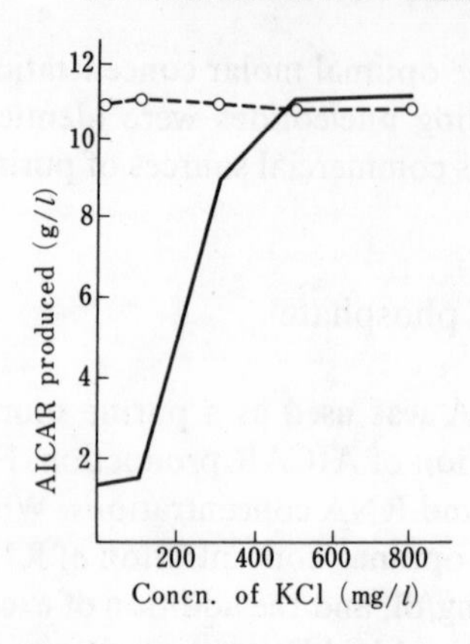

Fig. 11.5. Influence of potassium ion concentration: ---, neutralized with KOH;—, neutralized with NH_4OH. (Source: ref. 21. Reproduced by kind permission of J. Wiley and Sons, Inc., U.S.A)

however, it is more convenient to neutralize media with liquid ammonia, so the amount of potassium ions necessary for AICAR production had to be established. It is evident from Fig. 11.5 that this amount was 600–700 mg/l. With less than 600–700 mg/l, AICAR production was significantly reduced in spite of ordinary growth and consumption of glucose.

D. Concentration of sulfate

Sulfate ions were essential for growth, and the amount necessary for AICAR production was 40 mg/dl.

11.1.6 Chemical Synthesis of GMP from AICAR

AICAR was obtained from culture liquid by the following processes: adsorption on a cation-exchange resin, elution, concentration, and drying. The yield from AICAR in the culture broth was approximately 90%. The process[22] for synthesizing GMP from AICAR is shown in Fig. 11.6. The synthesis of guanosine was begun by ring closure of AICAR with sodium methylxanthate prepared *in situ* from carbon disulfide and sodium hydroxide in methanol. Namely, AICAR was allowed to react with 5 moles of sodium methylxanthate at 180° C for 3 h in an autoclave, giving 2-mercaptoinosine in nearly quantitative yield. The 2-mercapto derivative was then oxidized with 3 moles of hydrogen peroxide at 5° C for 1 h to give inosine-2-sulfonic acid. After the introduction of ammonia gas into the reaction solution, the compound, without isolation, was aminated at 120° C for 2 h in an autoclave to afford guanosine. The resulting guanosine was

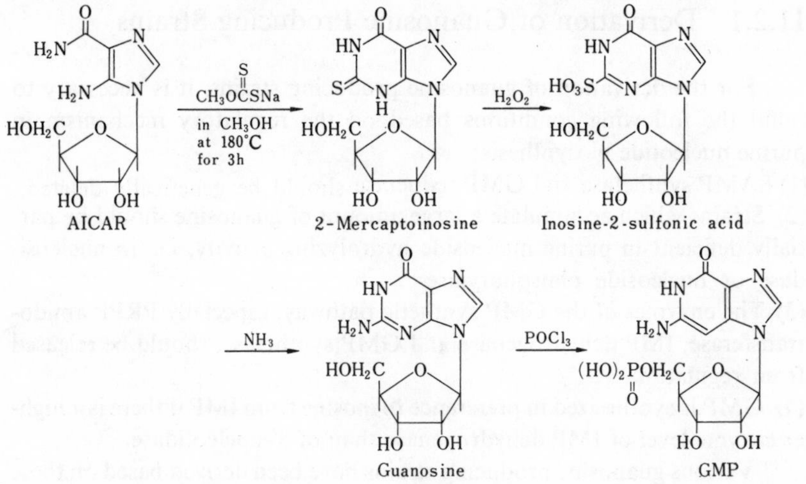

Fig. 11.6 Synthesis of GMP from AICAR.

easily obtainable as crystals on removal of the solvent by evaporation. The yield of guanosine thus obtained was about 80% based on AICAR. Guanosine was phosphorylated with phosphoryl chloride to give GMP in a yield of 90% or more. For the phosphorylation of nucleosides to give the corresponding nucleotides, see Chapter 14. GMP was finally neutralized with sodium hydroxide to give disodium guanylate.

11.2 Production of GMP from Guanosine

Guanosine can be produced in high yield by a fermentative procedure. The synthesis of GMP from guanosine is a simple and economical process involving chemical phosphorylation. In this section, the fermentative production of guanosine will be described; the phosphorylation of guanosine is dealt with in Chapter 14.

The fermentative production of guanosine was investigated using various mutants of *Bacillus subtilis*,[29] *B. pumilus*, *B. licheniformis*,[30] *Corynebacterium petrophilum*,[31] *C. guanofaciens*[32] and *Streptomyces griseus*.[33] Among these bacteria, guanosine biosynthesis in *B. subtilis* has been reported in detail, and its mechanism has also been clarified to some extent. Therefore, the fermentative production of guanosine using *B. subtilis* will be mainly considered here.

11.2.1 Derivation of Guanosine Producing Strains

For the derivation of guanosine producing strains, it is necessary to fulfill the following conditions based on the regulatory mechanism in purine nucleotide biosynthesis:

(1) SAMP synthetase and GMP reductase should be genetically deleted.

(2) Strains which accumulate a large amount of guanosine should be partially deficient in purine nucleoside hydrolyzing activity, i.e. in nucleosidase or nucleoside phosphorylase.

(3) The enzymes of the GMP synthetic pathway, especially PRPP amidotransferase, IMP dehydrogenase and GMP synthetase, should be released from regulation.

(4) GMP is synthesized in preference to inosine from IMP if there is a higher enzyme level of IMP dehydrogenase than of 5′-nucleotidase.

Various guanosine producing strains have been derived based on these principles.

Generally, IMP dehydrogenase is known to be regulated by GMP. Konishi *et al.*[34] investigated the possibility of inducing mutants in which the enzyme activity might not be regulated by GMP from inosine producing mutants of *B. subtilis* by obtaining mutants resistant to purine analogs: 8-azaguanine (8AG), 6-thioguanine (6TG), 6-mercaptopurine (6MP) and 2-thio-6-oxypurine (TOP). Among the colonies resistant to 8AG, two xanthosine producing mutants (AJ-1994 and P-246) and a guanosine producing mutant (AJ-1993) were found (Table 11.3). The IMP dehydrogenase activity of the guanosine producing strain AJ-1993 was three times higher than that of the parent strain AJ-1987, and 60% of that of the xanthosine producing mutant AJ-1994, when assayed at the final stage of the logarithmic phase (24 h), while its activity was similar to that of AJ-1994 at 42 h (Table 11.4). The GMP reductase activity of AJ-1993 was slight, while AJ-1994 showed some activity. These findings support the suggestion that acquisition of 8AG resistance caused metabolic alteration of inosine productivity to xanthosine or guanosine productivity, although the ability of these mutants to produce xanthosine or guanosine depends on the presence or absence of GMP reductase.

Momose *et al.*[35] investigated the effects of 8AG resistance of GMP reductase positive or negative strains on purine nucleoside production. GMP reductase negative, adenine requiring mutants were derived from strain 38–3, an adenine requiring inosine producer, according to the procedure shown in Table 11.5. Tryptophan requirement was used as a genetic marker throughout the isolation processes. Strain TD-9 (ade^{+}) was obtained from strain TR-101 by transduction. A mutant which grew on adeno-

TABLE 11.3

Characteristics of *Bacillus subtilis* AJ-1987, AJ-1994 and AJ-1993

Strain	Nutritional requirement†1	Resistance to 8AG†2	Ability to ribosylate 8AG†3	Accumulation of nucleosides†4		
				Inosine	Xanthosine (mg/ml)	Guanosine
AJ-1987	Adenine	−	−	7.2	0	trace
AJ-1994	Adenine	+	+	1.9	5.2	trace
(P-246)	Adenine	+	+	trace	6.9	trace
AJ-1993	Adenine	+	+	3.1	not detected	4.3

Basal medium: glucose 1.0%, KH_2PO_4 0.846%, $MgSO_4 \cdot 7H_2O$ 0.02%, $(NH_4)_2SO_4$ 0.1%, sodium citrate 0.05%, adjusted to pH 7.0 with KOH.

†1 Adenine, 50 μg/ml, was added to the basal medium.

†2 8AG resistance of the mutants was tested in medium 1, which was composed of the following components: glucose 1.0%, KH_2PO_4 0.8%, $MgSO_4 \cdot 7H_2O$ 0.02%, casein hydrolyzate 0.5%, sodium citrate 0.05%, adenine 0.01%, analog 0.03–0.08% and agar 2.0%, adjusted to pH 7.0 with KOH.

†3 8AG was added after culture for 24 h in medium 2, which was composed of the following components: glucose 7.0%, NH_4Cl 1.5%, KH_2PO_4 0.1%, $MgSO_4 \cdot 7H_2O$ 0.04%, $MnSO_4 \cdot 4H_2O$ 0.001%, $FeSO_4 \cdot 7H_2O$ 0.001%, soybean hydrolyzate (as total nitrogen) 0.048%, adenine 0.03% and $CaCO_3$ 2.5%, adjusted to pH 7.0 with KOH, and cultivation was continued for 42 h.

†4 Each mutant was aerobically cultured in a 500 ml Sakaguchi flask containing 20 ml of medium 2 at 30°C for 3 days.

TABLE 11.4

Comparison of Activities of IMP Dehydrogenase and GMP Reductase

Strain	IMP dehydrogenase (specific activity)		GMP reductase activity
	(24 h)	(42 h)	(24 h)
AJ-1987	0.7	not detected	++
AJ-1994	3.8	1.2	++
AJ-1993	2.3	1.0	−

IMP dehydrogenase and GMP reductase activities were assayed according to the modified method of Magasanik.[43]

The reaction mixture for IMP dehydrogenase activity, containing 0.15 ml of 1.5 M Tris-HCl buffer (pH 8.2), 0.15 ml of 0.8 M KCl, 0.15 ml of 0.03 M GSH, 0.15 ml of 0.04 M NAD, 0.3 ml of 0.01 M IMP, 0.18 ml of 0.1 M EDTA, 8 mg of acetone powder and water to a final volume of 1.8 ml, was incubated at 30°C for 15 min.

The reaction mixture for GMP reductase activity, containing 0.3 ml of 1.5 M Tris-HCl buffer (pH 7.4), 0.3 ml of 0.06 M cysteine, 0.2 ml of 0.016 M GMP, 0.5 ml of 0.004 M $NADPH_2$, 0.3 ml of crude enzyme extract and water to a final volume of 2.4 ml, was incubated at room temperature.

A similar mixture from which IMP or GMP had been omitted served as a control.

TABLE 11.5

Derivation of GMP Reductase-negative Mutants from Strain 38–3

Mutant	Genetic procedure	Growth response on				Genotype
		Adenosine	Inosine	Guanosine	Basal†2	
38–3		+	–	–	–	*ade⁻*
↓	NG⁺ treatment†1					
TR–101		+	–	+	+	*ade⁻ try⁻*
↓	*ade⁺*-transduction					
TD–9		+	+	–	–	*try⁻*
↓	NG treatment†1					
PU–17		+	+	–	+	*try⁻ pur⁻*
↓	NG treatment†1					
RD–37		+	+	+	+	*try⁻ pur⁻ red⁻*
↓	*pur⁺*-transduction					
TP–3		+	+	–	–	*try⁻ red⁻*
↓	NG treatmen†1					
30–12		+	–	–	–	*try⁻ red⁻ ade⁻*

†1 Treatment with *N*-methyl-*N*-nitro-*N*-nitrosoguanidine.
†2 The MCT plate was used as a basal plate. 100 μg/ml of adenosine, inosine, and guanosine were respectively supplemented to the basal plate.

sine or inosine but not on guanosine was isolated from strain PU-17, which was a non-exacting purine auxotroph derived from strain TD-9. This mutant, strain RD-37, was shown to be devoid of GMP reductase activity. Several adenine requiring mutants lacking SAMP synthetase were finally derived from this GMP reductase negative strain via a pur⁺ transductant, TP-3. All of the adenine auxotrophs obtained in this way produced inosine in amounts similar to that of the original strain 38-3 or strain TR-101 without accumulation of any by-product. Attempts were made to isolate mutants resistant to purine analogs from two types of inosine producing strains: a GMP reductase positive, adenine requiring mutant (strain 38-3) and a GMP reductase negative, adenine requiring mutant (strain 30-2). The accumulation patterns of nucleosides were measured in mutants resistant to 1 mg/ml of 8AX derived from strains 38-3 and 30-12 (Table 11.6). About 60% of the resistant mutants from strain 38-3 were found to produce xanthosine as well as inosine, while 70% of the resistant mutants from strain 30-12 produced guanosine as well as inosine. Among these mutants, strain 30-12-3 accumulated up to 5 mg/ml of guanosine after cultivation for four days.

Nogami *et al.*[36,37] showed that the essential genetic characteristics of mutants showing guanosine productivity were an adenine requirement, lack of GMP reductase and adenine or adenosine resistance from adenine and adenosine sensitivity. As shown in Fig. 11.7, a GMP reductase negative, ade⁻ mutant (strain Nt 1203) was derived from *Bacillus sp.* No. 102, and an adenosine sensitive mutant Nt 508 derived from strain Nt 422 was unable to grow on basal medium containing 0.1 mM adenosine. Strain

TABLE 11.6

Accumulation of Purine Nucleosides in 8AX-resistant Mutants Derived from Strains 38–3 and 30–12

Accumulation of purine nucleoside†	Number of mutants	
	Derived from	
	38–3 (GMP reductase+)	30–12 (GMP reductase−)
Inosine	109 (43%)	120 (30%)
Inosine+xanthosine	147 (57%)	0 (0%)
Inosine+guanosine	0 (0%)	280 (70%)
Total number	256 (100%)	400 (100%)

† Detected after cultivation for 4 days at 30°C in the AA medium (for mutants from 38–3) or AAT medium (for mutants from 30–12).

The AA medium contained 70 g glucose, 15 g NH_4Cl, 0.4 g $MgSO_4 \cdot 7H_2O$, 2 ppm Fe^{2+} (as $FeSO_4 \cdot 7H_2O$), 2 ppm Mn^{2+} (as $Mn^{2+}SO_4 \cdot 4H_2O$), 1.5 g KCl, 1g KH_2PO_4, 2 g amino acid mixture, 300 mg adenine and 25 g $CaCO_3$ in 1000 ml of distilled water. The pH was adjusted to 7.2. The AAT medium was prepared by adding 300 μg/ml of L-tryptophan to the AA medium.

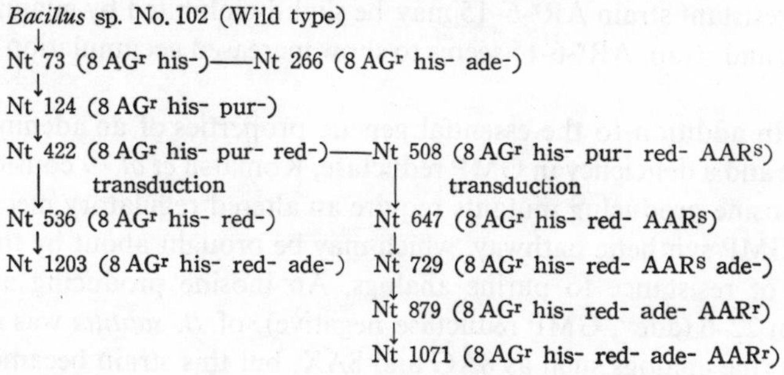

Bacillus sp. No. 102 (Wild type)
↓
Nt 73 (8 AG^r his$^-$)——Nt 266 (8 AG^r his$^-$ ade$^-$)
↓
Nt 124 (8 AG^r his$^-$ pur$^-$)
↓
Nt 422 (8 AG^r his$^-$ pur$^-$ red$^-$)——Nt 508 (8 AG^r his$^-$ pur$^-$ red$^-$ AAR^S)
| transduction ↓ transduction
Nt 536 (8 AG^r his$^-$ red$^-$) Nt 647 (8 AG^r his$^-$ red$^-$ AAR^S)
↓ ↓
Nt 1203 (8 AG^r his$^-$ red$^-$ ade$^-$) Nt 729 (8 AG^r his$^-$ red$^-$ AAR^S ade$^-$)
↓
Nt 879 (8 AG^r his$^-$ red$^-$ ade$^-$ AAR^r)
↓
Nt 1071 (8 AG^r his$^-$ red$^-$ ade$^-$ AAR^r)

Fig. 11.7 Derivation of guanosine producing mutants.

Nt 647 was derived from Nt 508 by transduction. The ade$^-$ mutant Nt 729 was then obtained from Nt 647. Furthermore, many mutants which could grow on basal medium containing adenosine at various concentrations were obtained from Nt 729. Mutant Nt 1071, which became more resistant to adenosine, accumulated more than 5 mg/ml of guanosine on cultivation for three days. Nogami *et al.*,[38] also studied the biological significance of XMP aminase inhibition by adenosine. XMP aminase produced by the adenosine sensitive strain AR^s-1-II and the resistant strains AR^r-6-7 and AR^r-6–15, as well as the parent strain W-171, was partially purified. The activity of XMP aminase in the sensitive strain was strongly inhibited by adenosine, but that in the resistant and parent strains was not (Fig. 11.8). Thus the adenosine sensitivity of the mutant may be attributed

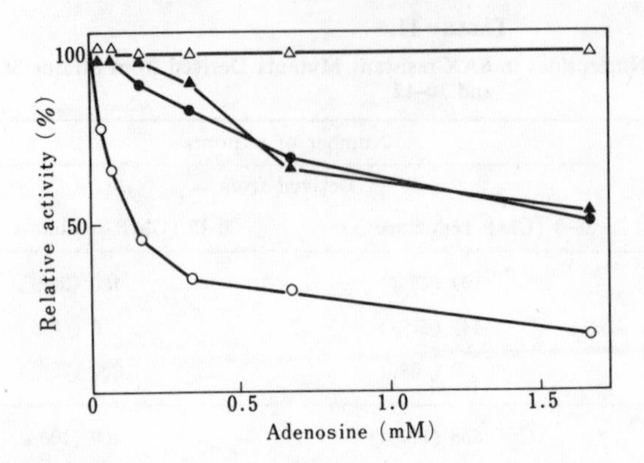

Fig. 11.8. Inhibition of XMP aminase by adenosine and psicofuranine. ⊙, W 171–5; ○, AR^s-1-II; ▲, AR^r-6-7; △, AR^r-6-15.

to the inhibition of its XMP aminase by adenosine. This was in agreement with the finding that the growth inhibition by adenosine was restored only by guanine or guanine derivatives. Therefore, XMP aminase of the adenosine resistant strain AR^r-6–15 may be slightly inhibited by guanine derivatives, and strain AR^r-6-15 seems to show increased accumulation of guanosine.

In addition to the essential genetic properties of an adenine requirement and a deficiency in GMP reductase, Komatsu *et al.*[39] considered that guanosine producing mutants require an altered regulatory mechanism in the GMP synthetic pathway, which may be brought about by the acquisition of resistance to purine analogs. An inosine producing auxotroph, strain 22–6 (ade$^-$, GMP reductase negative), of *B. subtilis* was insensitive to purine analogs such as 8AG and 8AX, but this strain became sensitive when its adenine deaminating activity was lost by further mutation. From this adenase deficient mutant S-1, mutants resistant to 8AX and 8AG were derived. A large number of strains resistant to 8AX accumulated xanthosine, but those resistant to 8AG accumulated guanosine. Among these mutants, strain TG-252 produced 2.5 mg/ml of guanosine on cultivation for four days (Fig. 11.9). Further, two mutants, GnR-44 and GnR-176, which were assumed to be partially deficient in purine nucleoside hydrolyzing activity, were isolated from one of the GMP reductase deficient mutants.[40] These mutants were found to accumulate large amounts of guanosine as a major product. It was presumed that the partially defective purine nucleoside phosphorylase of the mutants played an important role in the accumulation of guanosine.

Matsui *et al.*[41] derived guanosine producing mutants, No. 14119 and AG-169, resistant to methionine sulfoxide from an inosine producing

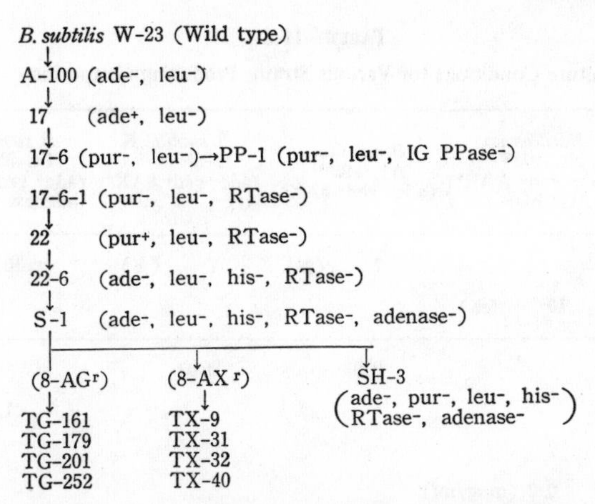

Fig. 11.9. Derivation of purine nucleoside producing mutants.
RTase⁻, GMP reductase lacking; IG PPase⁻, IMP (GMP) pyrophosphorylase lacking; leu⁻, his⁻, leucine or histidine requiring; pur⁻, purine requiring.

strain, No. 1411, of *B. subtilis*. The activities of IMP dehydrogenase and 5′-nucleotidase were assayed in these mutants and in the parent strain (Table. 11.7). The IMP dehydrogenase activity of these mutants was increased slightly compared to that of the parent strain, but the 5′-nucleotidase activity was decreased greatly compared to that of the parent strain. It seems to be necessary for GMP to be synthesized in preference to inosine from IMP, due to a higher level of IMP dehydrogenase than of 5′-nucleotidase.

TABLE 11.7

Relationship between Guanosine Productivities and IMP Dehydrogenase and 5′-Nucleotidase Activities of Strains No. 1411, No. 14119 and AG-169

Strain	Productivity (mg/ml)			IMP dehydrogenase activity[†1] (ΔOD/3mg/cells)		5′-Nucleotidase activity[†2] (ΔOD/15mg/cells)		
	Inosine	Xanthosine	Guanosine	None	GMP (0.55 mM)	IMP	XMP	GMP
No. 1411 (Parent)	11.0	0	5.5	130	128	400	40	200
No. 14119 (MSOr)[†3]	4.8	0	9.6	170	168	165	20	110
AG 169 (MSOrr)	0	6.0	8.0	190	190	95	15	65

†1 IMP dehydrogenase activities were determined according to Konishi *et al.*
†2 5′-Nucleotidase activities were determined according to Heppel *et al.*
†3 MSO = methionine sulfoxide.

TABLE 11.8

Culture Conditions for Various Strains Producing Guanosine

Strain (Genetic marker) / Culture condition	*Bacillus* sp. Nt 1071[37] (ade⁻ red⁻ AARʳ 8 AGʳ his⁻)	*B. subtilis* AJ-1993[34] (ade⁻ red⁻ 8AXʳ)	*B. subtilis* K 12-3[35] (ade⁻ red⁻ 8AXʳ try⁻)	*B. subtilis* TG-252[39] (ade⁻ red⁻ 8AGʳ adenase⁻ leu⁻ his⁻)
Glucose		7 (%)	7 (%)	10 (%)
Maltose	10 (%)			
Adenine		0.03	0.03	
Yeast extract				1.0
Dry yeast	2			
Biotin	0.2 (mμg/ml)			
Casamino acid				0.4
Amino acid mixture			0.2	
Soybean protein hydrolyzate		0.048 (N%)		
NH_4Cl		1.5	1.5	1.0
$(NH_4)_2SO_4$	2			
KH_2PO_4		0.1	0.1	0.5
$CaHPO_4$	0.5			
$Ca_3(PO_4)_2$	0.5			
$MgSO_4 \cdot 7H_2O$	0.2	0.04	0.04	0.1
$FeSO_4 \cdot 7H_2O$		0.001	Fe^{2+} 2ppm	
$MnSO_4 \cdot 7H_2O$		0.001	Mn^{2+} 2ppm	
$CaCO_3$	2	2.5	2.5	2.0
Temperature(°C)	37	30	30	32
Accumulation of nucleoside				
Guanosine (g/l)	5.4	4.3	5	2.5
Inosine (g/l)	3.3	3.1	10	3.1
Hypoxanthine (g/l)	1.5			

In order to obtain higher guanosine productivity, it seems to be necessary to increase the synthetic activities of the source compounds for purine nucleotides; PRPP, glycine, glutamine, asparagine, etc. For these reasons, the derivation of guanosine producing strains requires further investigation.

11.2.2 Culture Conditions

In inosine fermentation, systematic studies of culture conditions have been reported, but this is not yet the case for guanosine fermentation.

Genetic properties, culture conditions and guanosine accumulation by various guanosine producing strains are summarized in Table 11.8. Amino acids were required for the growth of the mutant, but they did not have any significant effect on guanosine accumulation. Adenine derivatives also were required for good growth of the guanosine producing strains. Guanosine accumulation, however, seemed to be affected primarily by the concentration of adenine derivatives in the medium. Generally, guanosine accumulation is inhibited when an excess of adenine derivatives is added to the culture medium. Recently, it was shown that guanosine accumulation is not inhibited in the presence of an excess of adenine derivatives if the growth is limited by the addition of chloramphenicol to the culture medium.[42]

11.2.3 Isolation and Chemical Phosphorylation of Guanosine

A procedure for the isolation of guanosine from culture broth is shown in Fig. 11.10. The culture broth was adjusted to pH 12 with NaOH.

Culture broth 5 l (Guanosine 40 g, Inosine 41 g)
-adjusted to pH 12 with NaOH
-separated by centrifugation
↓
Supernatant ⟶ Cells
-concentrated
↓
3 l
-adjusted to pH 6.8
-5°C, 15 h
↓
Guanosine 36 g
Inosine 8 g

Fig. 11.10. Isolation of guanosine accumulated in culture broth.

The supernatant fraction was separated from the cells by centrifugation and was concentrated under vacuum, then adjusted to pH 6.8. Crude guanosine was separated as crystals after 15 h at 5° C.

Phosphorylation of guanosine was carried out using phosphorus oxychloride. After phosphorylation, the product was hydrolyzed and neutralized to GMP·Na_2·$7H_2O$. The chemical phosphorylation of guanosine is described in more detail in Chapter 14.

11.3 Production of GMP by Fermentation

Strains capable of producing GMP require the following characteristics, based on the regulatory mechanism of purine nucleotide biosynthesis:

(1) The enzymes in the GMP synthetic pathway must not be regulated by GMP in GMP producing strains.
(2) GMP producing strains should have increased permeability for GMP excretion.
(3) Strains lacking 5′-nucleotidase and alkaline phosphatase activities (which degrade GMP) are able to produce higher amounts of GMP.

Screening of GMP producing bacteria was carried out from natural sources by Aoki *et al.*[45,46] GMP producing bacteria of *Pseudomonas aeruginosa*, *B. subtilis* and *Brevibacterium helvolum* accumulated 2.7–4.0 mg/ml of GMP in the culture medium after cultivation for four days.

Momose *et al.*[47] obtained GMP producing strains from mutants of *E. coli* No. 9 which were altered to have very weak 5′-nucleotidase activities (Table 11.9). The strain U-181, deficient in 5′-nucleotidase, was selected from mutants having weak phosphate releasing activities from *p*-nitro-

Table 11.9

IMP-Degrading Activities in Various Mutants

Strain	Specific activity of IMP degradation†
E. coli No. 9 (wild type)	1.58
E. coli U-18	0.52
E. coli U-181	0.24

The reaction mixture contained 1 mg of washed cells, 5 μmole of IMP, 1 μmole of $MgCl_2$ in 1 ml of 0.5 M Tris-buffer (pH 8). Incubation was carried out at 37°C without shaking.
† Orthophosphate released (μmole/mg dry cells/h).

phenylphosphate. These mutants were derived from a uracil auxotroph strain. The strain U-181 was converted to an adenine auxotroph by mutation using 2-aminopurine as a mutagen, and then reverted to uracil$^+$ by transduction (UA-28). The GMP producing strain, UA-667, was selected from the pool of mutants having weak 5′-nucleotidase activities by two-step mutation, using 2-aminopurine and UV. This strain, UA-667, accumulated 0.92 mg/ml of GMP and a small amount of guanosine.

Demain *et al.*[48] derived a strain deficient in 5′-nucleotidase, *B. subtilis* NRRL B-2911. This strain accumulated 0.48 mg/ml of GMP in the culture broth after cultivation for four days.

Abe *et al.*[49–51] derived adenine requiring mutants, which accumulated GMP as well as IMP in the culture broth, from various strains of *Brevibacterium*, *Arthrobacter*, *Corynebacterium* and *Micrococcus*, which are generally known to have comparatively weak activities of alkaline phosphatase and 5′-nucleotidase. Among these GMP producing bacteria, *Brevibacterium ammoniagenes* ATCC 6871 (ade$^-$) produced 5.1 mg/ml of GMP in the culture broth after cultivation for four days.

As for alteration of the regulatory mechanisms in the GMP synthetic pathway, feedback inhibition of IMP dehydrogenase prevents high accumulation of GMP and enhances the excretion of IMP. Demain *et al.*[52] considered the modification of this enzyme by means of revertant mutation in order to by-pass this control mechanism. The adenine requiring strain of *M. glutamicus* ATCC 13761, which excreted IMP, was further mutated to a xanthine dependent strain. The mutant strain was reverted to xanthine independent in an attempt to obtain a strain with an altered IMP dehydrogenase which would be less sensitive to feedback inhibition by GMP. A revertant was obtained which produced both GMP and IMP, each at 0.5 mg/ml.

Concerning the carbon source for GMP production, it was shown by Oosawa *et al.*[53,54] that mutants derived from *Candida tropicalis* and *Pseudomonas aeruginosa* produced 1.8–2.7 mg/ml of GMP from hydrocarbon as a carbon source.

11.4 Production of GMP from XMP

In addition to the above processes, the production of guanylic acid from xanthylic acid-related compounds (XMP or xanthosine), which can be accumulated rather easily by guanine auxotrophs of various strains, has been studied by the following two methods: (1) two-step fermentation in which an XMP-related compound is accumulated and isolated, then converted to GMP; (2) mixed culture with an XMP accumulating strain and an XMP converting strain.

The most important point in XMP and xanthosine fermentation is to cultivate a guanine auxotroph (lacking GMP synthetase) with a limiting supply of guanine, causing the release of feedback regulation (inhibition and repression) of IMP dehydrogenase by GMP. The accumulation of an XMP-related compound can be enhanced markedly by increasing the *de novo* synthesis of purine nucleotide by means of, for example, limiting adenine as well as guanine with a guanine-adenine double auxotroph.

The following strains have been reported as parents of xanthosine and XMP producers: for xanthosine producers, *Aerobacter aerogenes*,[55,56] *Salmonella typhimurium*,[57] *B. subtilis*,[58,59] *E. coli*,[60] etc., and for XMP producers, *Micrococcus glutamicus*,[61,62] *Brevibacterium ammoniagenes*,[63] coryneform bacteria,[64] *B. subtilis*,[65] *Bacillus sp.*,[66] etc. Whether a guanine auxotroph accumulates xanthosine or XMP depends on the 5′-phosphate cleaving activity of the mutant.

The derivation of xanthosine accumulating mutants was studied extensively with mutants of *B.subtilis*. Fujimoto *et al.*[58] found that guanine-adenine lacking double mutants accumulate inosine as well as xanthosine, and that the ratio of the two nucleotides accumulated varies depending on the adenine-guanine balance in the medium. They also found that the maximum amount of xanthosine accumulated by the double mutants is about twice that in guanineless mutants.

Ishii *et al.*[59] found improved inosine producers in high frequency among mutants resistant to a low concentration of 8AG derived from an adenine auxotroph of *B. subtilis*, in which GMP reductase, AMP deaminase and SAMP synthetase were lacking, and hence AMP could not be converted to GMP or vice versa. Since the formation of PRPP amidotransferase and SAMP lyase, both involved in IMP biosynthesis, was not repressed by adenosine in these mutants, they derived further guanine auxotrophs from these mutants, obtaining large accumulations of xanthosine. Fig. 11.11 shows the growth and accumulation patterns of xanthosine and inosine by a typical mutant, No. 75-13, cultured for 96 h at 31.5° C, in a medium containing 8% glucose, 2% ammonium sulfate, various concentrations of adenine and guanosine, etc. The optimum accumulation of xanthosine (in the presence of adenine 100 μg/ml, guanosine 150 μg/ml) was 17–18 mg/ml, 70–80% higher than that of the 8AG sensitive double mutant.

XMP fermentation is a more direct method than xanthosine fermentation for the production of GMP. Misawa *et al.*[63] reported that a guanine auxotroph of *B. ammoniagenes* produced 6.5 mg/ml of XMP from 10% glucose on cultivation at 30° C for 6 days. Exogenously supplied xanthine was not converted to XMP by the growing cells and the activity of XMP

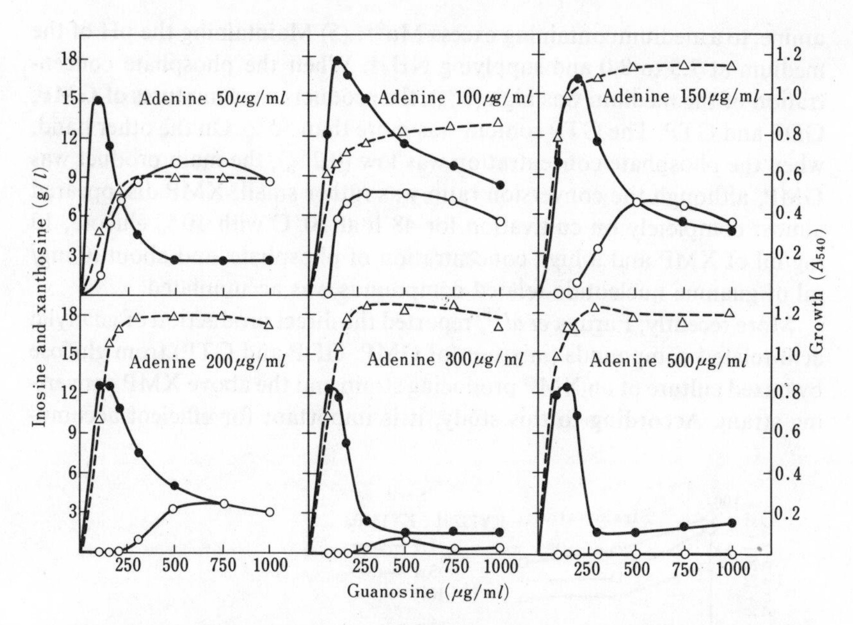

Fig. 11.11. Effect of adenine and guanosine concentration on the accumulation of inosine and xanthosine by a genetically derepressed adenine-guanine double auxotroph, No. 75–13. △---△, Growth (72 h); ○—○, inosine (96 h); ●—●, xanthosine (96 h).

pyrophosphorylase was very low. These facts suggest that XMP accumulation by the mutant is probably due to direct excretion of *de novo* synthesized nucleotide from the cells. In this fermentation, excess levels of Mn^{2+} in the medium markedly stimulated XMP accumulation, in striking contrast with IMP fermentation by adenine auxotrophs of the same species.

Beppu *et al.*[67] and Komuro *et al.*[68] showed independently that the accumulation of XMP-related compounds could also be achieved by adding angustomycin (also known as psicofuranine) to the culture medium with wild type *E. coli* and *B. ammoniagenes*, but not with guanine auxotrophs of these strains.

As for the conversion of xanthylic acid-related compounds to GMP, the conversion of XMP to GMP-related compounds was studied by Furuya *et al.*,[69] using mutants of *B.ammoniagenes*. They concluded that the following conditions were required for efficient conversion: (1) The use of a parent strain which is practically devoid of 5′-nucleotide cleaving activity. (2) Derivation of a psicofuranine- or decoynine-resistant mutant to promote GMP synthetase activity. (3) Limiting Mn^{2+} concentration in the medium. (4) Adding a surfactant, for example polyoxyethylene stearyl amine, to a medium containing excess Mn^{2+}. (5) Maintaining the pH of the medium at 7.5 to 8.0 and supplying NH_4^+. When the phosphate concentration of the medium was high (2%), the products were mixtures of GMP, GDP and GTP. The GTP content was more than 50%. On the other hand, when the phosphate concentration was low (0.2%), the main product was GMP, although the conversion ratio was rather small. XMP disappeared almost completely on cultivation for 48 h at 30° C with 10% glucose, 13 mg/ml of XMP and a high concentration of phosphate, and about 10 mg/ml of guanine nucleotide-related compounds was accumulated.

More recently, Furuya *et al.*[70] reported the direct production of guanylic acid-related compounds (mixtures of GMP, GDP and GTP) from glucose by mixed culture of an XMP producing strain and the above XMP converting strain. According to this study, it is important for efficient accumu-

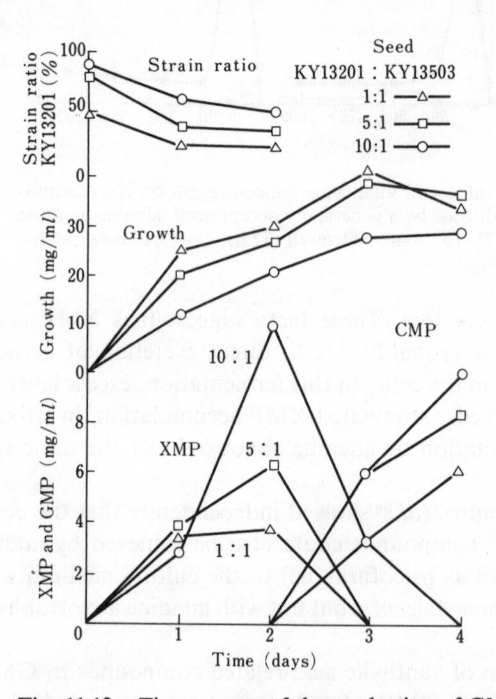

Fig. 11.12. Time courses of the production of GMP by mixed cultivation. Strain KY 13201, an XMP accumulator, and strain KY 13503, an XMP converter, were inoculated into a fermentation medium which contained, per liter, 130 g glucose, 10 g KH_2PO_4, 10 g K_2HPO_4, 10 g $MgSO_4 \cdot 7\ H_2O$, 0.1 g $CaCl_2 \cdot 2\ H_2O$, 10 mg $FeSO_4 \cdot 7\ H_2O$, 1 mg $ZnSO_4 \cdot 7H_2O$, 1 mg Mn^{2+}, 100 mg adenine, 100 mg guanine, 20 mg cysteine, 5 mg thiamine•HCl, 10 mg Ca pantothenate, 5 mg nicotinic acid, 30 μg biotin, 5 g yeast extract and 6 g urea, at the strain ratio indicated, and cultivated at 30°C. The starting pH was adjusted to 8.3 with NaOH.

lation of the nucleotides to allow the former strain to accumulate a large amount of XMP in the medium at an early stage of the cultivation and to allow the converting strain to grow appropriately in the later stage. As shown in Fig. 11.12, these conditions were achieved by making the inoculum ratio of the XMP accumulator to the XMP converter 10:1, because the growth rate of the former strain was lower than that of the latter strain, and by additional feeding of glucose and urea as well as the addition of a surfactant during the cultivation (after 48 h).

REFERENCES

1. C. L. Fox, Jr., *Proc. Soc. Exptl. Biol. Med.*, **51**, 102 (1942).
2. M. R. Stetten and C. L. Fox, Jr., *J. Biol. Chem.*, **161**, 333 (1945).
3. W. Shive, W. W. Ackermann, M. Gordon, M. E. Getzendaner and R. E. Eakin, *J. Am. Chem. Soc.*, **69**, 725 (1947).
4. J. M. Ravel, R. E. Eakin and W. Shive, *J. Biol, Chem.*, **172**, 67 (1948).
5. D. W. Wooley and P. B. Pringle, *J. Am. Chem. Soc.*, **72**, 634 (1950).
6. J. S. Gots, *Arch. Biochem.*, **29**, 222 (1950).
7. G. R. Greenberg, *J. Am. Chem. Soc.*, **74**, 6307 (1952).
8. R. C. Stewart and M. G. Sevag, *Arch. Biochem. Biophys.*, **41**, 9 (1952).
9. M. G. Sevag and R. C. Stewart, *ibid.*, **41**, 14 (1952).
10. J. S. Gots, *Nature*, **172**, 256 (1953).
11. J. S. Gots and S. H. Love, *J. Biol. Chem.*, **210**, 395 (1954).
12. S. H. Love and J. S. Gots, *ibid.*, **212**, 647 (1955).
13. I. J. Slotnick and M. G. Sevag, *Arch. Biochem. Biophys.*, **57**, 491 (1955).
14. J. S. Gots and E. G. Gollub, *J. Biol. Chem.*, **228**, 57 (1957).
15. H. R. Alimchandani and A. Sreenivasan, *J. Bact.*, **74**, 175 (1957).
16. J. S. Gots and E. G. Gollub, *Proc. Soc. Exptl. Biol. Med.*, **101**, 641 (1959).
17. J. S. Franzen and S. B. Binkley, *J. Biol. Chem.*, **236**, 515 (1961).
18. T. Shiro, *Agr. Biol. Chem.*, **25**, 350 (1961).
19. T. Shiro, *ibid.*, **26**, 75 (1962).
20. T. Shiro, *ibid.*, **26**, 269 (1962).
21. K. Kinoshita, T. Shiro, A. Yamazaki, I. Kumashiro, T. Takenishi and T. Tsunoda, *Biotechnol. Bioeng.*, **9**, 329 (1967); K. Kinoshita, T. Nishiyama, H. Tsuri, K. Konishi, T. Shiro and H. Okada, *Nippon Nogeikagaku Kaishi* (Japanese), **42**, 523 (1968); *Amino Acid and Nucleic Acid*, no. 17, 150 (1968).
22. A. Yamazaki, I. Kumashiro and T. Takenishi, *J. Org. Chem.*, **32**, 1825 (1967); I. Kumashiro, A. Yamazaki, T. Meguro, T. Takenishi and T. Tsunoda, *Biotechnol. Bioeng.*, **10**, 303 (1968).
23. T. Shiro, A. Yamanoi, S. Konishi, S. Okumura and T. Takahashi, *Agr. Biol. Chem.*, **26**, 785 (1962).
24. H. Shirafuji, A, Imada, S. Yashima and M. Yoneda, *Agr. Biol. Chem.*, **32**, 69 (1968).
25. K. Kinoshita, S. Sakai, M. Yasunaga, H. Sasaki and T. Shiro, *Nippon Nogeikagaku Kaishi* (Japanese), **43**, 404 (1969).
26. K. Kinoshita, H. Yasunaga, S. Sakai and T. Shiro, *ibid.*, **43**, 473 (1969).
27. K. Kinoshita, M. Aoki, A. Yamanoi and T. Shiro, *ibid.*, **42**, 529 (1968).
28. K. Kinoshita, H. Tsuri, S. Sakai, H. Yasunaga, H. Okada and T. Shiro, *ibid.*, **43**, 400 (1969).
29. T. Shiro, S. Okumura, Y. Tamagawa, T. Tsunoda, M. Takahashi and S. Motozaki, *Japanese Patent Publication* No. 39–16347 (1964).

30. I.Nogami, M. Yoneda, S. Eijima and M. Yoneda, *Japanese Patent Publication* No. 48–33392 (1973).
31. K. Takeda, T. Iguchi, T. Ooka and T. Kawamura, *Japanese Patent Publication* No. 46–20758 (1971).
32. J. Ishiyama, S. Sugiyama and T.Yokotsuka, *Japanese Patent Publication* No. 46–22550 (1971).
33. T. Ooka, S. Hayakawa, T. Kawamura, T. Iguchi and K. Takeda, *Japanese Patent Publication* No. 43–19840 (1968).
34. S. Konishi and T. Shiro, *Agr. Biol. Chem.*, **32**, 396 (1968).
35. H. Momose and I. Shiio, *J. Gen. Appl. Microbiol.*, **15**, 399 (1969).
36. I. Nogami and M. Yoneda, *Kagaku to Seibutsu* (Japanese), **7**, 371 (1969).
37. I. Nogami, M. Kida, I. Iijima and M. Yoneda, *Agr. Biol. Chem.*, **32**, 144 (1968).
38. M. Kida, F. Kawashima, A. Imada, I. Nogami, I. Suhara and M. Yoneda, *J. Biochem.* (Tokyo), **66**, 487 (1969).
39. K. Komatsu, A. Saijyo, R. Kodaira and H. Oosawa, *Amino Acid and Nucleic Acid*, no. 22, 54 (1970).
40. K. Komatsu and R. Kodaira, *J. Gen. Appl. Microbiol.*, **19**, 263 (1973).
41. H. Matsui, K. Satou, Y. Anzai, H. Enei and Y. Hirose, Proc. Symp. Amino Acid and Nucleic Acid (Japanese), p. 20, 1974.
42. K. Komatsu and R. Kodaira, *Japanese Patent Published Application* No. 47–19091 (1972).
43. B. Magasanik, *Methods in Enzymology*, vol. VI, p. 106–117, Academic, 1963.
44. R. Suzuki, T. Nomura, Y. Suzuki, T. Hirahara, *Japanese Patent Publication* No. 47–38199 (1972).
45. S. Okumura, A. Yamanoi, K. Komagata, T. Shiro, N. Matsumura, R. Aoki, T. Tsunoda, M. Takahashi and S. Motozaki, *Japanese Patent Publication* No. 42–6158 (1967).
46. R. Aoki, S. Okumura, N. Matsumura, Y. Tamagawa, T. Tsunoda and S. Motozaki, *Japanese Patent Publication* No. 39–16344 (1964).
47. H. Momose, *Japanese Patent Publication* No. 43–11760 (1968).
48. A. L. Demain, *Japanese Patent Publication* No. 41–12680 (1966).
49. S. Abe and K. Udagawa, *Japanese Patent Publication* No. 41–12673 (1966).
50. S. Abe and K. Udagawa, *Japanese Patent Publication* No. 41–12796 (1966).
51. S. Abe and K. Udagawa, *Japanese Patent Publication* No. 41–13797 (1966).
52. A. L. Demain, M. Jackson, R. A. Vitali, D. Hendlin and T. A. Jacob, *Appl. Microbiol.*, **14**, 821–825 (1966).
53. T. Oosawa, S. Seto, S. Watanabe and T. Ishida, *Japanese Patent Publication* No. 42–3836 (1967).
54. T. Oosawa, S. Seto, S. Watanabe and T. Ishida, *Japanese Patent Publication* No. 42–3839 (1967).
55. B. Magasanik and M. B. Brook, *J. Biol. Chem.*, **206**, 83 (1954).
56. K. Nakayama, T. Suzuki, K. Satou and S. Kinoshita, *Amino Acid and Nucleic Acid*, no. 8, 88 (1963).
57. B. Magasanik, H. S. Moyed and L. B. Gehring, *J. Biol. Chem.*, **226**, 339 (1957).
58. M. Fujimoto, K. Uchida, M. Suzuki and H. Yoshino, *Agr. Biol. Chem.*, **30**, 605 (1966).
59. K. Ishii and I. Shiio, *ibid.*, **37**, 287 (1973).
60. G. Powell, K. V. Rajagopalan and P. Haudler, *J. Biol. Chem.*, **244**, 4793 (1969).
61. M. Misawa, T. Nara, K. Udagawa. S. Abe and S. Kinoshita, *Agr. Biol. Chem.*, **28**, 690 (1964).
62. M. Misawa, T. Nara and S. Kinoshita, *ibid.*, **28**, 694 (1964).
63. M. Misawa, T. Nara, K. Udagawa, S. Abe and S. Kinoshita, *ibid.*, **33**, 370 (1969).
64. A.L. Demain, M. Jackson, R.A. Vitali, D. Hendlin and T.A. Jacob, *Appl. Microbiol.*, **13**, 757 (1965).

65. T. Akiya, Y. Midorikawa, A. Kuninaka, H. Yoshino and Y. Ikeda, *Agr. Biol. Chem.*, **36**, 227 (1972).
66. K. Imada, I. Nogami, Y. Nakao, M. Igarashi, *Ann. Rept. Takeda Res. Lab.* (Japanese), **23**, 54 (1964).
67. T. Beppu, M. Nose and K. Arima, *Agr. Biol. Chem.*, **32**, 197 (1968).
68. T. Komuro, T. Nara, M. Misawa and S. Kinoshita, *ibid.*, **33**, 230 (1969).
69. A. Furuya, S. Abe and S. Kinoshita, *Biotechnol. Bioeng.*, **13**, 229 (1971).
70. A. Furuya, R. Okachi, K. Takayama and S. Abe, *ibid.*, **15**, 795 (1973).

CHAPTER 12

Production of Other Nucleic Acid-related Substances

12.1 OROTIC ACID

The pyrimidine nucleotides such as UMP, CMP and TMP, and their di- and triphosphates are formed in living cells, and various kinds of sugar

*1 Akira FURUYA, Kyowa Hakko Kogyo Co. Ltd.
*2 Jiro ISHIYAMA, Kikkoman Shoyu Co. Ltd.
*3 Ken'ichi SASAJIMA, Institute for Fermentation, Osaka.
*4 Akira KUNINAKA, Yamasa Shoyu Co. Ltd.

nucleotides containing pyrimidine bases are also widely distributed. Considerable amounts of UMP and CMP are produced as by-products in the production of IMP and GMP from microbial RNA by enzymatic hydrolysis, and so the utilization of these compounds has been intensively investigated. CDP-choline which is synthesized from CMP and choline or choline phosphate is used in the medical field as a psychostimulant. Among pyrimidine derivatives formed by the *de novo* pathway, orotic acid is used as a hepatic drug (see Chapter 21). Orotic acid was first discovered in cows' milk by Biscaro and Belloni[1] in 1905, and was later found to be accumulated as an intermediate in the biosynthesis of pyrimidine nucleotides by various kinds of mutants.[2–7] In this section, the microbial accumulation of orotic acid by the *de novo* pathway will be described.

12.1.1 Orotic Acid Producing Microorganisms

Microorganisms which accumulate pyrimidine derivatives through the *de novo* pathway are listed in Table 12.1. In 1948, accumulation of orotic acid was first reported by Mitchell *et al.*[2] using a pyrimidine auxotrophic mutant of *Neurospora.* They examined the conditions of accumulation and obtained a maximal accumulation of 1.3 mg of orotic acid per ml. Thereafter, accumulations of orotic acid by *Aerobacter aerogenes,*[3] *Escherichia coli*[4] and *Serratia marinorubra*[7] were reported but these studies focused on the biosynthetic pathways of pyrimidine derivatives and the amounts accumulated were very low. Later, in order to produce orotic acid by fermentation, Tanaka *et al.*[8] studied its accumulation by a uracil requiring mutant induced from *Micrococcus glutamicus* (*Corynebacterium glutamicum*), a glutamic acid producing microorganism, and obtained a maximal orotic acid accumulation of 14 mg per ml in a medium containing 10% glucose. Nakayama *et al.*[12] examined the accumulation of orotic acid and orotidine by uracil requiring mutants derived from the genera *Micrococcus, Corynebacterium*, *Brevibacterium*, *Bacillus* and *Escherichia.* It was found that *M. glutamicus* 9824 strain ($ade^- + ura^-$) and *B. ammoniagenes* 7349 strain (ura^-) accumulated large amounts of orotic acid; the maximal amount was 6.0 mg per ml by strain 9824. Škodovă *et al.*[14,15] also induced a uracil requiring mutant of *B. ammoniagenes* and reported an accumulation of 5.8 mg of orotic acid per ml in a medium containing 5% glucose. Nakayama *et al.*[18] observed the accumulation of orotidine 5′-monophosphate (OMP) (about 1 mg/ml) or UMP (2.5 mg/ml) by a wild strain, *B. ammoniagenes* ATCC 6872, in the presence of 6-azauracil (6AU). The time of addition of 6AU determined which nucleotide was accumulated. Watanabe *et al.*[13] observed the accumulation of orotic acid by an adenine or

TABLE 12.1
Accumulation of Pyrimidine Derivatives by Microorganisms

Accumulated substance	Microorganism	Genetic character	Ref.
Orotic acid	*Neurospora*	pyr⁻	2)
Orotic acid	*Aerobacter aerogenes*	pyr⁻	3)
Orotic acid	*Escherichia coli*	pyr⁻	4)
Orotic acid, Uracil (6-Azauridine)	*E. coli*	(6-Azauridine added)	5)
Orotic acid, OMP, UMP (6-Azauridine)	*E. coli*	(6-Azauridine added)	6)
Orotic acid	*Serratia marinorubra*	ura	7)
Orotic acid	*Micrococcus glutamicus*	ura	8)
Orotic acid	*Penicillium commune*	(NaF added)	9)
UDP	*Zygosaccharomyces sojae*	(riboflavin producing mutant)	10)
Orotidine	*Bacillus subtilis*	ura	11)
Orotic acid, Orotidine	*Micrococcus glutamicus*	ade+ura	12)
Orotic acid, Orotidine	*M. glutamicus*	ade+pyr	12)
Orotic acid, Orotidine	*Corynebacterium rathayi*	ura	12)
Orotic acid, Orotidine	*Brevibacterium ammoniagenes*	ura	12)
Orotic acid, Orotidine	*B. subtilis*	ura	12)
Orotic acid	*E. coli*	amino acid	12)
Orotic acid	*Candida tropicalis*	ade or hypoxanthine	13)
Orotic acid	*B. ammoniagenes*	ura	14, 15)
Orotic acid	*Corynebacterium* sp.	ura	14)
Orotic acid	*E. coli*	no requirement	16, 17)
OMP, UMP	*B. ammoniagenes*	(6 AU added)	18)
Orotic acid, Orotidine	*Arthrobacter paraffineus*	ura (hydrocarbon assimilating)	19)
Orotic acid, Orotidine	*Streptomyces showdoensis*	ura	20, 21)

hypoxanthine requiring mutant induced from *Candida tropicalis* and obtained a maximal accumulation of 7 mg per ml in a medium containing 2% glucose, 1% aspartate and 0.1% yeast extract. Machida and Kuninaka[16,17] observed the excretion of orotic acid by a wild K strain of *E. coli*, and found that excretion occurred only with the K strain, not with other strains of *E. coli*. After examination of the conditions, a maximal accumulation of 0.45 mg per ml was obtained. In addition, Kawamoto *et al.*[19] observed the accumulation of orotic acid and orotidine by a uracil requiring mutant, KY 7122, induced from *Arthrobacter paraffineus*, which is able to assimilate hydrocarbon as a carbon source; the maximal accumulations of orotic acid and orotidine were 6.0 and 3.5 mg per ml, respectively. Ozaki *et al.*[20,21] found that a uracil leaky mutant induced from *Streptomyces showdoensis*, which produces showdomycin, accumulated orotic acid and orotidine; the accumulations were 1.2 and 0.8 mg per ml, respectively.

12.1.2 Conditions and Mechanisms of Orotic Acid Accumulation

As shown in Table 12.1, most orotic acid accumulating strains require uracil for growth, so the concentration of uracil in media severely affects orotic acid accumulation as well as cell growth. Fig. 12.1(a) and (b) show the considerable effects of uracil on the accumulation of orotic acid by *M. glutamicus* and *B. ammoniagenes*, respectively.[12] Maximal accumulations were obtained at concentrations of uracil which allowed the microorganisms to grow to about half-maximal levels. Similar results were obtained with uracil requiring mutants of other microorganisms.[3,8,11,19,20] It has been shown in *E. coli*, *Salmonella typhimurium* and *Neurospora crassa* that the biosynthesis of pyrimidine nucleotides is finely controlled by

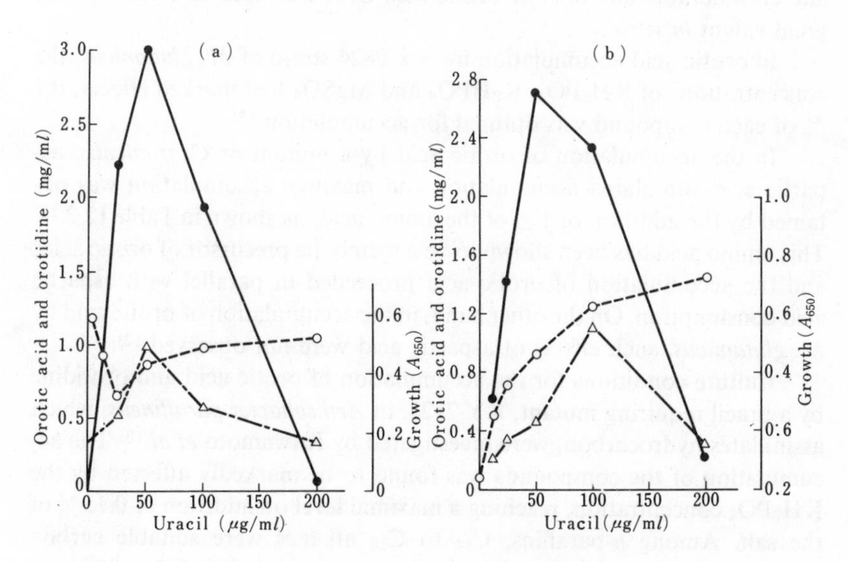

Fig. 12.1. Effect of uracil on orotic acid accumulation. (a) *M. glutamicus* 9824. (b) *B. ammoniagenes* 7349. ●, Orotic acid; ▲, orotidine; ○, growth.

TABLE 12.2

Effects of Nitrogen Source on Orotic Acid Accumulation†

Nitrogen source	Concn. (%)	Orotic acid accumulated (mg/ml)
$NaNO_3$	0.3	0.26
〃	1.0	0.30
$(NH_4)_2HPO_4$	0.3	2.22
〃	1.0	0.32
NH_2CONH_2	0.3	0.94
〃	1.0	1.69
Aspartic acid	0.5	4.19
〃	1.0	7.15
Casamino acids	0.5	1.80
〃	1.0	3.44
Polypeptone	0.5	2.50
〃	1.0	3.92
(Control)		0.32

† Cultivation was carried out at 27°C for 90 h.

uracil and cytosine nucleotides through regulatory mechanisms such as feedback inhibition and repression,[22,23] and it appeared that similar regulatory mechanisms were working in the accumulation of orotic acid. However, the mechanisms in microorganisms having the ability to accumulate considerable amounts of orotic acid have not been analyzed to any great extent *in vitro*.

In orotic acid accumulation by No. 9824 strain of *M. glutamicus*, the concentrations of KH_2PO_4, K_2HPO_4 and $MgSO_4$ had marked effects; 0.1 % of each compound was optimal for accumulation.[12]

In the accumulation of orotic acid by a mutant of *C. tropicalis*, aspartic acid stimulated accumulation and maximal accumulation was obtained by the addition of 1% of the amino acid, as shown in Table 12.2.[13] This amino acid has been shown to be a metabolic precursor of orotic acid, and the accumulation of orotic acid proceeded in parallel with aspartic acid consumption. On the other hand, in the accumulation of orotic acid by *M. glutamicus*, such effects of aspartic acid were not observed.[12]

Culture conditions for the accumulation of orotic acid and orotidine by a uracil requiring mutant, KY 7122, of *Arthrobacter paraffineus*, which assimilates hydrocarbon, were investigated by Kawamoto *et al.*[19] The accumulation of the compounds was found to be markedly affected by the KH_2PO_4 concentration, reaching a maximal level on addition of 0.15% of the salt. Among *n*-paraffins, C_{14} to C_{16} alkanes were suitable carbon sources, as shown in Table 12.3. Furthermore, it was found that fructose, sorbitol and mannitol were also suitable, and more orotic acid and oroti-

TABLE 12.3

Orotic Acid and Orotidine Production from Various *n*-Paraffins†

Carbon source	pH	Orotic acid	Orotidine (mg/ml)
n-Paraffin mixture	6.2	2.49	2.33
n-Tridecane	5.0	1.81	1.09
n-Tetradecane	5.8	2.50	2.80
n-Pentadecane	5.8	2.81	2.80
n-Hexadecane	5.6	2.81	3.16
n-Heptadecane	5.8	2.41	1.84

† Strain: KY7122. Each *n*-paraffin (5%) was added to the basal medium before inoculation. Analyses were made on 4-day culture broth.

dine were accumulated with these carbohydrates than with *n*-paraffins, as shown in Fig. 12.2.[19]

Since pyrimidine nucleotides are synthesized *in vivo* through the pathway shown in Fig. 12.3, accumulation of orotic acid by uracil requiring mutants can be ascribed to two genetic blocks caused by mutation. One is the deletion of OMP decarboxylase and the other is the deletion of OMP pyrophosphorylase. Since nucleotides permeate poorly through cell membranes, it is assumed that OMP is excreted only after being dephosphorylated to orotidine or further degraded to orotic acid. Therefore, when

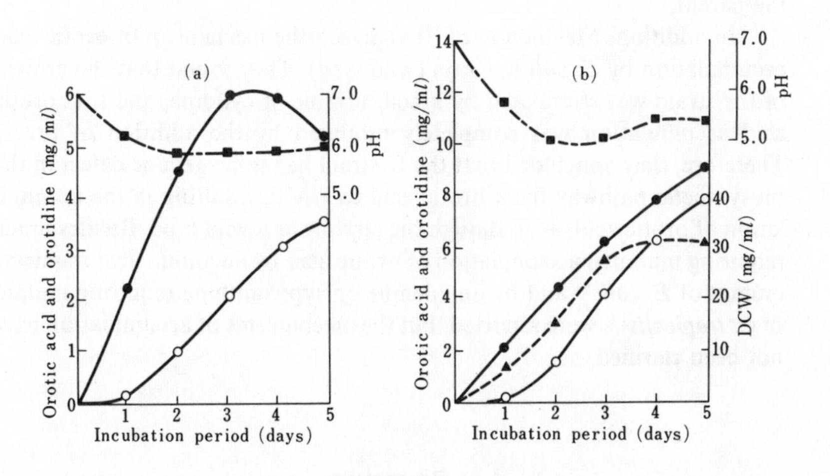

Fig. 12.2. Time course of orotidine accumulation. *A. paraffineus* KY7130 was cultivated at 30°C. (a) 10% *n*-paraffin mixture was added. (b) 10% sorbitol was added. DCW = dry cell weight. ■, pH; ●, orotic acid; ○, orotidine; ▲, growth.

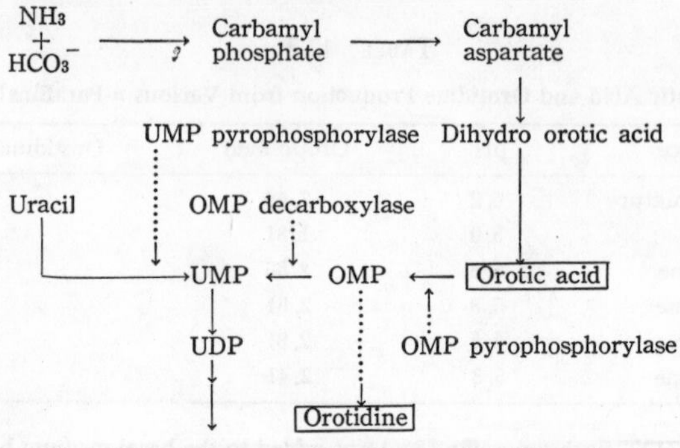

Fig. 12.3. Biosynthetic pathway of uracil nucleotides.

accumulation of orotidine in addition to orotic acid is observed, it is reasonable to assume the deletion of OMP decarboxylase. However, when accumulation of orotic acid alone is observed, it should be confirmed which enzyme of the two is deleted. This point has been covered in few papers, even though most orotic acid and orotidine accumulations were carried out with uracil requiring mutants, as shown in Table 12.1. Škodovă and Škoda[15] analyzed the accumulation of orotic acid by *B. ammoniagenes* and demonstrated the deletion of OMP pyrophosphorylase. Ozaki *et al.*[21] measured OMP decarboxylase in a uracil requiring mutant accumulating orotidine and orotic acid, derived from *S. showdoensis*, and observed a considerable decrease of the enzyme activity in the mutant compared to the parent.

In addition, Machida *et al.*[17] analyzed the mechanism of orotic acid accumulation by *E. coli* K strain (wild type). They found that the growth of the strain was stimulated by uracil, uridine or cytidine, and that orotic acid accumulation was completely inhibited by the addition of uracil. Therefore, they concluded that the K strain has some genetic defect in the biosynthetic pathway from orotic acid to UMP, resulting in the accumulation of orotic acid, even though the strain was a wild type. Besides uracil requiring mutants, accumulation of orotic acid by an amino acid requiring mutant of *E. coli*[12] and by an adenine or hypoxanthine requiring mutant of *C. tropicalis*[13] were reported, but the mechanisms of accumulation have not been clarified.

References

1. G. Biscaro and E. Belloni, *Chem. Zentr.*, **2**, 64 (1905).
2. H. K. Mitchell, M. B. Houlahan and J. F. Mye, *J. Biol. Chem.*, **172**, 525 (1948).
3. M. S. Brooke, D. Ushiba and B. Magasanik, *J. Bact.*, **68**, 534 (1954).

4. R. E. Yates and A. B. Pardee, *J. Biol. Chem.*, **221**, 743 (1956).
5. J. Škoda and F. Sŏrm, *Biochim. Biophys. Acta*, **28**, 659 (1958).
6. R. E. Handschumacher, *Nature*, **182**, 1090 (1958).
7. W. L. Belser, *Biochem. Biophys. Res. Commun.*, **4**, 56 (1961).
8. K. Tanaka, Y. Nakajima and S. Kinoshita, *Ann. Mtg. Agr. Chem. Soc.* Japan, 1961. Abstracts, p. 14.
9. H. Sugimoto, T. Iwasa and J. Ishiyama, *Nippon Nogeikagaku Kaishi* (Japanese), **36**, 690 (1962).
10. Y. Tsukada and T. Sugimori, *Ann. Mtg. Agr. Chem. Soc.* Japan, 1963. Abstracts, p. 126.
11. S. Konishi, T. Shiro and M. Takahashi, *ibid.*, p. 80.
12. K. Nakayama, Z. Sato, H. Tanaka and S. Kinoshita, *Nippon Nogeikagaku Kaishi* (Japanese), **39**, 118 (1965).
13. A. Watanabe, K. Tani and Y. Sasaki, *Amino Acid and Nucleic Acid*, no. 18, 9 (1968).
14. H. Škodova, H. Šolinova, J. Škoda and J. Dyr, *Folia Microbiol.*, **14**, 145 (1969).
15. H. Škodova and J. Škoda, *Appl. Microbiol.*, **17**, 188 (1969).
16. H. Machida and A. Kuninaka, *Agr. Biol. Chem.*, **33**, 868 (1969).
17. H. Machida, A. Kuninaka and H. Yoshino, *ibid.*, **34**, 1129 (1970).
18. K. Nakayama and H. Tanaka, *ibid.*, **35**, 518 (1971).
19. I. Kawamoto, T. Nara, M. Misawa and S. Kinoshita, *ibid.*, **43**, 1142 (1970).
20. M. Ozaki, S. Tagawa and T. Kimura, *Amino Acid and Nucleic Acid*, no. 26, 24 (1972).
21. M. Ozaki and T. Kimura, *ibid.*, no. 26, 31 (1972).
22. G. A. O'Donovan and J. Neuhard, *Bact. Rev.*, **34**, 278 (1970).
23. I. Shiio, *Seikagaku* (Japanese), **44**, 7 (1972).

12.2 Cyclic AMP

In 1957, cyclic AMP (cAMP) was discovered independently by Sutherland *et al.*[1] and Lipkin *et al.*[2]. The former demonstrated it to be an activating factor of phosphorylase from dog liver; the latter isolated it from a barium hydroxide digest of ATP. Since then, many studies have been carried out on the physiological roles of cAMP in organisms and it is now known to participate in various regulatory mechanisms with multiple functions.[3–5] In 1963, Okabayashi *et al.*[6] and Makman and Sutherland[7] found that some microorganisms can form cAMP. Their discoveries led to attempts to produce large amounts of cAMP with microorganisms. Many attempts have also been made to utilize cAMP for the medical treatment of diabetes, asthma and cancer, and as an agent stimulating the differentiation of higher plants.

12.2.1 Microbial Strains

Makman *et al.* detected about 1.1 ng of cAMP per ml in cell extracts of *Escherichia coli* Crooke's strain.[7] Okabayashi *et al.* reported that

Brevibacterium liquefaciens ATCC 14929 accumulated cAMP to the extent of 64 μg per ml in the culture fluid,[6] and that the accumulation was strikingly enhanced to about 1.2 mg per ml when *dl*-alanine was fed as a sole source of nitrogen.[8] Hayaishi *et al.*, who first purified adenylate cyclase (ATP pyrophosphate-lyase, EC 4.6.1.1) from *B. liquefaciens* and characterized the properties of the enzyme, described a method for preparing cAMP from ATP with the enzyme.[9] In 1971, Ishiyama *et al.* succeeded in producing 3–6 mg of cAMP per ml of the culture fluid by salvage synthesis from exogenously supplied precursor (such as hypoxanthine or its derivatives) with *Microbacterium* sp. No. 205 bio$^-$(ATCC 21376) and some other coryneform bacterial strains.[10,11] They found it possible to accumulate more than 2 mg of cAMP per ml of the broth from glucose by *de novo* synthesis with certain mutant strains, which were resistant to drugs such as 6-mercaptopurine (6MP), 8-azaguanine (8AG) and methionine sulfoxide (MSO), derived from the parent strain *Microbacterium* sp. No. 205.[11,12] Suzuki *et al.* reported in 1973 that cAMP was produced from *n*-tetradecane to the extent of about 1.4 mg per ml of the culture broth with *Arthrobacter roseoparaffineus* ATCC 15584.[13] It should be noted that microorganisms capable of producing a larger amount of cAMP all belonged to the family Corynebacteriaceae.

There are many other patented methods for the processing of cAMP by microbial techniques.[14–19]

12.2.2 Procedures

Processes for cAMP production using bacterial or enzymatic techniques can be classified into four groups as illustrated in Table 12.4.

TABLE 12.4

Processes of Microbial cAMP Production

Process	Microorganism†	Yields (mg/ml)	Ref.
1. Fermentative			
A. salvage synthesis	*Corynebacterium murisepticum* No. 7 (ATCC 21374)		
	Arthrobacter sp. No. 11 (ATCC 21375)	3–6	10, 11)
	Microbacterium sp. No. 205 bio$^-$ (ATCC 21376)		
B. *de novo* synthesis	*Brevibacterium liquefaciens* (ATCC 14929)	1.2	8)
	Microbacterium sp. No. 205 bio$^-$ 6MP^r 8AG^{r}MSO^r	2.0	11, 12)

TABLE 12.4—*Continued*

Method	Strain	Yields(mg/ml)	Ref.
	Arthrobacter roseoparaffineus (ATCC 15584)	1.4	13)
	Others		18, 19)
2. Enzymatic			
C. adenylate cyclase	*B. liquefaciens* (ATCC 14929)		9)
	Streptococcus salivarius (ATCC 25975)		25)
	Neurospora crassa (NIII 8-FGSC)		26)
	Others		27, 28)
D. hydrolysis of RNA	Microbes producing RNase which hydrolyze RNA to 2′:3′-cyclic nucleotides:		
	Rhizopus niveus		16)
	Bacillus subtilis K		15)
	Escherichia coli K 12-W 3350 (CI 857)		17)
	Escherichia coli (ATCC 10798)		14)
	Others		14)

† 6MP = 6-Mercaptopurine, 8AG = 8-azaguanine, MSO = methionine sulfoxide.

A. Salvage synthesis

At present, the procedure proposed by Ishiyama *et al.* gives the highest yields in cAMP production, as described below.[10]

Microbacterium sp. No. 205 was grown aerobically at 30°C in a seed medium comprising 2% glucose, 1% peptone, 0.5% yeast extract and 0.3% NaCl at pH 7.0. At the late exponential phase of growth, 5% (v/v) inocula were transferred into the cAMP production medium containing 5% glucose, 1% KH_2PO_4, 1% K_2HPO_4, 1% $MgSO_4 \cdot 7H_2O$, 0.01% $ZnSO_4 \cdot 7H_2O$, 1% peptone, 0.5% yeast extract and 0.3% hypoxanthine in distilled water at pH 7.5, and were incubated aerobically at 30°C for 40 h. Three to 6 mg of cAMP per ml was accumulated in the culture fluids. The broth was centrifuged and the supernatant solution was treated with charcoal, from which cAMP was then eluted with 50% ethanol containing 1% ammonia. After concentration to about 100 mg of cAMP per ml under reduced pressure, ethanol or isopropanol was added to the solution and the resulting precipitate was discarded by decantation. On adjusting the pH of the solution to 2.0 or below, cAMP crystals of 90% or more purity appeared. Recovery of cAMP from the culture broth was about 90%.[20]

In this procedure, such precursors as adenine, adenosine, AMP, inosine or IMP were found to substitute for hypoxanthine.[11]

B. *de novo* synthesis

This procedure seems to be the most suitable for commercial production at present, although the yield of cAMP by this method is not yet satisfactory.

With *B. liquefaciens*:[8] The bacterium was grown aerobically at 27°C for 91 h in a medium comprising 4% *dl*-alanine, 2% glucose, 0.3% KH_2PO_4, 0.7% K_2HPO_4 and 0.02% $MgSO_4 \cdot 7H_2O$ in distilled water at pH 7.0; 1.2 mg of cAMP was accumulated per ml of the culture broth. When *dl*-alanine was not fed, the yield was markedly reduced.

With *Microbacterium* sp. No. 205 bio^- $6MP^r8AG^rMSO^r$:[11,12] When the bacterium was incubated under aerobic conditions at 28°C for 48 h in a medium similar to that used for salvage synthesis, except that hypoxanthine was omitted, a considerable amount of cAMP (more than 2 mg per ml) was produced. No effect of *dl*-alanine was observed in this case.[10]

With *A. roseoparaffineus*:[13] The microbe produced 1.4 mg of cAMP per ml of the broth when cultured aerobically at 30°C for six days in a medium containing 10% *n*-tetradecane, 0.5% $(NH_4)_2SO_4$, 0.2% KH_2PO_4, 0.2% Na_2HPO_4, 0.1% $MgSO_4 \cdot 7H_2O$, 0.1% $FeSO_4 \cdot 7H_2O$, 0.01% $MnSO_4 \cdot 4H_2O$, 0.01% $ZnSO_4 \cdot 7H_2O$, 0.03% $CuCl_2 \cdot 2H_2O$, 0.2% corn steep liquor, 2% yeast extract and 100 μg/l of biotin in tap water at pH 7.0.

C. Direct method using adenylate cyclase

Adenylate cyclase was isolated and purified from cellular extracts of *B. liquefaciens* in the presence of pyruvate.[9] cAMP could be prepared efficiently from ATP with this enzyme at pH 8 to 9 in the presence of 10 mM pyruvate.[9,21] Attempts were made to immobilize this enzyme.[22]

D. Production from RNA

Norimoto *et al.*[14] stated that 2′:3′-cyclic XMP was transformed to 3′:5′-cyclic XMP in a solution containing M/3 phosphate ions (pH 7.0 to 8.7). On the basis of this finding, RNA was hydrolyzed to 2′:3′-cyclic XMP with any suitable RNase[15–17] and the solution was treated as described above; thus, a mixture containing 3′:5′-cyclic XMP was produced from RNA. cAMP was isolated by separation from the mixture. Washed cells of certain bacteria or yeasts could be utilized as the RNase preparation.

12.2.3 Mechanisms of cAMP Production

The biosynthetic pathway leading to cAMP production and its control systems are illustrated in Fig. 12.4. Accumulation of cAMP in the culture medium involves the following processes: (1) incorporation of the substrate into cells, (2) participation of enzymes involved in the cAMP biosynthetic sequence, (3) excretion of cAMP from the cells, and (4) release from feedback controls and loss or inactivation of degradative sequences.

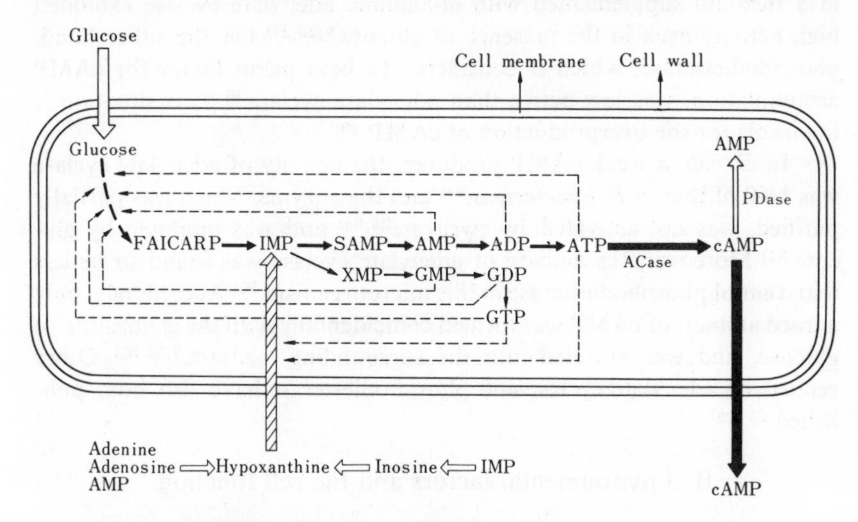

Fig. 12.4. Postulated biosynthetic pathway leading to cAMP production and its control systems. ACase, Adenylate cyclase; PDase, cAMP phosphodiesterase; ---feedback controls.

We will consider cAMP accumulation under the following headings: (A) specificities of the enzymes participating in cAMP accumulation, (B) environmental factors controlling the cellular activities, and (C) alteration of metabolic control mechanisms.

A. Specificities of enzymes involved in cAMP accumulation: adenylate cyclase (ATP pyrophosphate-lyase, EC 4.6.1.1) and cAMP phosphodiesterase (3′:5′-cyclic AMP 5′-nucleotido-hydrolase, EC 3.1.4.17).

It is clear that the rate limiting step in cAMP formation is the final reaction by adenylate cyclase: ATP $\rightleftharpoons$ cAMP + PPi.[21)] As already described, adenylate cyclase was isolated from *B. liquefaciens*, purified to homogeneity and crystallized by Hayaishi *et al.*[9] The molecular weight of the enzyme was estimated to be 9.24×10^4; the K_m value for ATP was 0.4 mM; and V_{max} was 1.6 μM/min.mg of protein.[9] The mature enzyme was composed of two ellipsoidal subunits. Interestingly, this enzyme was activated about 100-fold by the addition of pyruvate,[9,21,23,29] and was not subject to glucose suppression, which had been observed in *E. coli*.[29,31] Tracer experiments showed that this bacterium converted ^{14}C-*d*-alanine to ^{14}C-pyruvate *in vivo*.[29] Consequently, when the bacterium was grown in a medium supplemented with *dl*-alanine, adenylate cyclase exhibited high activity even in the presence of glucose.[9,24,29] On the other hand, phosphodiesterase, which is considered to be a *minus* factor for cAMP accumulation, was less active than adenylate cyclase.[9] This situation is favorable for the overproduction of cAMP.[29]

In *E. coli*, a weak cAMP producer, the activity of adenylate cyclase was 1/60 of that in *B. liquefaciens*,[24] and the enzyme, which was partially purified, was not activated by pyruvate[23,24] and was inhibited by glucose.[31] Moreover, the activity of adenylate cyclase was found to be less than that of phosphodiesterase in this microorganism.[30] Accordingly, only a trace amount of cAMP was formed coincidentally with the exhaustion of glucose, and was excreted into the surrounding medium.[7,30,31] Other reports on adenylate cyclase and phosphodiesterase have also been published.[25-28]

B. Environmental factors and the cell function

In medium containing high concentrations of both Mg^{2+} and phosphate ions, *Microbacterium* sp. No. 205 bio$^-$, wild strain, formed elongated, swollen and branched cells, took up exogenous hypoxanthine and excreted cAMP (salvage synthesis) when the amount of Mn^{2+} was limited.[12] If the Mn^{2+} content was increased, the bacterium formed short coccoid cells, could not take up hypoxanthine and produced no cAMP.[12] In the case of *de novo* synthesis with a mutant strain derived from this bacterium, the relations among Mn^{2+} concentration, cellular form, and cAMP productivity were similar. However, the glucose uptake was not affected by the Mn^{2+} concentration.[37] The effects of Mn^{2+} on cell structure and function are similar to those observed in IMP fermentation (see Chapter 10).

In the salvage synthesis of cAMP with *Microbacterium* sp. No. 205, exogenous adenine or related compounds were first decomposed to hypoxanthine in the medium, then taken up into the cells.[11,36] Tracer experiments with (8-^{14}C)-hypoxanthine showed that cAMP was entirely formed from hypoxanthine at the early stage of growth, but at the later stage *de novo* synthesis participated in cAMP production. Overall, 75% of the accumulated cAMP was derived from exogenously supplied hypoxanthine and 25% from glucose.[36] The effect of *dl*-alanine was described previously.

C. Alteration of metabolic control mechanisms

In an ordinary bacterium such as *E. coli* or *Bacillus subtilis*, the uptake of exogenous hypoxanthine into the cells is inhibited by ATP and GTP,[38,39] and *de novo* synthesis of IMP from glucose is subjected to feedback inhibition and repression by AMP, GMP and related compounds.[38] In excellent cAMP producers, however, it seems likely that the inhibition of hypoxanthine uptake is released in bacteria such as *Corynebacterium murisepticum*, *Microbacterium* sp. and *Arthrobacter* sp., which are capable of producing cAMP from hypoxanthine by salvage synthesis, and that the feedback inhibition and repression of IMP synthesis are released in bacteria such as *B. liquefaciens*, *A. roseoparaffineus* and drug-resistant mutant strains derived from *Microbacterium* sp., which are capable of producing cAMP from glucose directly by *de novo* synthesis. No detailed study of the regulatory mechanisms in cAMP producers has yet appeared. Glucose suppression and pyruvate activation of adenylate cyclase were noted.

In conclusion, the overproduction of cAMP by bacteria may be understood to result from (1) high activity of adenylate cyclase and low activity of phosphodiesterase, (2) induction and stimulation of cellular activities by limiting Mn^{2+} or adding *dl*-alanine, and (3) genetic release from feedback control of the cAMP biosynthetic sequence.

12.2.4 Remarks

The physiological roles of cAMP in organisms have been elucidated in detail. There have been many attempts to utilize cAMP in the medical and agricultural fields, but few practical uses have yet been developed. The yields of cAMP which can be obtained by fermentative processes are not yet satisfactory for commercial production. As for the mechanisms permitting the overproduction of cAMP in bacterial cells, many problems remain—for instance the detailed properties of adenylate cyclase and related enzymes, correlations between the presence of metal ions and cellular metabolic activities, and alteration of the control mechanisms. It is noteworthy that cyclic nucleotides such as deoxy cAMP,[40] cGMP,[41] cIMP,[42,43] cCMP[43] and cUMP[43] have also been accumulated from their respective precursors.

References

1. T. W. Rall, E. W. Sutherland and J. Berthet, *J. Biol. Chem.*, **224**, 463 (1957); *ibid.*, **232**, 1065 (1958); E. W. Sutherland and T. W. Rall, *J. Am. Chem. Soc.*, **79**, 3608 (1957); *J. Biol. Chem.*, **232**, 1077 (1958).
2. W. H. Cook, D. Lipkin and R. Markham, *J. Am. Chem. Soc.*, **79**, 3607 (1957); *ibid.*, **81**, 6075 (1959).
3. K. Yoshitoshi *et al.*, *Taisha* (Japanese), vol. 8, November 1971; *ibid.*, vol. 9, August 1972.
4. G. A. Robison *et al.*, *Cyclic AMP*, p. 17–398, Academic, 1971.
5. P. G. Greengard *et al.*, *Advances in Cyclic Nucleotide Research*, vol. 1–6, Raven Press, 1972–75.
6. T. Okabayashi, M. Ide and A. Yoshimoto, *Arch. Biophys.*, **100**, 158 (1963); T. Okabayashi, A. Yoshimoto and M. Ide, *J. Bact.*, **86**, 930 (1963).
7. R. S. Makman and E. W. Sutherland, *Fed. Proc.*, **22**, 470 (1963); *J. Biol. Chem.*, **240**, 1309 (1965).
8. T. Okabayashi, M. Ide and A. Yoshimoto, *Amino Acid and Nucleic Acid*, no. 10, 117 (1964).
9. M. Hirata and O. Hayaishi, *Biochem. Biophys. Res. Commun.*, **21**, 361 (1965); A. Hirata and O. Hayaishi, *Biochim. Biophys. Acta*, **149**, 1 (1967); K. Takai, Y. Kurashina, C. Hori, H. Okamoto and O. Hayaishi, *J. Biol. Chem.*, **249**, 1965 (1974).
10. J. Ishiyama, T. Yokotsuka and N. Saito, *Ger. Offen.* No. 1926072 (1969); *Agr. Biol. Chem.*, **38**, 507 (1974); J. Ishiyama, T. Yokotsuka and N. Saito, Mtg. Kanto Branch Agr. Chem. Soc. Japan, 1971; J. Ishiyama, T. Yokotsuka, F. Yoshida and M. Kato, *Japanese Patent Publication* No. 49–11437 (1974).
11. J. Ishiyama, T. Yokotsuka, N. Saito, M. Kato and F. Yoshida, Ann. Mtg. Agric. Chem. Soc. Japan, 1973. Abstracts, p. 306.
12. J. Ishiyama, T. Yokotsuka, N. Saito, F. Yoshida and M. Kato, *Japanese Patent Application Laid-Open* No. 49–61391 (1974); J. Ishiyama, M. Kato, F. Yoshida, T. Kidima and T. Yoshida, *ibid.*, No. 49–75791 (1974); J. Ishiyama, *ibid.*, No. 49–109586 (1974); J. Ishiyama, *personal communication.*
13. T. Suzuki and F. Tomita, *Japanese Patent Application Laid-Open* No. 48–67492 (1973); F. Tomita and T. Suzuki, *Agr. Biol. Chem.*, **38**, 71 (1974).
14. U. Norimoto, Y. Shimizu, S. Omura and T. Tatano, *Japanese Patent Application Laid-Open* No. 50–14684 (1975); U. Norimoto, S, Omura, Y. Shizu and T. Tatano, *ibid.*, No. 49–102890 (1974); *ibid.*, no. 49–118893 (1974).
15. M. Yamasaki and K. Arima, *Biochim. Biophys. Acta*, **139**, 202 (1967).
16. S. Mantani, J. Fukumoto and T. Yamamoto, *Agr. Biol. Chem.*, **36**, 242 (1972).
17. T. Ando, E. Hayase and S. Ekawa, *Japanese Patent Application Laid-Open* No. 49–335577 (1974).
18. K. Haneda, I. Tsukuda, I. Takeda, O. Kanemitsu and M. Tuzuki, *Japanese Patent Application Laid-Open* No. 49–69891 (1974).
19. I. Kihata, J. Kato, K. Watanabe and Y. Uchida, *Japanese Patent Application Laid-Open* No. 48–85792 (1973).
20. J. Ishiyama, T. Yokotsuka, M, Kato, N. Yamagi and F. Yoshida, *Japanese Patent Application Laid-Open* No. 49–80095 (1974).
21. O. Hayaishi, P. Greengard and S. P. Colowick, *J. Biol. Chem.*, **246**, 5840 (1971); K. Takai, Y. Kurashina, C. Suzuki, H. Okamoto, A. Ueki and O. Hayaishi, *ibid.*, **246**, 5843 (1971); Y. Kurashina, K. Takai, C. Hori, H. Okamoto and O. Hayaishi, *ibid.*, **249**, 4824 (1974).
22. Y. Kurashina, K. Takai, K. Umezawa, C. Hori and O. Hayaishi, *Seikagaku* (Japanese), **45**, 622 (1973).

23. M. Ide, *Biochem. Biophys. Res. Commun.*, **36**, 42 (1969); H. Brana, *Folia Microbiol.*, **14**, 185 (1969); M. Tao, *Proc. Natl. Acad. Sci. U.S.A.*, **63**, 86 (1969).
24. M. Ide, *Arch. Biochem. Biophys.*, **144**, 262 (1971).
25. R. L. Khandelwal and I. R. Hamilton, *J. Biol. Chem.*, **246**, 3297 (1971); *Arch. Biochem. Biophys.*, **151**, 75 (1972).
26. M. M. Flawia and H. N. Torres, *Biochim. Biophys. Acta*, **289**, 428 (1972); *J. Biol. Chem.*, **247**, 6873 (1972).
27. W. S. M. Wold and I. Suzuki, *Can. J. Microbiol.*, **20**, 1567 (1974); E. F. Rossomando, *Fed. Proc.*, **33**, 1362 (1974).
28. J. C. Londesborouch and T. Nurmin, *Acta Chem. Scand.*, **26**, 3396 (1972); G. E. Wheeler, A. B. Schibeci, R. M. Epand, J.B.M. Rattray and D. K. Kidby, *Biochim. Biophys. Acta*, **372**, 15 (1974).
29. M. Ide, A. Yoshimoto and T. Okabayashi, *J. Bact.*, **94**, 317 (1967); K. Umezawa, K. Takai, S. Tsuzi, Y. Kurashina and O. Hayaishi, *Proc. Natl. Acad. Sci. U.S.A.*, **71**, 4597 (1974).
30. A. Peterkofsky and C. Gazdar, *Proc. Natl. Acad. Sci. U.S.A.*, **68**, 2794 (1971); *ibid.*, **70**, 2149 (1973); M. J. Buetner, E. Spitz and H. V. Rickenberg, *J. Bact.*, **14**, 1068 (1973).
31. A. Peterkofsky and C. Gazdar, *Proc. Natl. Acad. Sci. U.S.A.*, **71**, 2324 (1974); *J. Cyclic Nucleotide Research*, **1**, 11 (1975).
32. T. Okabayashi and M. Ide, *Biochim. Biophys. Acta*, **220**, 116 (1970).
33. L. D. Nielsen, D. Monard and H. N. Rickenberg, *J. Bact.*, **116**, 857 (1973).
34. K. Iwakura, A. Ichikawa and K. Tomita, *Seikagaku* (Japanese), **43**, 583 (1971).
35. V. Riedel and D. Malchew, *Biochem. Biophys. Res. Commun.*, **46**, 279 (1972); G. Gerisch, D. Malchew, V. Riedek, E. Muller and M. Every, *Nature New Biol.*, **235**, 90 (1972); B. M. Chassy, *Science*, **175**, 1016 (1972).
36. J. Ishiyama, M. Kato and T. Yokotsuka, Ann. Mtg. Agr. Chem. Soc. Japan, 1972. Abstracts, p. 410.
37. J. Ishiyama, *personal commucation.*
38. M. Momose, *Tanpakushitsu Kakusan Koso* (Japanese), **13**, 78 (1968).
39. J. H. Ozer and M. Cashel, *J. Biol. Chem.*, **247**, 7067 (1972).
40. J. Ishiyama, F. Yoshida, M. Kato and T. Yokotsuka, Ann. Mtg. Agr. Chem. Soc. Japan, 1974. Abstracts, p. 3.
41. J. Ishiyama, *Agr. Biol. Chem.*, **39**, 1329 (1975).
42. J. Ishiyama and T. Yokotsuka, *Japanese Patent Application Laid-Open* No. 50–5594 (1975); *ibid.*, No. 50–6782 (1975).
43. J. Ishiyama, *Biochem. Biophys. Res. Commun.*, **65**, 286 (1975); J. Ishiyama and Yokotsuka, *T. W. Germany-OLS* No. 2427484 (1974); J. Ishiyama and T. Yokotsuka, *Japanese Patent Application Laid-Open* No. 50–18691 (1975).

12.3 D-Ribose

D-Ribose (referred to as ribose below) is an aldopentose which was first isolated from yeast nucleic acid and characterized as one of its constituents by Levene and Jacobs in 1909.[1)] Since then, the sugar has been found to be a constituent of various ribonucleic acids such as mRNA, tRNA, rRNA and 5S RNA, and of various nucleotide coenzymes.

Ribose has been produced in large quantities from glucose as a raw material for the chemical synthesis of vitamin B_2. So far, the microbial production of ribose has been described by Simonart and Godin,[2] Suzuki *et al.*,[3] Saito and Sugiyama,[4] and Sasajima and Yoneda.[5] The last authors, who used a transketolase mutant of a *Bacillus* sp., subsequently improved their method.[6] Ribose can now be produced cheaply and in high yield by the microbial method. Among methods of chemical synthesis of ribose is a process involving electrolytic reduction with a mercury electrode. However, mercury pollution is a social problem in Japan as well as in other countries, and chemical methods of synthesis of ribose are now being replaced by microbial methods.

12.3.1 Ribose Production using Wild type Strains

Except in the method described by Sasajima and Yoneda, all the microorganisms used are wild type strains. The published reports will be reviewed briefly below.

In 1951, Simonart and Godin[2] found ribose in the culture medium of *Penicillium brevicompactum*. Only a trace amount of ribose was accumulated. In 1963, Suzuki *et al.*[3] found that an unidentified bacterium formed 5 mg of ribose per ml in a glucose medium containing large amounts of Mn^{2+}, Fe^{2+} or Zn^{2+}. Without the metal ion, D-ribose-5-phosphate and PRPP were accumulated instead of ribose. Although the mechanism has not been determined, this phenomenon is very interesting. In 1966, Saito and Sugiyama[4] reported that *Pseudomonas reptilivola* accumulated 0.72 mg of ribose per ml of the medium.

The yields using wild type strains are relatively poor. Therefore, the method described by Sasajima and Yoneda, in which a transketolase mutant of *Bacillus* sp. is used and in which ribose can be produced cheaply and in good yield, will be described in more detail.

12.3.2 Ribose Production using a Transketolase Mutant of *Bacillus* sp.

A. Isolation of transketolase mutants

On the basis of the pentose phosphate pathway (Fig. 12.5), a transketolase mutant may be isolated as a non-utilizer of D-gluconate or L-arabinose or as a shikimic acid requiring mutant. The reason is that transketolase also catalyzes the reaction forming D-erythrose-4-phosphate, a precursor of aromatic compounds such as L-tryptophan, L-tyrosine, L-phenylalanine, coenzyme Q, vitamin K, folic acid, etc. In fact, Sasajima *et al.* succeeded in isolating transketolase mutants as shikimic acid requir-

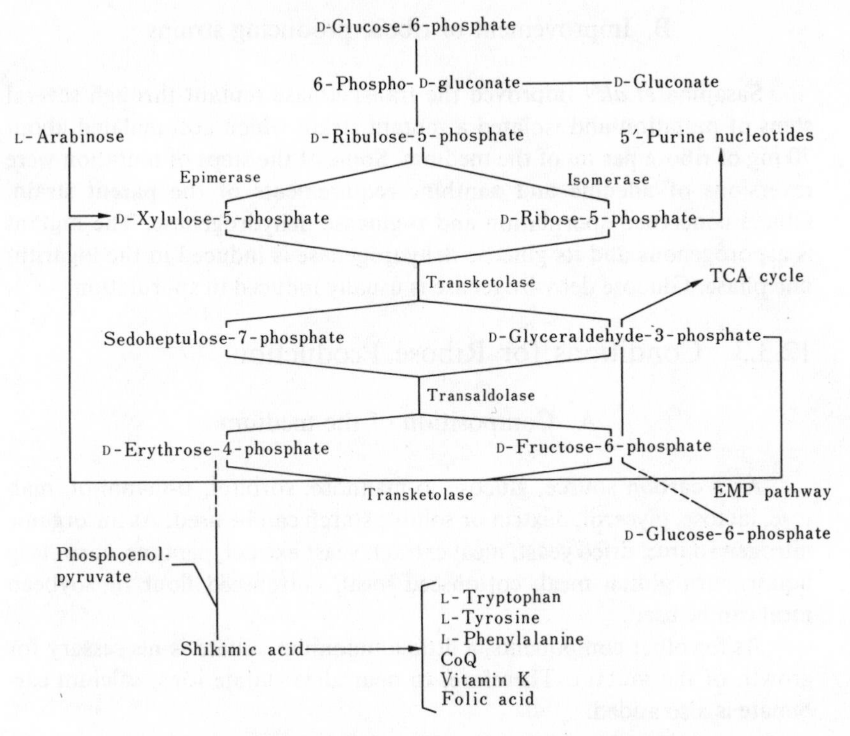

Fig. 12.5. Metabolic map of the pentose phosphate pathway and related pathways.

ing mutants.[7] At the same time, a D-ribulose-5-phosphate 3-epimerase mutant was also isolated as a non-utilizer of D-gluconate. The deleted enzymes of these mutants were determined.[8] The transketolase mutants accumulated only ribose but the D-ribulose-5-phosphate 3-epimerase mutant accumulated both ribose and D-ribulose.[5] Thus, the transketolase mutant appeared promising for ribose production. The amount of ribose accumulated in the culture medium was 35 mg per ml.

To investigate whether pentose production by the transketolase or D-ribulose-5-phosphate 3-epimerase mutant is general in the genus *Bacillus* or not, mutants of another strain, *Bacillus subtilis* IFO 3026, were isolated.[6] The transketolase and D-ribulose-5-phosphate 3-epimerase mutants both accumulated ribose and D-ribulose. Thus, the products varied according to the parent strain used, but pentose production appears to be general in the *Bacillus* genus. Josephson and Fraenkel[9] isolated transketolase mutants of *Escherichia coli* independently. However, ribose was not produced.[10] Eidels and Osborn[11] also isolated a transketolase mutant of *Salmonella typhimurium* but they did not describe pentose production.

B. Improvement of ribose producing strains

Sasajima *et al.*[6] improved the transketolase mutant through several steps of mutation and isolated a mutant strain which accumulated about 70 mg of ribose per ml of the medium. Some of the steps of mutation were reversions of adenine and xanthine requirements of the parent strain. Others concerned sporulation and D-glucose dehydrogenase. The mutant is asporogenous and its glucose dehydrogenase is induced in the logarithmic phase. Glucose dehydrogenase is usually induced in sporulation.

12.3.3 Conditions for Ribose Production

A. Composition of the medium

As a carbon source, glucose, D-mannose, sorbitol, D-mannitol, maltose, lactose, glycerol, dextrin or soluble starch can be used. As an organic nitrogen source, dried yeast, meat extract, yeast extract, peptone, corn steep liquor, corn gluten meal, cottonseed meal, cottonseed flour or soybean meal can be used.

As for other components, a little ammonium sulfate is neccessary for growth of the mutant. Therefore, to neutralize sulfate ions, calcium carbonate is also added.

B. Physical conditions

Three dehydrogenases participate in ribose production from glucose, i.e. glucose dehydrogenase, D-glucose-6-phosphate dehydrogenase and 6-phospho-D-gluconate dehydrogenase. Therefore, aerobic conditions are necessary. The optimum temperature is around 37° C. An example of a typical fermentation time course is shown in Fig. 12.6.

12.3.4 Purification of Ribose

After incubation, the culture medium was purified through general procedures such as removal of the cells and other solid debris, exclusion of cations and anions, and decolorization. Ribose can then be crystallized by adding methanol or ethanol to the concentrated solution.

12.3.5 Mechanism of Ribose Formation

Ribose may be formed from glucose through the oxidative part of the

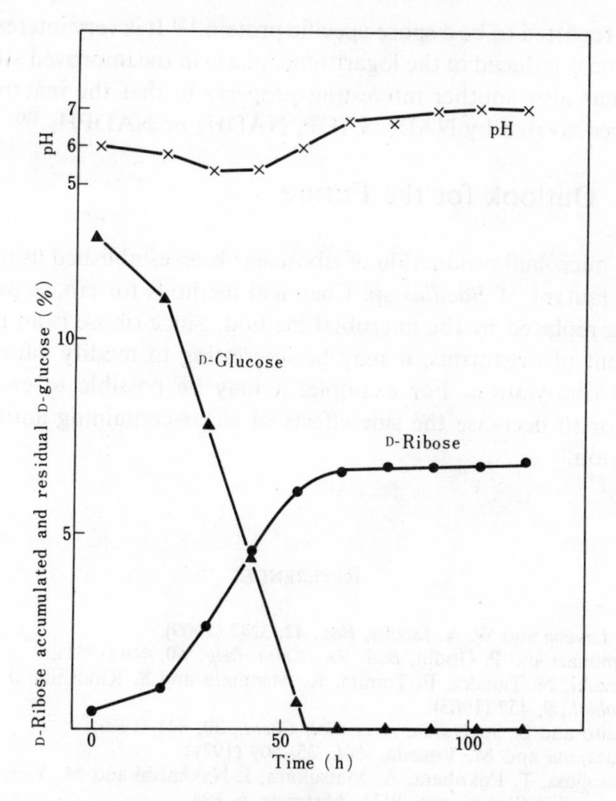

Fig. 12.6. Fermentation time course of D-ribose production.

pentose phosphate pathway. In the improved ribose producing strain, a by-pass via D-gluconate, in which glucose dehydrogenase and gluconokinase are included, also functions. The pathways through which ribose is formed are shown in Fig. 12.7. Glucose dehydrogenase of *Bacillus* sp.

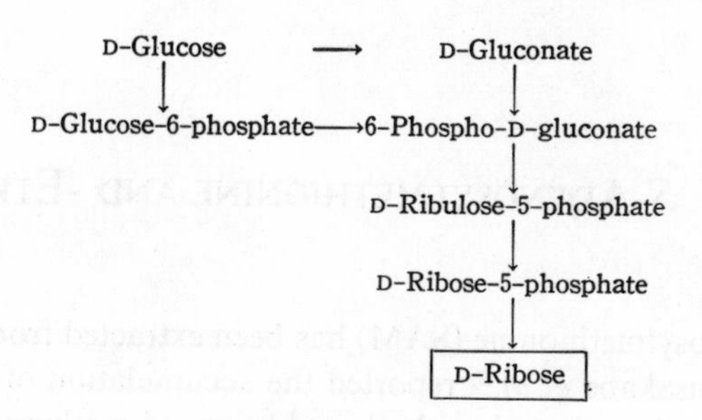

Fig. 12.7. The pathways of D-ribose formation.

has been reported to be a spore-specific protein.[12] It is very interesting that the enzyme is induced in the logarithmic phase in the improved strain. The enzyme has also another interesting property in that the inactivated enzyme is reactivated by NAD, NADP, $NADH_2$ or $NADPH_2$.[13]

12.3.6 Outlook for the Future

The microbial production of ribose has been established using a transketolase mutant of *Bacillus* sp. Chemical methods for ribose production are being replaced by the microbial method. Since ribose is an important constituent of organisms, it may be interesting to modify various medicines by ribosylation. For example, it may be possible to enhance the efficacy or to decrease the side-effects of sugar-containing antibiotics by ribosylation.

References

1. P. A. Levene and W. A. Jacobs, *Ber.*, **42**, 3247 (1909).
2. P. Simonart and P. Godin, *Bull. Soc. Chim. Belg.*, **60**, 446 (1951).
3. T. Suzuki, N. Tanaka, F. Tomita, K. Mizuhara and S. Kinoshita, *J. Gen. Appl. Microbiol.*, **9**, 457 (1963).
4. N. Saito and S. Sugiyama, *Agr. Biol. Chem.*, **30**, 841 (1966).
5. K. Sasajima and M. Yoneda, *ibid.*, **35**, 509 (1971).
6. K. Sasajima, T. Fukuhara, A. Matsukura, I. Nakanishi and M. Yoneda, 4th Int. Fermentation Symposium, 1972. Abstracts, p. 194.
7. K. Sasajima, I. Nogami and M. Yoneda, *Agr. Biol. Chem.*, **34**, 381 (1970).
8. K. Sasajima and M. Yoneda, *ibid.*, **38**, 1305 (1974).
9. B. L. Josephson and D. G. Fraenkel, *J. Bact.*, **100**, 1289 (1969).
10. D. G. Fraenkel, *personal communication.*
11. L. Eidels and M. J. Osborn, *Proc. Natl. Acad. Sci. U.S.A.*, **68**, 1673 (1971).
12. J. Bach and H. L. Sadoff, *J. Bact.*, **83**, 699 (1962).
13. A. Yokota, K. Sasajima and M. Yoneda, Ann. Mtg. Jap. Biochem. Soc., 1974. Abstracts, p. 504.

12.4 *S*-Adenosylmethionine and -Ethionine

S-Adenosylmethionine (SAM) has been extracted from yeast cells.[1,2] Recently, Kusakabe *et al.*[3] reported the accumulation of SAM in fungal cells secondarily cultured with the addition of methionine. The fungal cells, which were cultured in a medium composed of 5% glucose, 0.5%

peptone, 0.1% KH_2PO_4, 0.1% K_2HPO_4, 0.04% $CaCl_2 \cdot 2H_2O$ and 0.04% $MgSO_4 \cdot 7H_2O$, at pH 5.8 for 3 days at 28° C, were harvested and washed twice with water. Then, 3 g of the wet mycelia was transferred to a 500 ml Sakaguchi flask containing 100 ml of fresh medium supplemented with more than 10 mM of L-methionine or D,L-ethionine. After further cultivation at 28° C for 24 h, the amounts of SAM and *S*-adenosylethionine (SAE) reached about 20 μmoles and 4 μmoles, respectively, per g of dry cells of *Aspergillus tamarii* (Fig. 12.8). SAM accumulating ability was

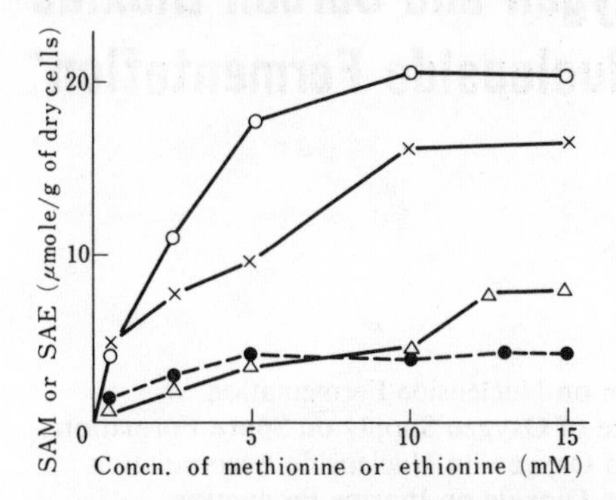

Fig. 12.8. Effect of methionine and ethionine concentration on the accumulation of SAM and SAE. ○—○, L-Methionine → SAM; ×—×, DL-methionine → SAM; △—△, D-methionine→SAM; ●—●, DL-ethionine → SAE.

widely distributed in molds, especially in *Mucor jansseni*, *M. circinelloides*, *Penicillium concavorugulosum*, *A. tamarii* and *Rhizopus pseudochinensis*. Among them, *M. jansseni* had the highest ability; it accumulated 45 μmoles of SAM per g of dry cells.

References

1. F. Schlenk and R. E. Depalma, *J. Biol. Chem.*, **229**, 1037, 1051 (1957).
2. L. J. Gawel, J. R. Turner and L. W. Parks, *J. Bact.*, **83**, 497 (1962).
3. H. Kusakabe, A. Kuninaka and H. Yoshino, *Agr. Biol. Chem.*, **38**, 1669. (1974).

CHAPTER **13**

Effects of Oxygen and Carbon Dioxide in Nucleoside Fermentation*

In order to ensure successful scale-up and stable production at commercial plants, the problems of agitation and aeration require extensive investigation. Cells which biosynthesize purine nucleotides require large amounts of energy. Therefore, the system for purine nucleotide biosynthesis, as well as the energy producing system within the cell, must be efficient for the microbial production of AICAR or inosine. It is especially important to determine quantitatively the features of oxygen and carbon dioxide transfer in culture systems for nucleoside fermentation.

13.1 Effect of Oxygen on Nucleoside Fermentation

13.1.1 Influence of Oxygen Supply on Spore Formation

In AICAR fermentation using *Bacillus megaterium* No. 336 (purine-less mutant) as an AICAR producer, the fermentation results are in-

* Yoshio Hirose, Ajinomoto Co. Inc.

fluenced markedly by the degree of spore formation (spores/(spores+ vegetative cells)).[1] Only vegetative cells can biosynthesize AICAR from glucose. Therefore, it is essential to restrict spore formation in order to obtain satisfactory production of AICAR. The conditions of agitation and aeration have a close relationship to the degree of spore formation in that vegetative cells do not form spores under conditions where the oxygen demand of the cells is not satisfied due to limited oxygen supply. However, AICAR accumulation is markedly depressed in oxygen-deficient cultures. On the other hand, a large amount of spores appears when the oxygen demand of the cells is satisfied. However, it is only in the period 8–12 h after the beginning of fermentation that spore formation is absolutely influenced by oxygen supply. When the oxygen demand of the cells is satisfied during that period of the exponential growth phase, a large amount of spores is formed at the stationary phase of cell growth. When the oxygen demand of the cells is not satisfied at that time, on the other hand, spores are not formed at the stationary phase of cell growth, and the vegetative cells accumulate a large amount of AICAR. It is considered that an energy requiring step exists in the process where the vegetative cells change to spores. Vegetative cells remain unchanged until the end of the fermentation as a result of the inhibition of spore formation caused by insufficient oxygen supply. When the oxygen supply is controlled during the critical period at a level between 0.85 and 1.0 (expressed in terms of the degree of satisfaction of the oxygen demand of cells, i.e. the rate of cell respiration in the culture system/the rate of oxygen demand of cells), the degree of spore formation is 0.05. Under these conditions, satisfactory accumulation of AICAR occurs (Table 13.1).

The effect of oxygen on AICAR production by vegetative cells will be described below.

TABLE 13.1
Effect of Oxygen Supply on Spore Formation in AICAR Fermentation

Liquid-phase oxygen(atm)	$r_{ab}^{\dagger 1}/K_rM^{\dagger 2}$	Spore formation (%)	AICAR accumulated (g/l)
0.06	1.0	35	10.6
0.02	1.0	25	11.6
nearly zero	0.85–1.0	5	14.3
nearly zero	0.6–0.7	3	14.0
nearly zero	0.2–0.4	0	13.2

The conditions of oxygen supply were changed in the period 8 to 12 h after the beginning of fermentation.

†1 Rate of cell respiration in the culture system (moles of O_2/ml.min).

†2 Maximum oxygen demand of the cells with a sufficient oxygen supply (moles of O_2/ml.min).

13.1.2 Effect of Oxygen on Nucleoside Accumulation

The effect of oxygen on nucleoside production was investigated in AICAR fermentation using a mutant of *B. megaterium*,[1] and in inosine fermentation using a mutant of *B. subtilis*.[2,3] In these investigations, the liquid-phase oxygen tension of the culture medium was employed as an index of oxygen supply. The accumulation of both nucleosides was markedly inhibited when the liquid-phase oxygen tension of the culture medium was nearly zero, as recorded by a membrane-coated oxygen electrode. On the other hand, large amounts of both nucleosides were accumulated when the liquid-phase oxygen tension of the culture medium was measured as being above zero (sufficient oxygen supply). Fig. 13.1 shows the relationship between product accumulation and the liquid-phase oxygen tension in inosine fermentation. Extremely low dissolved oxygen levels, less than 0.01 atm, are measured as zero by a membrane-coated oxygen electrode, so it is impossible to determine the critical values of the liquid-phase oxygen tension for nucleoside formation; these are considered to be smaller than 0.01 atm. For industrial purposes, however, the oxygen electrode can be usefully employed to control the practical operation of these fermentations, assuming that the critical level of dissolved oxygen for nucleoside fermentation is slightly above zero.

The inosine-producing mutant of *B. subtilis* accumulates acetoin, 2,3-butyleneglycol and lactic acid at dissolved oxygen levels of less than 0.01 atm.[3] Under these oxygen-deficient conditions, the redox potential of the culture medium is a useful index of the oxygen supply in place of the dissolved oxygen tension of the medium as measured by an oxygen electrode.[4,5] Under conditions of oxygen deficiency in inosine fermentation, acetoin accumulates at a redox potential of −160 mV, 2,3-butyleneglycol at −190 mV and lactic acid at −220 mV. Inosine is completely replaced by these products (Fig. 13.2)[5]

In aerobic fermentation, where it is essential to satisfy the oxygen demand of the cells to ensure product formation, an oxygen electrode is useful for controlling the fermentation. In anaerobic fermentation (e.g. the microbial production of acetone-butanol), on the other hand, the redox potential of the medium is closely related to product formation as well as the physiological changes in the cells.[6–9] Therefore, measurement of the redox potential of the medium offers an effective method of controlling the fermentation.

The redox potential is mainly affected by changes in the liquid-phase oxygen tension in aerobic fermentation. However, in anaerobic fermenta-

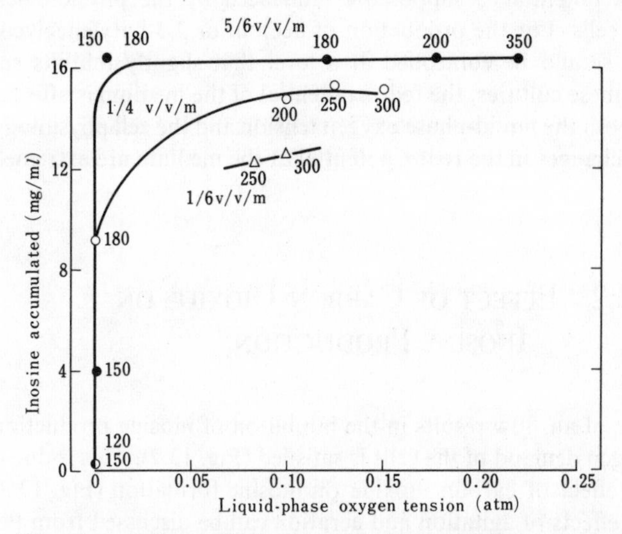

Fig. 13.1. Relationship between inosine accumulation and liquid-phase oxygen tension in jar fermenter culture. Air flow rate was 5/6(●), 1/4 (○), or 1/6 (△) v/v/m, respectively; numbers indicate the agitation speed (rpm).

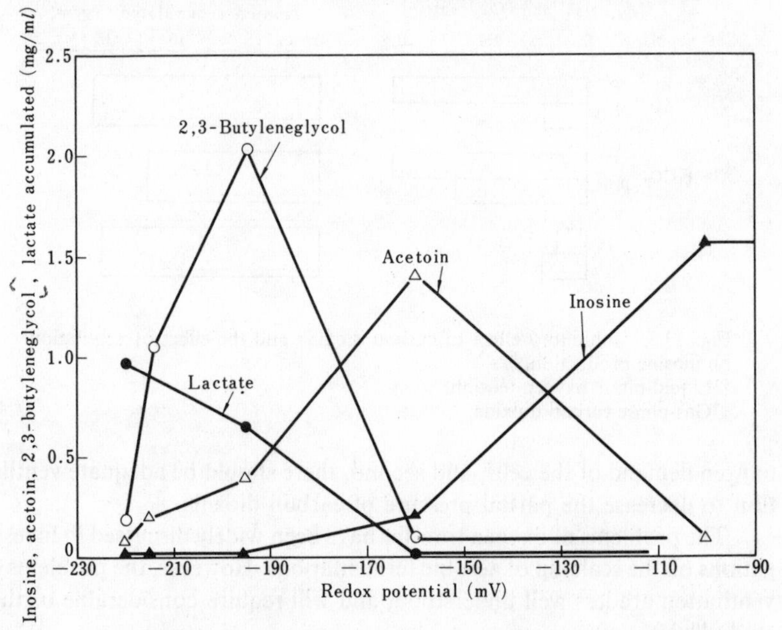

Fig. 13.2. Conversion of microbial products in relation to oxygen supply in inosine fermentation.

tion, the redox potential is supposedly influenced by the physiological changes in the cells. For the production of acetoin or 2,3-butyleneglycol, oxygen supply should be controlled at a level that slightly inhibits cell respiration. In these cultures, the redox potential of the medium is affected by changes in both the liquid-phase oxygen tension and the cell physiology. Therefore, the changes in the redox potential of the medium are extremely complicated.

13.2 Effect of Carbon Dioxide on Inosine Production

A low rate of air flow results in the inhibition of inosine production, even if the oxygen demand of the cells is satisfied (Fig. 13.2). This is due to the inhibitory effect of carbon dioxide on inosine formation (Fig. 13.3). Therefore, the effects of agitation and aeration can be discussed from two standpoints: first, there should be sufficient oxygen supply to satisfy the

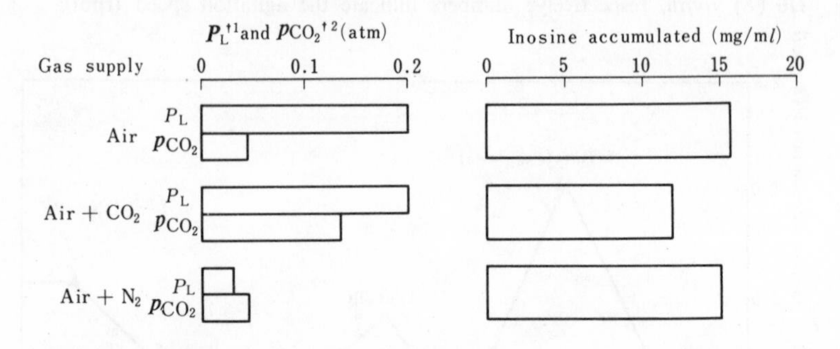

Fig. 13.3. Inhibitory effect of carbon dioxide and the effect of ventilation on inosine production.
†1 Liquid-phase oxygen tension.
†2 Gas-phase carbon dioxide.

oxygen demand of the cells, and second, there should be adequate ventilation to decrease the partial pressure of carbon dioxide.

The problems of oxygen transfer have been widely discussed in investigations on the scale-up of aerobic fermentations. However, the problems of ventilation are less well understood, and will require considerable further study.[10-14]

The biosynthesis of arginine and purine nucleotides includes a process of carbonate fixation. Therefore, carbon dioxide has a promoting effect on the initial growth of *Penicillium chrysogenum*[15] and on the initiation of growth of germinated spores of *Bacillus stearothermophilus*. Carbon dioxide is considered to promote the rapid biosynthesis of both protein and nucleic acid. For the growth of methane assimilating strains, carbon dioxide is reported to be essential.[17] On the other hand, carbon dioxide is inhibitory to the production of antibiotics such as penicillin,[18] tetracycline,[19] oleandmycin[19] and streptomycin.[19] Charles *et al.*[20] described a CO_2-inhibited mutant, the growth of which was inhibited by carbon dioxide at concentrations which did not inhibit the parent strain. The growth inhibition of the mutant *Neurospora crassa* caused by carbon dioxide is reversed by the addition of purines. This implies that carbon dioxide is inhibitory to purine biosynthesis in this mutant.

In addition to the physiological changes mentioned above, carbon dioxide affects the acid-base equilibrium of the culture medium, influencing the fermentation results. High partial pressure of carbon dioxide brings the pH of the culture medium to the acid side, resulting in the production of high amylase activity by *B. subtilis*.[21] The behavior of carbon dioxide in aerobic fermentation is complicated, depending on the strains employed, the composition of the medium and the nature of the products. Therefore, it is important to understand the role of carbon dioxide in order to achieve successful scale-up and stable production at commercial plants.

13.3 CONCLUSION

Investigations on agitation and aeration have been carried out mainly from the viewpoint of chemical engineering, dealing with problems concerning oxygen and carbon dioxide transfer and liquid mixing. Studies on agitation and aeration in nucleoside fermentation have shown that the conditions of agitation and aeration are closely related to the fermentation results, influencing cell physiology (such as spore formation), ATP yield and CO_2 inhibition. From such studies, extremely important and basic data can be obtained for the scale-up and practical operation of commercial plants. Problems concerning gas transfer and liquid mixing are all related to the microbial activities in the culture system. Therefore, it is expected that further investigations on agitation and aeration will be carried out from the viewpoint of both microbiology and chemical engineering in order to obtain data applicable to aerobic tank culture.

References

1. K. Kinoshita, K. Niwa, H. Sasaki, S. Sakai and Y. Hirose, *Nippon Nogeikagaku kaishi* (Japanese), **47**, 793 (1973).
2. H. Shibai, A. Ishizaki, H. Mizuno and Y. Hirose, *Agr. Biol. Chem.*, **37**, 91 (1973).
3. H. Shibai, A. Ishizaki and Y. Hirose, *ibid.*, **37**, 2083 (1973).
4. A. Ishizaki, H. Shibai and Y. Hirose, *ibid*, **38**, 2399 (1974).
5. H. Shibai, A. Ishizaki, K. Kobayashi and Y. Hirose, *ibid.*, **38**, 2407 (1974).
6. M. Hongo, *Nippon Nogeikagaku Kaishi* (Japanese), **31**, 731 (1957).
7. M. Hongo, *ibid.*, **31**, 735 (1957).
8. M. Hongo, *ibid.*, **31**, 811 (1957).
9. M. Hongo, *ibid.*, **31**, 816 (1957).
10. A. Ishizaki, H. Shibai, Y. Hirose and T. Shiro, *Agr. Biol. Chem.*, **35**, 1733 (1971).
11. A. Ishizaki, Y. Hirose and T. Shiro, *ibid.*, **35**, 1852 (1971).
12. A. Ishizaki, Y. Hirose and T. Shiro, *ibid.*, **35**, 1860 (1971).
13. A. Ishizaki, H. Shibai, Y. Hirose and T. Shiro, *ibid.*, **37**, 99 (1973).
14. A. Ishizaki, H. Shibai, Y. Hirose and T. Shiro, *ibid.*, **37**, 107 (1973).
15. N. S. Golding, *J. Dairy Sci.*, **23**, 879 (1940).
16. D. W. Cook, L. R. Brown and R. G. Tischer, *Develop. Ind. Microbiol.*, **5**, 326 (1963).
17. S. Fukuoka, *Sekiyu to Biseibutsu* (Japanese), **4**, 15 (1970).
18. L. Nyiri and L. Lengyel, *Biotechnol. Bioeng.*, **7**, 343 (1965).
19. E. S. Bylinkina, T. S. Nikitina, V. V. Biryukov and O. N. Cherkasova, *Biotechnol. Bioeng. Symp.*, no. 4, 197 (1973).
20. G. A. Roberts and H. P. Charles, *J. Gen. Microbiol.*, **63**, 21 (1970).
21. S. Ikeda, *personal communication.*

CHAPTER **14**

Phosphorylation of Nucleosides*1,2

14.1 Biochemical Phosphorylation of Nucleosides

Nucleoside phosphotransferase (NPTase), which catalyzes the phosphoryl transfer reaction from a phosphate donor to nucleosides, was first demonstrated by Chargaff and Brawerman[1,2] in mammalian and plant tissues. 5′-Nucleotides were synthesized by NPTase in carrot but NPTase in mammalian tissue was found to synthesize a small amount of 3′(2′)-isomers together with the 5′-isomers. The enzyme found in human prostate gland synthesized a large amount of the 2′- and 3′-isomers. Bacterial NPTases found in *Escherichia coli* and *Serratia marcescens*[3] were reported to

*1 Koji Mitsugi, Ajinomoto Co. Inc.
*2 Masaharu Yoshikawa, Ajinomoto Co. Inc.

synthesize the 5′-isomers, like NPTase in carrot. Glucose-1-phosphate phosphotransferase was also observed in *E. coli* to catalyze the phosphorylation of various kinds of nucleosides.[4)]

Mitsugi *et al.*[5)] studied the distribution of NPTases in microorganisms in detail and found that 5′-nucleotides were synthesized by NPTase in bacteria belonging to the genera *Pseudomonas*, *Alcaligenes*, *Achromobacter*, *Flavobacterium*, *Serratia* and *Staphylococcus*, while bacteria belonging to the Enterobacteriaceae, such as the genera *Aeromonas*, *Escherichia*, *Proteus*, *Aerobacter* and *Salmonella*, were found to synthesize mainly 3′(2′)-nucleotides. Purification of these bacterial NPTases has not yet been attempted, but these enzymes were intracellular[6)] and *p*-nitrophenyl phosphate (NPP) was employed as the best phosphate donor, with an optimum pH of 4 to 5.[7)]

14.1.1 Production of 5′-Nucleotides

On the basis of the above findings, biochemical phosphorylation of nucleosides was studied for the industrial production of IMP or GMP. A reaction mixture containing 20 mM inosine or guanosine, 70 mM NPP, 1 mM $CuSO_4$ and culture broth or intact cells of *Pseudomonas trifolii* (later re-identified as *Erwinia herbicola*[8)]) was incubated at pH 4.0 and 37°C for 20 h. IMP or GMP was synthesized in a molar yield of more than 85%. The reaction mixture was centrifuged to remove the bacterial cells, and the supernatant was extracted with an organic solvent to remove *p*-nitrophenol (NP), followed by chromatography to obtain 5′-nucleotide fractions. 5′-Nucleotides were crystallized by adding alcohol.

14.1.2 Distribution of NPTases in Microorganisms

All bacteria having NPTase activities, other than the genus *Staphylococcus*, are gram-negative bacteria and can be divided into two groups, depending on the nucleotide isomers synthesized. Komagata *et al.*[9)] proposed a relationship between gram-negative bacteria and their flagellation or physiological characteristics from the taxonomic viewpoint. As shown in Fig. 14.1, bacteria having NPTases were distributed in the fermentative group of gram-negative bacteria; the 3′(2′)-nucleotide producers belonged to the Enterobacteriaceae; and the 5′-nucleotide producers were distributed between oxidative and more fermentative bacterial groups. Thus, NPTase activity is a useful criterion for taxonomic studies on gram-negative bacteria.[10)] NPTases were found to be distributed in various fungi or actinomycetes, in addition to bacteria, but their activities were very low.[11)]

Polar flagellation	Peritrichous flagellation	Non-motile
Pseudomonas (Fluorescent group)	*Agrobacterium*	
Pseudomonas (Achromogenic group)	*Achromobacter*	
	Alcaligenes	
Pseudomonas (Chromogenic group)	*Flavobacterium*	
	Serratia	
	Escherichia	*Escherichia*
	Aerobacter	*Aerobacter*
Aeromonas	*Proteus*	
	Salmonella	

Fermentative ⇌ oxidative

Fig. 14.1. Distribution of nucleoside phosphotransferase in gram-negative rod bacteria. ———: 5′-Nucleotide producer; ═══: 3′(2′)-Nucleotide producer

14.1.3 Acceptor and Donor Specificities

Acceptors in this phosphoryl transfer reaction included various ribonucleosides,[12] deoxyribonucleosides,[13] and their analogs,[14] and mononucleotides.[15] As shown in Table 14.1, *Ps. trifolii* or *S. marcescens*, as a

TABLE 14.1

Acceptor and Donor Specificities in Nucleoside Phosphate Transfer Reactions

Enzyme Source	Donor	Acceptor	Nucleotide formed
5′-Nucleotide producers†1	NPP or 5′-Nucleotide	5′-OH(primary)	
		Nucleoside	5′-Nucleotide
		2′-Nucleotide	Nucleoside-2′,5′-diphosphate
3′(2′)-Nucleotide producers†2	NPP or 3′(2′)-Nucleotide	3′(2′)-OH(secondary)	
		Nucleoside	3′(2′)-Nucleotide
		5′-Nucleotide	Nucleoside-3′(2′,5′)-diphosphate

†1 5′-Nucleotide producers: *Ps. trifolii, S. marcescens.*
†2 3′(2′)-Nucleotide producers: *E. coli, Proteus vulgaris.*

typical 5′-nucleotide producer, also catalyzed phosphoryl transfer from a phosphate donor to 2′-nucleotides to synthesize nucleoside 2′,5′-diphosphates, while *E. coli* or *Proteus vulgaris*, a typical 3′(2′)-nucleotide producer, synthesized nucleoside 3′(2′),5′-diphosphates from 5′-nucleotides and a phosphate donor. In order to confirm the acceptor specificities of these reactions, angustmycin A (decoyinine) and angustmycin C (psicofuranine) were used as subjects for phosphorylation. As shown in Table 14.2,[16] NPTase in 5′-nucleotide producers was confirmed to phosphorylate the primary hydroxyl group at the C-5′ position specifically but not to phosphorylate the primary hydroxyl group at the C-1′ position of the ribose moiety. This phosphorylation, however, was sterically prohibited by amino and phosphate substituents at C-3′ of the moiety, as observed in the phosphorylation of 3′-AMP, puromycin (6-dimethylamino-9-[3′-deoxy-3′-(*p*-methoxy-L-phenylalanylamino)-β-D-ribofuranosyl]-β-purine), or aminonucleoside (6-dimethylamino-9-[3′-amino-3′-deoxyribosyl]purine).

On the other hand, phosphorylation of the secondary hydroxyl group at C-2′ or C-3′ of the ribose moiety by 3′(2′)-nucleotide producers was also observed in the phosphorylation of NAD. *Proteus mirabilis* was found to phosphorylate NAD to NADP (2′-phosphate in the adenosine moiety of NAD) and NADP phosphate (3′(2′)-phosphate in the nicotinamide ribose moiety of NADP).[17,18] This bacterium also phosphorylated nicotinamide monophosphate to give nicotinamide riboside diphosphate.[19] This phosphorylation at the secondary hydroxyl group at C-2′ or C-3′ of the ribose moiety was prevented by the presence of a primary hydroxyl group at C-1′ in the case of angustmycin A and C,[16] or by the 4-carboxyl group of the pyrimidine base in the case of orotidine.

Both types of bacteria phosphorylated formycin[20] or pseudouridine,

TABLE 14.2

Phosphorylation of Nucleoside Derivatives[16]

Acceptor	Phosphorylated Product	
	Ps. trifolii	*Proteus mirabilis*
	(μmoles/ml)	
Adenosine	9.1	5.8
3′-AMP	0.0	—
Puromycin	0.0	0.0
Aminonucleoside	0.0	0.0
Angustmycin A	0.0	0.0
Angustmycin C	7.3	0.3
Uridine	8.5	7.0
Orotidine	7.2	0.0
Pseudouridine	7.2	5.4

which have a C–C linkage between the purine or pyrimidine base and the ribose moiety.

As phosphate donors, aromatic phosphates can be employed. Among these, NPP was the best donor, and phenyl or benzyl phosphate was also effective in the acidic pH range.[6] In addition, mononucleotides were also employed as phosphate donors subject to the specificities of the NPTases. As shown in Table 14.1, 5′-nucleotide producers catalyze reversible phosphoryl transfer between 5′-nucleotides and nucleosides to synthesize 5′-nucleotides corresponding to the nucleoside acceptors. Interconversion between 3′(2′)-nucleotides and nucleosides was catalyzed only by 3′(2′)-nucleotide producers.[6,21] The apparent optimal pH of this reversible phosphoryl transfer reaction catalyzed by 5′-nucleotide producers, however, was at 8 to 9,[21] which was different from the value for NPP as a donor.

Phosphoryl transfer in the alkaline pH range was observed in crude extracts of chick embryo, especially between deoxythymidylic acid and deoxypyrimidine nucleosides.[22] The apparent optimal pH's of bacterial NPTases were not always the same, as mentioned above, but these reactions require further investigation in detail through the purification of NPTase, separating it from phosphomonoesterase and related enzymes in crude extracts of bacterial cells. Further, the reaction mechanism and physiological significance of NPTase should be elucidated in detail.

14.2 Chemical Phosphorylation of Nucleosides

Synthesis of 5′-nucleotides by the chemical phosphorylation of nucleosides has been the object of great interest in laboratory work and many studies have been reported. Since the production of nucleosides by fermentation appeared feasible, industrial methods for the phosphorylation of nucleosides have been extensively investigated.

The early studies were focused on the development of phosphorylating agents. Phosphoryl oxychloride, which is the most common phosphorylating agent, has three active P–Cl bonds and often yields a secondary phosphate or a chlorinated compound as a by-product, depending on the reaction conditions. To avoid this, phosphorochloridates (Fig. 14.2) such

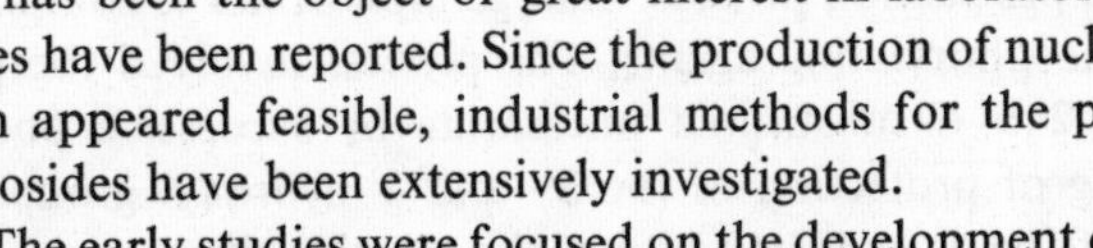

$(C_6H_5O)_2P(O)Cl$ (1) $(C_6H_5CH_2O)_2P(O)Cl$ (2) $NCCH_2CH_2OP(O)(OH)_2$ (3)

Fig. 14.2

as diphenyl phosphoro chloridate (1)[23] and dibenzyl phosphorochloridate (2),[24] prepared by substituting chlorine atoms of phosphoryl oxychloride, have been developed as phosphorylating agents. The substituents in these agents can be removed from the phosphorylated nucleosides by catalytic hydrogenation. Later, dicyclohexyl carbodiimide came into use as a dehydrating agent in the phosphorylation of nucleosides, and cyanoethyl dihydrogen phosphate (3),[25] the substituent of which is easily removable by alkaline hydrolysis, has also been used successfully.

Effective phosphorylating agents have also been found among phosphoric anhydrides. Examples include polyphosphoric acid[26] (a condensed product of 85% phosphoric acid and phosphoric pentoxide obtained by heating) and tetra-*p*-nitrophenyl pyrophosphate (4)[27] (Fig. 14.3). Pyrophosphoryl tetrachloride (5) will be described later.

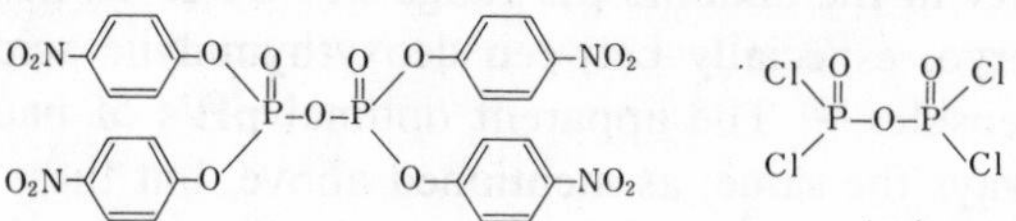

(4) (5)

Fig. 14.3

14.2.1 Production of 5′-Nucleotides

The main object of studies on the industrial production of 5′-nucleotides was to find conditions under which nucleosides could be phosphorylated with readily available or producible phosphorylating agents, and to simplify the overall production process. Halogenophosphorus compounds were chosen as phosphorylating agents. The process involved the phosphorylation of a 2′,3′-*O*-substituted nucleoside by successive one-step reactions consisting of protection of the 2′- and 3′-hydroxyl groups and phosphorylation of the 5′-hydroxyl group, though the ultimate goal was the selective phosphorylation of unprotected nucleosides.

A. Phosphorylation of 2′,3′-*O*-substituted nucleosides

It is necessary that the protective groups on the 2′- and 3′-hydroxyl groups should be easily removable after phosphorylation. The most common protective group is isopropylidene, formed by reaction with acetone in the presence of an acid catalyst such as hydrogen chloride,[28] zinc chloride,[29] or a mixture of phosphoryl oxychloride and water.[30] The latter mixture is also used for phosphorylation by the successive one-step reaction described later. Protective groups formed by coupling with arylaldehydes (e.g. benzylidene)[31] can be used.

The phosphorylation of a 2′,3′-*O*-isopropylidene nucleoside was carried out for the first time with phosphoryl oxychloride by Levene *et al.*[29,32] The method (Fig. 14.4) consists of dissolving the 2′,3′-*O*-isopropylidene nucleoside in pyridine and adding the solution, with cooling, to phosphoryl oxychloride dissolved in pyridine. Subsequent studies showed that very effective removal of water from the reaction mixture and an excess of phosphoryl oxychloride with respect to nucleoside are essential in the reaction.[33] Insufficient phosphoryl oxychloride caused an increase in the formation of secondary phosphates as by-products (e.g. inosinyl (5′–5′) inosine).[34] IMP can be obtained in a good yield (about 83 %) by this method, while GMP is scarcely produced due to the lower solubility of 2′, 3′-*O*-isopropylidene guanosine in pyridine.

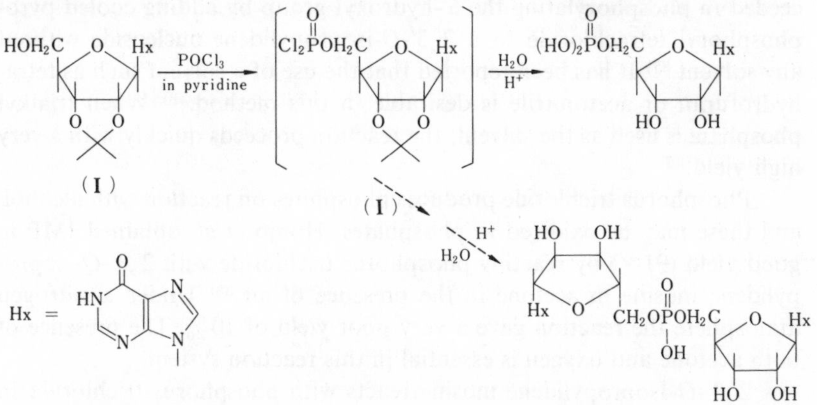

Fig. 14.4

Another effective method is to dissolve 2′,3′-*O*-isopropylidene nucleoside in a large excess of cooled phosphoryl oxychloride.[35] The addition of a small amount of water to the reaction mixture is a characteristic feature of this method and increases the phosphorylation yield to more than 80 % with most nucleosides. The exact mechanism of promotion by the addition of water is unknown, but it is known that the dichlorophosphoric acid formed promotes the phosphorylation.

The latest method, by Yoshikawa *et al.*, involves phosphorylation in a phosphoryl oxychloride-trialkyl phosphate system.[36] Suitable trialkyl phosphates are trimethyl phosphate and triethyl phosphate. The ratio of phosphoryl oxychloride and trialkyl phosphate in this system can be chosen with comparatively little restriction, and it is preferable to use an easily recoverable trialkyl phosphate as a solvent. The phosphorylation yield by this method is very high, due to the higher solubility of nucleosides in these phosphates than in pyridine. It is also considered that trialkyl phosphate plays some part in activating phosphoryl oxychloride.

Among phosphoric anhydride derivatives, pyrophosphoryl tetrachloride is one of the most active phosphorylating agents. Grunze *et al.* succeeded in phosphorylating the 5′-hydroxyl group by adding cooled pyrophosphoryl tetrachloride to a 2′,3′-*O*-isopropylidene nucleoside without any solvent.[37] It has been reported that the use of a solvent such as tetrahydrofuran or acetonitrile is desirable in this method.[38] When trialkyl phosphate is used as the solvent, the reaction proceeds quickly with a very high yield.[36]

Phosphorus trichloride produces phosphites on reaction with alcohol, and these may be oxidized to phosphates. Honjo *et al.* obtained IMP in good yield (91%) by reacting phosphorus trichloride with 2′,3′-*O*-isopropylidene inosine in acetone in the presence of air.[39] Under a nitrogen atmosphere the reaction gave a very poor yield of 10%. The presence of both acetone and oxygen is essential in this reaction system.

2′,3′-*O*-Isopropylidene inosine reacts with phosphorus trichloride in trialkyl phosphate to give the corresponding 5′-dichlorophosphite (Fig. 14.5). The 5′-dichlorophosphite can be partially hydrolyzed by the addi-

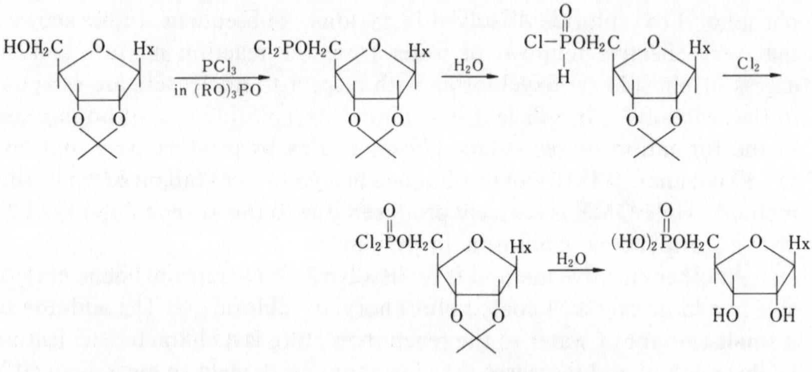

HOH₂C … Hx —[PCl_3 in $(RO)_3PO$]→ Cl_2POH_2C … Hx —[H_2O]→ Cl–P(=O)(H)–OH_2C … Hx —[Cl_2]→ $Cl_2P(=O)OH_2C$ … Hx —[H_2O]→ $(HO)_2P(=O)OH_2C$ … Hx (HO, OH)

Fig. 14.5

tion of water to yield the 5′-monochlorophosphite, which can then be oxidized with chlorine and hydrolyzed to produce IMP.[40]

B. Protection of the 2′- and 3′-hydroxyl groups and phosphorylation of the 5′-hydroxyl group by successive one-step reactions

After the acetonization of the 2′- and 3′-hydroxyl groups using excess phosphoryl oxychloride and water as a catalyst, the 5′-hydroxyl group can be phosphorylated with the excess phosphoryl oxychloride by adding an organic base such as pyridine to the reaction mixture.[41] Two successive one-step reactions (Fig. 14.6) are next explained here.

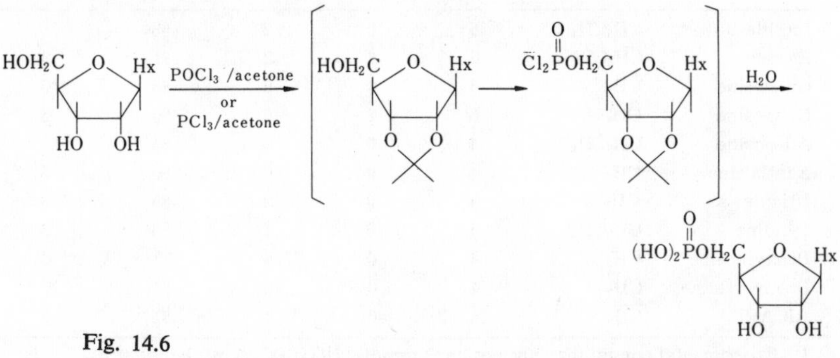

$$\text{HOH}_2\text{C-(inosine, 2',3'-OH)} \xrightarrow[\text{or } PCl_3/\text{acetone}]{POCl_3/\text{acetone}} \left[\text{HOH}_2\text{C-(2',3'-O-isopropylidene)} \longrightarrow Cl_2P(O)OH_2\text{C-(2',3'-O-isopropylidene)} \right] \xrightarrow{H_2O} (HO)_2P(O)OH_2\text{C-(2',3'-OH)}$$

Fig. 14.6

When the amount of phosphoryl oxychloride is sufficient in an acetone-phosphoryl oxychloride-water system for the acetonization of inosine, IMP can be obtained in a yield of more than 80%.[42] Careful hydrolysis of the reaction mixture gives 2′,3′-*O*-isopropylidene inosine 5′-phosphate. The exclusion of acetone from the system results in very limited phosphorylation. These two facts show that successive acetonization and phosphorylation take place.

A similar reaction occurs in an acetone-phosphorus trichloride-water system.[39] The presence of oxygen is also essential in this reaction, as in the reaction with 2′,3′-*O*-isopropylidene inosine.

C. Selective phosphorylation of unprotected nucleosides (Table 14.3)

An attempt has been made to phosphorylate selectively the primary hydroxyl group using phosphorus compounds carrying a large substituent, which sterically hinders reaction with the secondary hydroxyl groups.[43] There was also an attempt to produce selective reactivity of the primary hydroxyl group of nucleosides by converting them to alkoxytriphenylphosphonium salts.[44] However, two simple methods will be described here.

When unprotected nucleosides are suspended in trimethyl or triethyl phosphate and phosphoryl oxychloride is added to the suspension with cooling, naturally occurring nucleosides give the corresponding 5′-nucleotides in high yields (68–90%) together with small amounts of the 2′(3′),5′-diphosphates.[36] However, previous addition of water (about equimolar with respect to the nucleosides) decreases the production of 2′(3′),5′-

TABLE 14.3

Selective Phosphorylation of Unprotected Nucleosides in Trialkyl Phosphate†

Nucleoside	R group in $(RO)_3PO$	$POCl_3$ (mmole)	H_2O (mmole)	Time (h)	Yield of nucleoside phosphate (mole %) 5′-mono	2′(3′), 5′-di
Inosine	CH_3CH_2	6	0	2	68	18
Inosine	CH_3CH_2	6	2	2	91	8
Guanosine	CH_3	6	0	6	85	9
Guanosine	CH_3	6	2	6	90	5
Adenosine	CH_3CH_2	4	0	6	84	11
Xanthosine	CH_3	4	0	9	80	5
Uridine	CH_3	4	0	12	89	8
Cytidine	CH_3CH_2	4	0	1	88	8
Deoxyinosine	CH_3	4	0	5	73	6
Deoxycytidine	CH_3	4	0.5	6	64	19
AICAR	CH_3	4	0	4	91	5

† Experimental conditions: nucleoside, 2 mmole; $(RO)_3PO$, 5 ml; temp., 0°C.

diphosphates. In the case of inosine and guanosine, the yields of 5′-nucleotides reach more than 90%. This high selectivity is considered to be due to suppression of the reactivity of the 2′- and 3′-hydroxyl groups in the presence of acid. Since trialkyl phosphate is easily recovered by extraction with ethylene dichloride after hydrolysis of the reaction mixture, this method is one of the most convenient processes for industrial phosphorylation. This method may be used for various nucleosides. It is also possible to obtain a nucleoside 5′-phosphorodichloridate by extraction of the phosphorylating solution with a solvent such as ethyl ether. Therefore, the method also has very wide application for experimental purposes.

Pyrophosphoryl tetrachloride has much higher reactivity than phosphoryl oxychloride and phosphorylates not only the 5′-hydroxyl group but also the 2′- and 3′-hydroxyl groups. However, when *m*-cresol or *o*-chlorophenol is used as a solvent, pyrophosphoryl tetrachloride reacts selectively, and various naturally occurring nucleosides give the corresponding 5′-nucleotides in yields of 55–85%.[45] Though nitriles such as acetonitrile and benzonitrile or esters such as ethyl acetate and methyl acrylate can be used as solvents, some cleavage of the glycosyl bond in the case of the former and some formation of 2′(3′),5′-diphosphates in case of the latter are known to occur.

Another method has been developed, using a phosphoryl oxychloride-water-pyridine system in acetonitrile or nitromethane as a solvent[46] (see Chapter 9).

14.2.2 Future Prospects

When *ortho*-phosphoric acid reacts with alcohol in the presence of a tertiary amine and trichloroacetonitrile as a dehydrating agent, the alcohol is phosphorylated in good yield. It is considered that the intermediate, imidoyl phosphate (6) (Fig. 14.7), acts as the phosphorylating agent in the

$$\underset{(6)}{CCl_3-C(=NH)-O-P(=O)(OH)-O^-} \qquad \underset{(7)}{C_5H_5N^+-P(=O)(OH)-O^-}$$

Fig. 14.7

reaction medium.[47] NPP is transesterified with alcohol in pyridine on heating. The reaction proceeds by the formation of *N*-phosphoropyridinium (7) as an active phosphorylating agent.[48] When the tri-*n*-butylammonium salt of *ortho*-phosphoric acid and 2′,3′-*O*-isopropylidene inosine are heated in dimethyl formamide, the phosphorylation reaction is accelerated by the presence of malononitrile or acrylonitrile.[49] An active phosphorus compound is also presumed to be formed during this reaction.

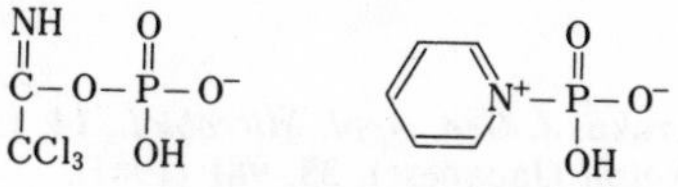

Phosphorous acid is formed as a by-product when phosphorus trichloride is used as a chlorine donor, and has some commercial value. Heating phosphorous acid, triethylamine and mercuric chloride with alcohol yields a monoester of phosphoric acid (Fig. 14.8).[50] *N*-Methylimidazole can also

$$(HO)_2P(=O)H + HgCl_2 + ROH \xrightarrow[\text{or } N\text{-methylimidazole}]{(C_2H_5)_3N} RO-P(=O)(OH)_2 + Hg + 2HCl$$

Fig. 14.8

be used instead of triethylamine in a similar phosphorylation of nucleosides.[51] When phosphorous acid reacts with iodine in a large excess of alcohol in the presence of a tertiary amine, the corresponding monoalkyl phosphate is obtained.[52]

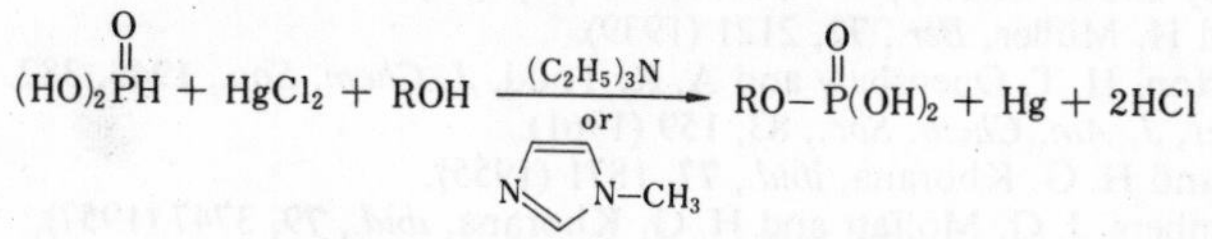

A future aim is the development of a closed process in which excess phosphorylating agents, reaction accelerators, and solvents can be recovered and reused in order to reduce environmental problems and to preserve resources. The above reactions suggest certain directions for future studies.

References

1. G. Brawerman and E. Chargaff, *J. Am. Chem. Soc.*, **75**, 4113 (1953).
2. G. Brawerman and E. Chargaff, *Biochim. Biophys. Acta*, **15**, 549 (1954).
3. G. Brawerman and E. Chargaff, *ibid.*, **16**, 524 (1955).
4. H. Katagiri, H. Yamada, and K. Imai, *J. Biochem.* (Tokyo), **46**, 1119 (1959).
5. K. Mitsugi, K. Komagata, M. Takahashi, H. Izuka and H. Katagiri, *Agr. Biol. Chem.*, **28**, 586 (1964).
6. K. Mitsugi, *ibid.*, **28**, 659 (1964).
7. K. Mitsugi, *ibid.*, **28**, 669 (1964).
8. K. Komagata, Y. Tamagawa and H. Izuka, *J. Gen. Appl. Microbiol.*, **14**, 19 (1968).
9. K. Komagata, *Nippon Nogeikagahu Kaishi* (Japanese), **35**, 981 (1961).
10. K. Komagata and Y. Tamagawa, *J. Gen. Appl. Microbiol.*, **12**, 191 (1966).
11. A. Kamimura, K. Mitsugi and S. Okumura, *Amino Acid and Nucleic Acid*, no. 16, 146 (1967).
12. K. Mitsugi, A. Kamimura, H. Nakazawa and S. Okumura, *Agr. Biol. Chem.*, **28**, 828 (1964).
13. K. Mitsugi, H. Nakazawa and S. Okumura, *ibid.*, **28**, 838 (1964).
14. K. Mitsugi, A. Kamimura, S. Okumura and N. Katsuya, *ibid.*, **28**, 849 (1964).
15. K. Mitsugi, H. Nakazawa, S. Okumura, M. Takahashi and H. Yamada, *ibid.*, **29**, 1051 (1965).
16. A. Kamimura, K. Mitsugi and S. Okumura, *ibid.*, **37**, 2037 (1973).
17. T. Tochikura, M. Kuwahara, S. Komatsubara, M. Fujisaki, A. Suga and K. Ogata, *ibid.*, **33**, 840 (1969).
18. T. Tochikura, M. Kuwahara, T. Tachiki, S. Komatsubara and K. Ogata, *ibid.*, **33**, 848 (1969).
19. M. Kuwahara, T. Tachiki, T. Tochikura and K. Ogata, *ibid.*, **34**, 983 (1970).
20. T. Sawa, Y. Fukagawa, Y. Shimauchi, K. Ito, M. Hamada, T. Takeuchi and H. Umezawa, *J. Antibiotics, Ser. A*, **18**, 259 (1969).
21. K. Mitsugi, A. Kamimura, M. Kimura and S. Okumura, *Agr. Biol. Chem.*, **29**, 1109 (1965).
22. G. F. Marley and F. Marley, *Arch. Biochem. Biophys.*, **101**, 342 (1963).
23. P. Brigl and H. Müller, *Ber.*, **72**, 2121 (1939).
24. F. R. Atherton, H. T. Openshaw and A. R. Todd, *J. Chem. Soc.*, **1945**, 382.
25. G. M. Tener, *J. Am. Chem. Soc.*, **83**, 159 (1961).
26. P. H. Hall and H. G. Khorana, *ibid.*, **77**, 1871 (1955).
27. R. W. Chambers, J. G. Moffatt and H. G. Khorana, *ibid.*, **79**, 3747 (1957).
28. O. Shimamura, T. Takenishi, T. Kato and H. Mori, *Japanese Patent* No. 41–476430 (1966).
29. P. A. Levene and R. S. Tipson, *J. Biol. Chem.*, **111**, 131 (1935).
30. Y. Tsuchiya, T. Takenishi, T. Kato and H. Mori, *Japanese Patent* No. 40–447419 (1965).
31. D. M. Brown, L. J. Haynes and A. R. Todd, *J. Chem. Soc.*, **1950**, 3299.
32. P. A. Levene and R. S. Tipson, *J. Biol. Chem.*, **106**, 113 (1934).
33. T. Kato, H. Mori, N. Muramatsu, T. Meguro, M. Yoshikawa, T. Ichikawa, T. Takenishi and Y. Tsuchiya, Ann. Mtg. Agr. Chem. Soc. Japan, 1963. Abstracts, p. 123.
34. N. Muramatsu and T. Takenishi, *J. Org. Chem.*, **30**, 3211 (1965).
35. M. Yoshikawa and T. Kato, *Bull. Chem. Soc. Japan*, **40**, 2849 (1967).
36. M. Yoshikawa, T. Kato and T. Takenishi, *ibid.*, **42**, 3505 (1969); *Tetr. Lett.*, **50**, 5065 (1967).
37. H. Grunze and W. Koransky, *Angew. Chem.*, **71**, 407 (1959).

38. S. Ouchi, K. Tsunoda and S. Senoo, *Japanese Patent* No. 42–506740 (1967).
39. M. Honjo, R. Marumoto, K. Kobayashi and Y. Yoshioka, *Tetr. Lett.*, **32**, 3851 (1966).
40. M. Yoshikawa, M. Sakuraba and K. Kusashio, *Bull. Chem. Soc. Japan*, **43**, 456 (1970).
41. M. Naruse and Y. Fujimoto, 84th Ann. Mtg. Pharm. Soc. Japan, 1964. Abstracts, p. 253.
42. (a) H. Mori, M. Yoshikawa, H. Takase, T. Kato and T. Takenishi, Ann. Mtg. Agr. Chem. Soc. Japan, 1964. Abstracts, p. 169; (b) M. Yoshikawa, H. Mori, K. Kusashio, T. Kato and T. Takenishi, *ibid.*, p. 170.
43. M. Ikehara, E. Ohtsuka and Y. Kodama, *Chem. Pharm. Bull.*, **11**, 1456 (1963).
44. O. Mitsunobu, K. Kato and J. Kimura, *J. Am. Chem. Soc.*, **91**, 6510 (1969).
45. K. Imai, S. Fujii, K. Takanohashi, Y. Furukawa, T. Masuda and M. Honjo, *J. Org. Chem.*, **34**, 1547 (1969).
46. S. Ouchi, H. Yamada, N. Kameyama, M. Kurihara, S. Senoo and T. Sowa, *Japanese Patent* No. 43–532649 (1968).
47. F. D. Cramer and G. Weimann, *Chem. Ind.*, **1960**, 46.
48. K. J. Chong and T. Hata, *Bull. Chem. Soc. Japan*, **44**, 2741 (1972).
49. Y. Sanno and A. Nohara, *Chem. Pharm. Bull.*, **16**, 2056 (1968).
50. T. Mukaiyama and T. Obata, *J. Org. Chem.*, **32**, 1063 (1967).
51. H. Takaku, Y. Shimada and H. Oka, *Chem. Pharm. Bull.*, **21**, 1844 (1973).
52. A. J. Kirby, *Chem. Ind.*, **1963**, 1877.

CHAPTER **15**

Interconversion of Nucleosides*

15.1 Nucleoside-*N*-Pentosyl Transfer

N-Ribosyl or *N*-deoxyribosyl transfer reactions between nucleosides and free bases occur widely in bacteria. From the results of numerous studies, *N*-pentosyl transfer reactions can be classified into two types based on their reaction mechanism, as shown in Fig.15.1. The first type of reaction is catalyzed by nucleoside phosphorylases, and phosphorolysis of the nucleoside results in the formation of free pentose-1-phosphate or enzyme-bound pentose-1-phosphate, which then reacts with a free base (B_2) to form a new nucleoside, effecting net pentosyl transfer. The second type of reaction is direct pentosyl transfer, involving a nucleoside or an enzyme-

* Takuo Sakai, College of Agriculture, University of Osaka Prefecture.

(1) Phosphorylase type reaction

$$B_1\text{-Pentose} + Pi \rightleftharpoons B_1 + \text{Pentose-1-phosphate}$$
$$\text{Pentose-1-phosphate} + B_2 \rightleftharpoons B_2\text{-Pentose} + H_3PO_4$$
$$\overline{\hspace{10em}}$$
$$B_1\text{-Pentose} + B_2 \overset{Pi}{\rightleftharpoons} B_1 + B_2\text{-Pentose}$$

(2) Transferase type reaction

$$B_1\text{-Pentose} + B_2 \rightleftharpoons B_1 + B_2\text{-Pentose}$$

Fig. 15.1. Mechanisms of nucleoside-*N*-pentosyl transfer in bacteria. B_1 and B_2 represent free bases and B_1- and B_2-pentose represent nucleosides.

pentosyl intermediate as the pentosyl donor to the free bases; enzymes catalyzing this reaction are known as nucleoside-*N*-glycosyl transferases.

15.1.1 *N*-Ribosyl Transfer

N-Ribosyl transfer among bases of ribonucleosides was first demonstrated by Stephenson and Trim[1] in *Escherichia coli*, which shows *N*-ribosyl transfer between inosine and adenine. Ott and Werkman[2–4] studied the formation of adenosine and hypoxanthine from inosine and adenine in *E. coli*, and suggested that *E. coli* might contain phosphorylases which are active toward adenosine and inosine, and that the phosphorylases might catalyze *N*-ribosyl transfer with ribose-1-phosphate as the intermediate. Tazuke and Yamada[5] found that highly purified nucleoside phosphorylase from *Aerobacter cloacae* can catalyze *N*-ribosyl transfer between purine nucleosides and bases. These studies showed that nucleoside phosphorylase catalyzes *N*-ribosyl transfer among bases of ribonucleosides in bacteria.

In contrast to *N*-ribosyl transfer among bases of purine ribonucleosides, there have been few studies of *N*-ribosyl transfer among bases of pyrimidine nucleosides in microorganisms. However, it is known that (1) uridine phosphorylase from animal sources catalyzes *N*-ribosyl transfer among bases of pyrimidine nucleosides,[6] and (2) uridine phosphorylase is present in various bacteria,[7] so *N*-ribosyl transfer among bases of pyrimidine nucleosides seems to occur widely in bacteria.

It is well known that purine and pyrimidine nucleosides have independent metabolic pathways, and *N*-ribosyl transfer between pyrimidine nucleoside and purine bases was not detected until 1956, when Koch[8] detected *N*-ribosyl transfer between pyrimidine nucleosides and purine bases in *E. coli*. He also found that the reaction was catalyzed by a transferase and named the enzyme, nucleoside ribosyltransferase [nucleoside purine (pyrimidine) ribosyltransferase, EC 2.4.2.5]. However, no details were reported.

In 1965, Sakai *et al.*[9] studied bacterial *N*-ribosyl transfer between pyrimidine nucleosides and purine bases, with the aim of utilizing pyrimidine nucleotides formed as by-products in the commercial production of flavor enhancing nucleotides by the enzymatic hydrolysis of RNA. They found that the reaction is distributed widely in bacteria, as shown in Table 15.1, and that it has the following characteristics. (1) It is an orthophosphate requiring reaction. (2) It is reversible, though the apparent equilibrium lies far on the side of purine nucleoside formation. (3) Various bases, including artificial bases, such as allopurinol, 8-azaguanine or 8-azaadenine, can act as substrates, as shown in Table 15.2. (4) Bacteria which show this activity can be divided into two groups, based on the pH dependency (neutral or alkaline) of the reaction.

A typical example of nucleoside synthesis by *N*-ribosyl transfer is as follows.[10] Allopurinol riboside was produced through *N*-ribosyl transfer from uridine to allopurinol, using cell-free extract of *Erwinia carotovora* as an enzyme source. The reaction was carried out at 30° C for 10 h in a reaction mixture containing 400 mg of NaF, Tris-HCl buffer (pH 7.0) and 2 g (as protein) of cell-free extract in a total volume of 1 liter. The forma-

TABLE 15.1

Distribution of Nucleoside *N*-Ribosyl Transfer Activity in Bacteria

Bacteria	Inosine formed (μmole/mg cell)†		
	pH 6.0	pH 7.0	pH 9.0
Aerobacter aerogenes	0.56	0.18	0.11
Erwinia carotovora	0.54	0.57	0.54
E. aroideae	0.66	0.74	0.55
Escherichia coli Crooks	0.26	0.39	0.34
E. coli 2bT	0.23	0.31	0.27
Proteus vulgaris	0.50	0.59	0.53
Flavobacterium arborescens	0.19	0.20	—
Serratia marcescens	0.23	0.31	0.20
Bacillus sphaericus	0.37	0.66	0.25
Pseudomonas polycolor	0.08	0.06	0.03
Bacillus megaterium	0.16	0.29	0.29
Bacillus subtilis	0.21	0.19	0.42
Bacillus pumilus	0.17	0.19	0.40
Bacillus thiaminolyticus	0.11	0.11	0.18
Sarcina lutea	0.04	0.04	0.06
Corynebacterium sepedonicum	0.35	0.43	0.88
Bacterium cadaveris	0.07	0.16	0.18

† The reaction mixtures contained 3 μmole each of uridine and hypoxanthine and 5 μmole of orthophosphate in a total volume of 0.5 ml. The reaction was carried out at 37°C for 3 h.

TABLE 15.2

Substrate Specificity of Nucleoside *N*-Ribosyl Transfer in *Erwinia carotovora*

N-Ribosyl donor	Acceptor	Relative activity (%)
Uridine	Hypoxanthine	100
	Adenine	138
	Guanine	8
	Allopurinol	170
	8-Azaguanine	9
	8-Azaadenine	30
	Xanthine	16
Inosine	Thymine	7
	5-Methylcytosine	5
	Uracil	40

tion of allopurinol riboside was accompanied by a decrease of uridine and allopurinol, and a yield of 500 mg was finally obtained. After the reaction, the reaction mixture was heated for 3 min at 100° C and adjusted to pH 2.0 with 6 N HCl. The solution was then centrifuged to remove the flocculated protein, 10 g of charcoal was added and the mixture was stirred for 30 min. The charcoal was collected and washed with water, and adsorbed materials were eluted with 1% ammoniacal methanol. The eluate was concentrated *in vacuo* at 35° C. The concentrated fluid was adjusted to pH 9.0 with NH_4OH, mixed with $K_2B_4O_7$ solution at a final concentration of 0.01 M, and applied to a column of Dowex-1 Cl^-. The column was washed successively with water and a solution containing 0.03 M KCl and 0.03 M $K_2B_4O_7$, then the adsorbed material was eluted with 0.01 N HCl. The fractions containing allopurinol riboside were combined and evaporated to dryness *in vacuo*; the dried material was dissolved in a minimum quantity of water, methanol containing 0.001 N HCl was added and the solution was again evaporated to dryness *in vacuo*. This treatment was repeated 5 times to remove borate as methyl borate, which is volatile. Finally, the dried material was dissolved in a minimum quantity of water, and insoluble material was removed by filtration. The transparent filtrate was mixed with 10 ml of absolute methanol, and ethyl ether was added gradually until the solution became faintly cloudy. After standing at 5° C for several days, the resulting needle-shaped crystals were collected, washed with ethyl ether, and dried *in vacuo*. The crystals were recrystallized three times in the same way, yielding 200 mg of allopurinol riboside (corresponding to a recovery of about 40%). Production of allopurinol ribotide was also reported by Tanaka and Nakayama, utilizing the salvage pathway in *Brevibacterium ammoniagenes* ATCC 6872.[11)]

The details of the above reaction mechanism have been studied by Sakai *et al.*,[12)] using *Erwinia carotovora* and *Corynebacterium sepedonicum*, and it was found that the reaction was catalyzed by the joint action of pyrimidine nucleoside phosphorylase and purine nucleoside phosphorylase. It appears that in the reaction, α-ribose-1-phosphate is formed in the initial phosphorolysis of the pyrimidine nucleoside and then reacts with the purine base to form β-purine riboside, effecting net *N*-ribosyl transfer (Fig. 15.2).

The metabolic role of *N*-ribosyl transfer between pyrimidine nucleosides and purine bases is considered to be not only the salvage synthesis of nucleosides but also a link between the metabolic pathways of pyrimidine nucleotides and purine nucleotides, as illustrated in Fig. 15.3.

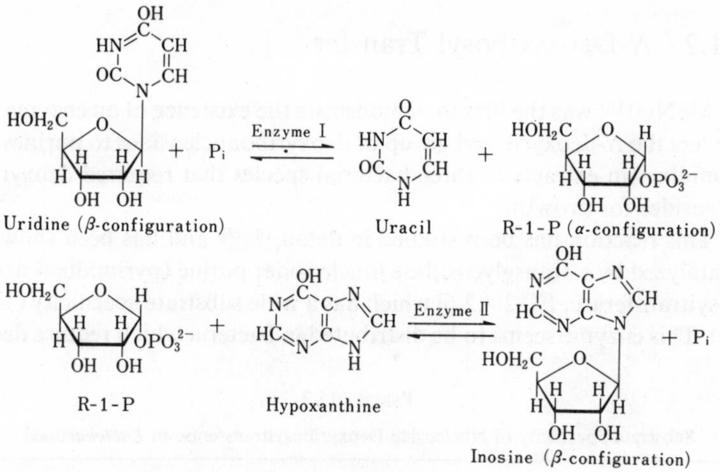

Uridine (β-configuration) + P_i ⇌ (Enzyme I) Uracil + R-1-P (α-configuration)

R-1-P + Hypoxanthine ⇌ (Enzyme II) Inosine (β-configuration) + P_i

Fig. 15.2. Mechanism of bacterial *N*-ribosyl transfer from uridine to hypoxanthine. Enzyme I: pyrimidine nucleoside phosphorylase. Enzyme II: purine nucleoside phosphorylase.

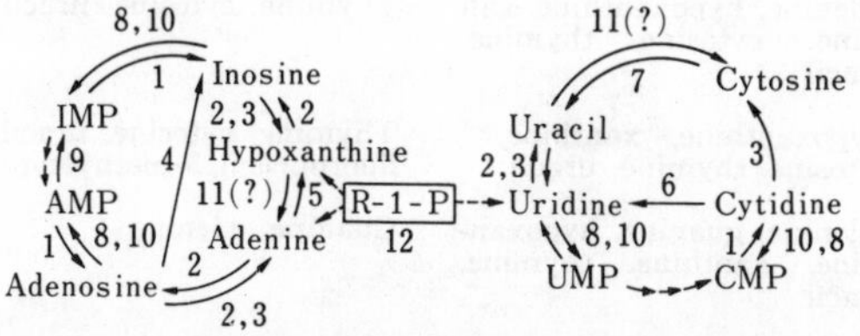

Fig. 15.3. Metabolism of nucleotides in bacteria.

1: Nucleotidase (5′- or 3′-ribonucleotide phosphohydrolase)
2: Nucleoside phosphorylase (purine or pyrimidine:orthophosphate ribosyltransferase)
3: Nucleoside hydrolase (*N*-ribosyl-purine or *N*-ribosyl-pyrimidine ribohydrolase)
4: Adenosine deaminase (adenosine aminohydrolase)
5: Adenine deaminase (adenine aminohydrolase)
6: Cytidine deaminase (cytidine aminohydrolase)
7: Cytosine deaminase (cytosine aminohydrolase)
8: Nucleoside kinase (ATP:nucleoside 5′-phosphotransferase)
9: Adenylic acid deaminase (adenylic acid aminohydrolase)
10: Phosphotransfer reaction (orthophosphoric monoester phosphohydrolase)
11: Transaminase (aminotransferase)
12: *trans*-*N*-Ribosylation (nucleoside:purine or pyrimidine ribosyltransferase, purine:orthophosphate ribosyltransferase and pyrimidine: orthophosphate ribosyltransferase)

15.1.2 *N*-Deoxyribosyl Transfer

McNutt[13] was the first to demonstrate the existence of an enzyme that transfers the *N*-deoxyribosyl group of deoxyribonucleosides to purines and pyrimidines in extracts of three bacterial species that required deoxyribonucleosides for growth.

This reaction has been studied in detail,[14–20] and has been shown to be catalyzed by a transglycosidase [nucleoside: purine (pyrimidine) deoxyribosyltransferase, EC 2.4.2.6] which has a wide substrate specificity (Table 15.3). This enzyme seems to be distributed in bacteria which require deoxy-

TABLE 15.3

Substrate Specificity of Nucleoside Deoxyribosyltransferase in *Lactobacillus*†

N-deoxyribosyl donor	Acceptor	
	L. helveticus	*L. leichmannii*
Deoxyinosine	Adenine, guanine, uric acid, azaguanine, cytosine. thymine, uracil	Thymine
Deoxyguanosine	Adenine, hypoxanthine, xanthine, cytosine, thymine, uracil	Thymine. cytosine, uracil
Deoxyadenosine	Hypoxanthine, xanthine, cytosine, thymine, uracil	Thymine, cytosine, uracil, fluorouracil, 5-methylcytosine
Deoxycytidine	Adenine, guanine, hypoxanthine, xanthine, thymine, uracil	Guanine, adenine
Deoxyuridine		Guanine, adenine
Deoxythymidine		Adenine, guanine, hypoxanthine, xanthine, 1-methylhypoxanthine, 6-methylaminopurine, 2-ethylamino-6-hydroxypurine, 6-methylpurine, 2, 6-diaminopurine

† Quoted from ref. 16 and 18 with modifications.

ribonucleoside for growth. From the fact that deoxyribonucleoside, which is an essential growth factor, can be replaced by vitamin B_{12} and the fact that the enzyme level in bacteria increases on limiting the supply of deoxyribonucleoside or vitamin B_{12} (Fig. 15.4), Beck and Levin[17] suggested that vitamin B_{12} is closely related to the biosynthesis of deoxyribonucleotides in bacteria, as shown in Fig. 15.5.

Phosphorylase-type *N*-deoxyribosyl transfer has also been demonstrated in bacteria. Manson and Lampen[21,22] and Hoffmann[23] have

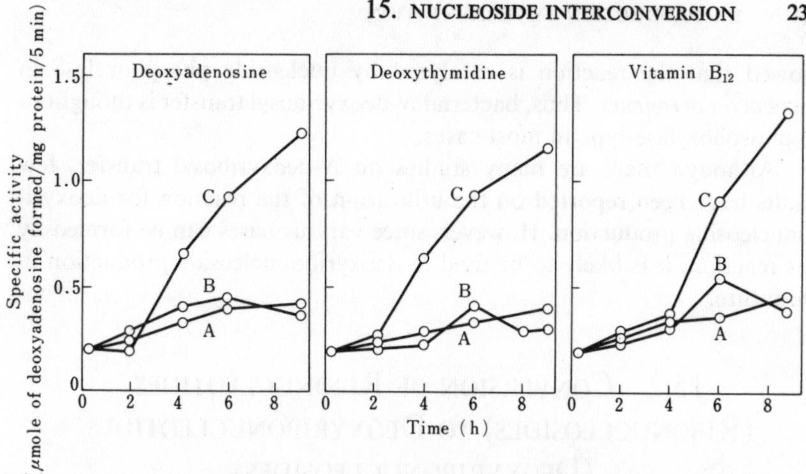

Fig. 15.4. Time course of specific activity of deoxyribosyl transferase of *Lactobacillus leichmannii* in various concentrations of deoxyadenosine, thymidine and vitamin B_{12} (quoted from ref. 17 with modifications). A, Nutrients in gross excess (50 μg, 50 μg and 5.0 μg/ml, respectively); B, nutrient concentrations optimal (2.5 μg, 2.5 μg and 0.1 mμg/ml, respectively); C nutrients limiting (0.05 μg, 0.05 μg and 0.005 mμg/ml, respectively).

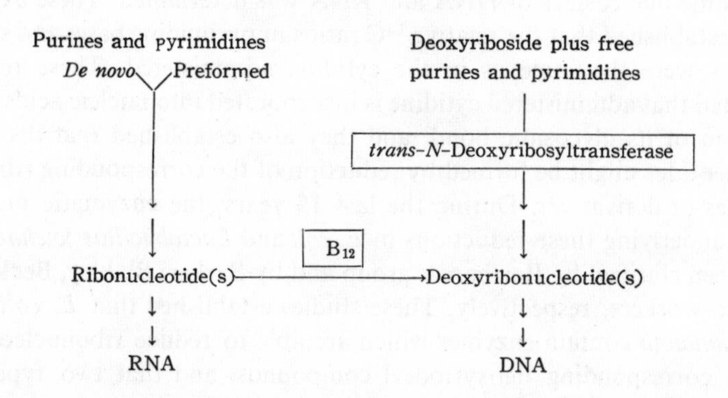

Fig. 15.5. Schematic diagram of postulated alternative pathways of deoxyribonucleotide synthesis in *Lactobacillus leichmannii* (quoted from ref. 17 with modifications).

found *N*-deoxyribosyl transfer in *E. coli*, and demonstrated that the reaction is a coupled nucleoside phosphorylase reaction. Sakai *et al.*[9] and Imada and Igarashi[24] found *N*-deoxyribosyl transfer between pyrimidine nucleosides and purine bases, and showed that the reaction is orthophosphate requiring, which is characteristic of nucleoside phosphorylase reactions. Further studies have been made by Imada and Igarashi,[24] who showed that the reaction is catalyzed by nucleoside phosphorylase in *Aerobacter aerogenes*. Thus, bacterial *N*-deoxyribosyl transfer is thought to be phosphorylase-type in most cases.

Although there are many studies on *N*-deoxyribosyl transfer, few results have been reported on the utilization of the reaction for deoxyribonucleoside production. However, since various bases can be formed by this reaction, it is likely to be used in deoxyribonucleoside production in the future.

15.2 Conversion of Ribonucleotides (Ribonucleosides) to Deoxyribonucleotides (Deoxyribonucleosides)

Our knowledge of deoxyribonucleotide biosynthesis was based on Rose and Schweigert's experiments in 1952,[25] in which rats were fed uniformly labeled cytidine and radioactivity appearing in the individual pyrimidine nucleosides of DNA and RNA was determined. These experiments established that the relative ^{14}C ratios in pyrimidine bases and sugar moieties were the same as in the cytidine administered. These results suggested that administered cytidine is incorporated into nucleic acids with retention of its glycosidic bond, and they also established that deoxyribonucleotides might be formed by reduction of the corresponding ribonucleotides or derivatives. During the last 15 years, the enzymatic mechanisms underlying these reductions in *E. coli* and *Lactobacillus leichmannii* have been clarified by Reichard's group and by Barker, Blakley, Beck and their co-workers, respectively. These studies established that *E. coli* and *L. leichmannii* contain enzymes which are able to reduce ribonucleotides to the corresponding deoxyribosyl compounds, and that two types of reductase (*E. coli* type and *Lactobacillus* type) exist in microorganisms. The direct reduction of ribonucleosides to the corresponding deoxyribonucleosides has been demonstrated by Grossman and Hawkins[26] in *Salmonella typhimurium*. However, the details are not known. It is now believed that deoxyribonucleotides are formed by the reduction of ribonucleoside diphosphates or ribonucleoside triphosphates.

15.2.1 Ribonucleotide Diphosphate Reduction in *E. coli*

During the last 15 years, the enzymatic mechanisms underlying ribonucleotide reduction in *E. coli* have been clarified by Reichard and his group. *E. coli* contains an enzyme system catalyzing the reduction of all four naturally occurring ribonucleotide diphosphates (ADP, GDP, UDP and CDP) to the corresponding 2′-deoxyribonucleotides. One of the characteristics of this reaction is that it requires a specific protein called thioredoxin as an electron donor *in vivo*. Thioredoxin is a low-molecular, heat-stable protein which contains a single disulfide bridge. The reduced form, thioredoxin-$(SH)_2$, is the immediate hydrogen donor for ribonucleotide reduction. The enzyme system contains ribonucleoside diphosphate reductase, which consists of two subunits (B_1 and B_2) and thioredoxin reductase, which catalyzes the reduction of the disulfide bridge of thioredoxin with NADPH as an electon donor[27–32]. The overall reaction is shown in Fig. 15.6.

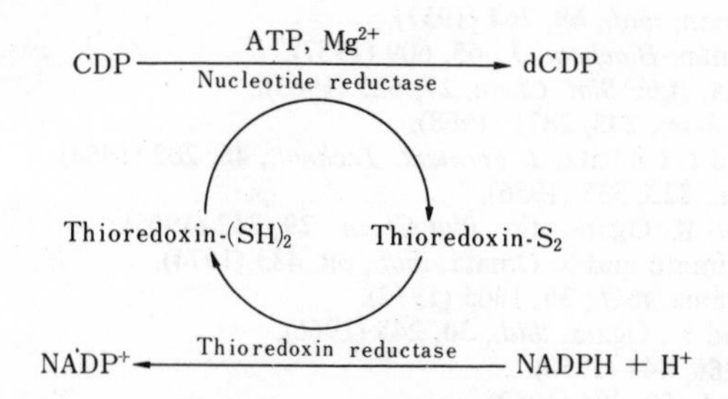

Fig. 15.6. Reduction of CDP to dCDP in *E. coli* (quoted from ref. 31 with modifications).

15.2.2 Ribonucleoside Triphosphate Reduction in *Lactobacillus leichmannii*

The biosynthesis of deoxyribonucleotides in *Lactobacillus leichmannii* has been studied in great detail by Barker, Blakley, Beck and their coworkers. Subsequent studies on the ribonucleotide reductase system

TABLE 15.4

Properties of Ribonucleotide Reductases from *Escherichia coli* and *Lactobacillus leichmannii*†

Properties	*E. coli*	*L. leichmannii*
Substrate	Nucleotide diphosphate	Nucleotide triphosphate
Cofactor	no	Deoxyadenosylcobalamine
Electron donor	Thioredoxin	Thioredoxin(?)
Mg^{2+} Requirement	yes	no
Inhibition by dATP	yes	no

† **Quoted from ref. 39 with modifications.**

revealed the following characteristics. (1) The substrates for reduction are ribonucleoside triphosphates.[33] (2) The nucleoside triphosphates are reduced in reactions requiring vitamin B_{12} as a coenzyme.[34–38] (3) All naturally occurring ribonucleoside triphosphates are reduced by one enzyme. These characteristics are different from those of *E. coli*. Some further apparent differences are shown in Table 15.4.

REFERENCES

1. M. Stephenson and A. R. Trim, *Biochem. J.*, **32**, 1740 (1938).
2. J. L. Ott and C. H. Werkman, *Arch. Biochem. Biophys.*, **48**, 483 (1954).
3. J. L. Ott and C. H. Werkman, *ibid.*, **69**, 264 (1957).
4. J. L. Ott and C. H. Werkman, *Biochem. J.*, **65**, 609 (1957).
5. Y. Tazuke and H. Yamada, *Agr. Biol. Chem.* **27**, 625 (1963).
6. T. A. Krenitsky, *J. Biol. Chem.*, **243**, 2871 (1968).
7. T. Sakai, T. Watanabe and I. Chibata, *J. Ferment. Technol.*, **46**, 202 (1968).
8. A. L. Koch, *J. Biol. Chem.*, **223**, 535 (1956).
9. T. Sakai, T. Tochikura and K. Ogata, *Agr. Biol Chem.*, **29**, 742 (1965).
10. T. Sakai, K. Ushio, I. Ichimoto and S. Omata, *ibid.*, **38**, 433 (1974).
11. H. Tanaka and K. Nakayama, *ibid.*, **36**, 1405 (1972).
12. T. Sakai, T. Tochikura and K. Ogata, *ibid.*, **30**, 245 (1966).
13. W. S. MacNutt, *Nature*, **166**, 444 (1950).
14. W. S. MacNutt, *Biochem. J.*, **50**, 384 (1952).
15. H. M. Kalckar, W. S. MacNutt and E. Hoff-Jørgensen, *ibid.*, **50**, 397 (1952).
16. A. H. Roush and R. F. Betz, *J. Biol. Chem.*, **233**, 261 (1958).
17. W. S. Beck and M. Levin, *Biochim. Biophys. Acta*, **55**, 245 (1962).
18. W. S. Beck and M. Levin, *J. Biol. Chem.*, **238**, 702 (1963).
19. J. C. Marsh and M. E. King, *Biochem. Pharmacol.*, **2**, 146 (1959).
20. M. Kanda and Y. Takagi, *J. Biochem* .(Tokyo), **46**, 725 (1959).
21. L. A. Manson and J. O. Lampen, *J. Biol. Chem.*, **191**, 87 (1951).
22. L. A. Manson and J. O. Lampen, *ibid.*, **193**, 539 (1951).
23. C. E. Hoffmann, *Fed. Proc.*, **11**, 231 (1952).
24. A. Imada and S. Igarashi, *J. Bact.*, **94**, 1551 (1967).
25. I. A. Rose and B. S. Schweigert, *J. Biol. Chem.*, **202**, 635 (1953).
26. L. Grossman and G. R. Hawkins, *Biochim. Biophys. Acta*, **26**, 657 (1957).
27. C. Biswas, J. Hardy and W. S. Beck, *J. Biol. Chem.*, **240**, 3631 (1965)
28. A. Larsson and P. Reichard, *Biochim. Biophys. Acta*, **113**, 407 (1966).
29. A. Larsson and P. Reichard, *J. Biol. Chem.*, **241**, 2533 (1966)
30. A. Larsson and P. Reichard, *ibid.*, **241**, 2540 (1966).
31. A. Holmgren, P. Reichard and L. Thelander, *Proc. Natl. Acad. Sci. U.S.A.*, **58**, 830 (1965).
32. H. Z. Sable, *Advan. Enzymol.*, **28**, 391 (1966).
33. R. L. Blakley, R. K. Ghambeer, P. F. Nixon and E. Vitols, *Biochem. Biophys. Res. Commun.*, **20**, 439 (1965).
34. R. Abrams, *J. Biol. Chem.*, **240**, 3697 (1965).
35. M. Goulian and W. S. Beck, *ibid.*, **241**, 4233 (1966).
36. W. S. Beck, M. Goulian, A. Larsson and P. Reichard, *ibid.*, **241**, 2177 (1966).
37. R. H. Abeles and W. S. Beck, *ibid.*, **242**, 3589 (1967).
38. H. A. Barker, *Biochem. J.*, **105**, 1 (1967).
39. A. Larsson and P. Reichard, *Progr. Nucl. Acid Res.*, **7**, 303 (1967).

PART IV

PRODUCTION OF NUCLEIC ACID-RELATED SUBSTANCES FROM ADDED PRECURSORS

CHAPTER 16

Production of Nucleosides and Nucleotides*

One of the biosynthetic pathways for nucleotides is the salvage pathway, as described by Lieberman *et al.*,[1,2] Kornberg *et al.*,[3,4] Buchanan *et al.*[5,6] and others, and reviewed in detail by Moat *et al.*[7] The salvage pathway may be generalized as follows:

(1) base + PRPP $\rightleftharpoons$ nucleotide + PPi

(2) base + ribose-1-P $\rightleftharpoons$ nucleoside + Pi

Reactions (1) and (2) are catalyzed by nucleotide pyrophosphorylase and nucleoside phosphorylase, respectively.

Many investigations have been done on the possiblity of 5′-nucleotide production by salvage synthesis since 1962.

16.1 Production of IMP from Adenine via Adenosine

Ott *et al.*[8] showed in 1954 that adenosine was formed from adenine in the presence of inosine by cell-free extracts of *Escherichia coli*.

---

* Koichi Ogata, Faculty of Agriculture, Kyoto University.

In 1962, Hara *et al.*,[9] Kojima *et al.*[10] and Kozae *et al.*[11,12] found that a large amount of adenosine was accumulated in cultures of a purine requiring strain 160-88 (st^r, try^-, pur^-) derived from *Bacillus subtilis* Marburg 160 (st^r, try^-) to which adenine synthesized chemically from malononitrile had been added. The use of glucose as the carbon source and the addition of a small amount of ribose increased the yield of adenosine. An increase was also observed with nitrogen sources such as polypepton, meat extract and casamino acids and on the addition of biotin. With 1–2 mg/ml of adenine, 1 mg/ml of adenosine was accumulated after cultivation for 40 h (Fig. 16.1). Later, the yield was increased to 80%. The adenosine thus formed from adenine was phosphorylated to AMP by cells of *Saccharomyces cerevisiae* according to the method of Ostern *et al.*[13,14] and then deamination with nitrite or mold AMP deaminase yielded IMP·Na. This process and the yield at each step are outlined in Fig. 16.2. The relation between adenosine synthesis and genetic markers in this strain was examined by back mutation. The results showed that the synthesis of adenosine is related to pur^- but not to try^- or st^r. Two different pathways were

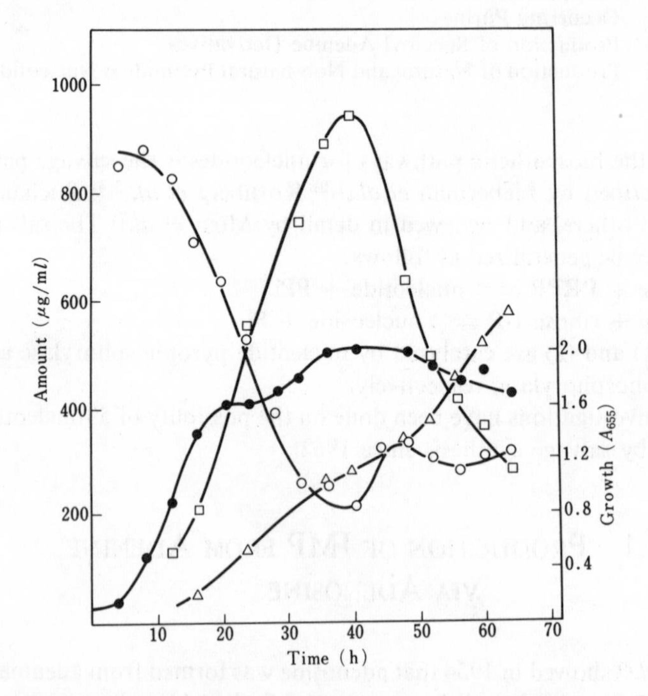

Fig. 16.1. Formation of adenosine during cultivation[9] in a jar fermenter. Temp., 37°C; agitation, 550 rpm; aeration, one volume per min. ○, Adenine; ●, growth; □, adenosine; △, hypoxanthine.

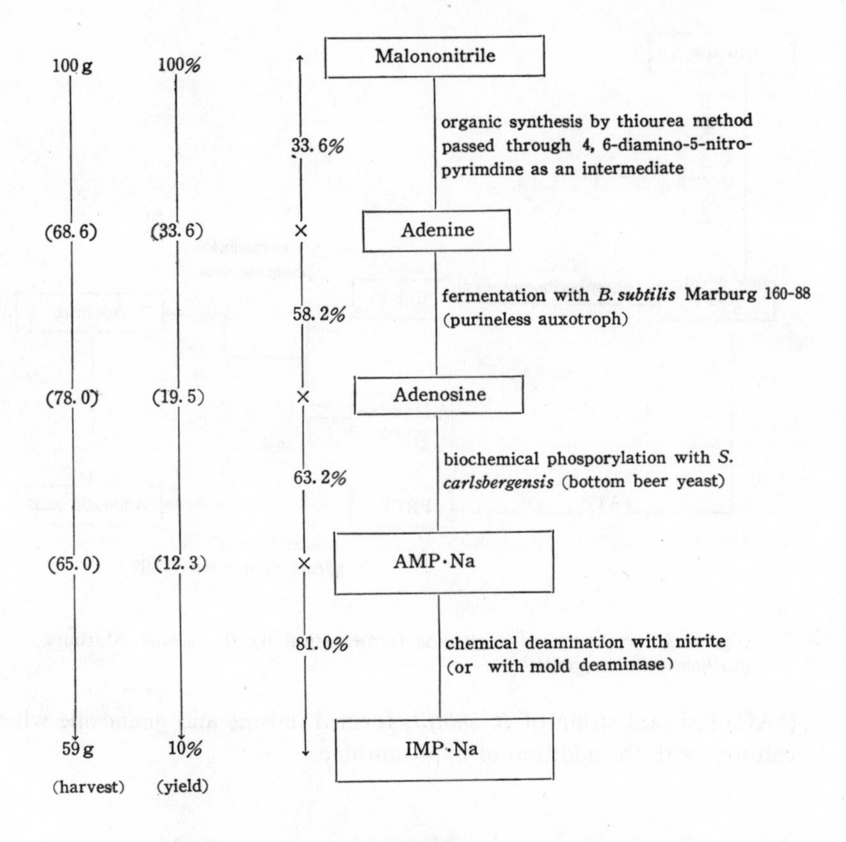

Fig. 16.2. Formation of IMP from malononitrile.[12]

thought to be involved in adenosine synthesis from adenine in this strain (Fig. 16.3).

16.2 SALVAGE SYNTHESIS ACCUMULATING INOSINE OR IMP

Yamanoi *et al.*[15,16] found that 5.3 mg of inosine per ml was accumulated in cultures of an adenine requiring strain of *B. subtilis* by the addition of hypoxanthine. The accumulation of IMP was demonstrated in cultures of *Micrococcus sodonensis*, *Arthrobacter citreus* and *Brevibacterium insectiphilum* to which hypoxanthine or adenine had been added.[17,18]

Kanamitsu[19] found that an amino acid requiring and 8-azaguanine

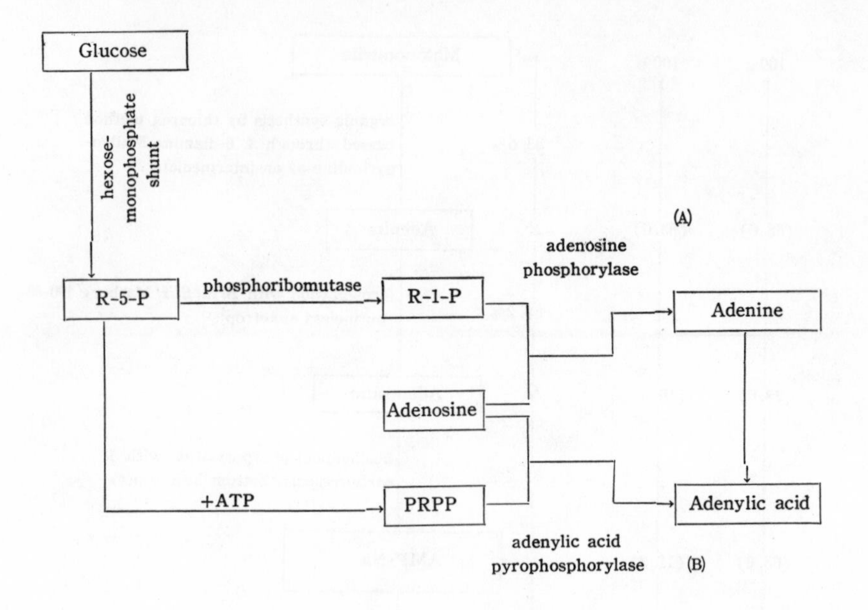

Fig. 16.3. Pathways of adenosine fermentation by *B. subtilis* Marburg, purineless auxotroph.[11)]

(8AG)-resistant strain of *B. subtilis* formed inosine and guanosine when cultured with the addition of hypoxanthine.

16.3 Production of Nucleosides and Nucleotides of Non-naturally Occurring Purines

Yamanoi *et al.*[20)] reported that various purine derivatives were formed by the addition of non-naturally occurring purine bases to cell suspensions of an inosine producer, *B. subtilis* No. 11023 (Table 16.1), and that ribosylation from hypoxanthine to inosine was strongly repressed by 8AG and 6-mercaptopurineriboside (6MP-riboside).

Tanaka *et al.*[21,22)] showed that *Brevibacterium ammoniagenes* ATCC 6872, which accumulated IMP, AMP, ADP, ATP, GMP, GDP and GTP from the corresponding purine bases, was able to produce non-naturally occurring nucleotides from the corresponding base analogs. The medium for cultivation of this strain was: glucose 10%, urea 0.6%, KH_2PO_4 1%, K_2HPO_4 1%, $MgSO_4 \cdot 7H_2O$ 1%, $CaCl_2 \cdot 2H_2O$ 0.01%, biotin 30 μg/1 and yeast extract 1%, at pH 7.4 before sterilizing. When non-naturally occur-

TABLE 16.1

Ribosylation of Non-natural Purine Bases by Washed Cell Suspensions[20]

Base (mM)	Chemical structure	Final pH	Inosine formed (mM)	Degree of ribosidation
2,6-Dichloropurine (6.0)		5.8	0.8	—
Dichloro-6-methoxypurine (3.0)		5.4	7.3	++
Isoguanine (6.0)		5.4	10.1	—
2-Methylhypoxanthine (3.0)		5.8	11.7	+++
N^2N^2-dimethylguanine (3.0)		5.6	12.1	+++
8-AG (3.0)		6.4	2.3	—
8-Azahypoxanthine (3.0)		5.6	11.7	—
6-Methoxypurine (6.0)		5.2	11.0	+++
6-Methylthiopurine (6.0)		5.2	14.0	+++
6-Chloropurine (6.0)		5.4	15.4	—
6-Mercaptopurine (6.0)		5.2	8.0	—
N^6-Acetyladenine (6.0)		5.6	14.0	+++
None —	—	5.4	13.0	

ring purine bases (2 mg/ml) were added to a 3-day culture of the medium and cultivation was continued until the 4th or 5th day, the nucleotides shown in Table 16.2 were formed. Ribotidation of 8-azahypoxanthine (8AHx), 8-azaadenine (8AA), 8AG and AICA to produce 8-aza-IMP, 8-aza-ATP, 8-aza-GMP and 8-aza-GDP, and AICARP was also confirmed. Further, allopurinol (4-hydroxypyrazolo [3,4-*d*]pyrimidine), an inhibitor of xanthine oxidase, and APP (4-aminopyrazolo[3,4-*d*]pyrimidine) were recognized as being converted to their ribotides and ribosides.

TABLE 16.2

Production of Non-natural Purine Ribosides by *B. ammoniagenes*[21]

Purine analog added (*Rf*)†	Ribotide produced	
	Compound (*Rf*)	Accumulation (mg/m*l*)
6-MP (0.32)	6-Thio-IMP (0.57)	3.27
6-Mercaptoguanine (0.55)	6-Thio-GMP (0.23)	1.46
6-Thioxanthine (0.46)	6-Thio-XMP (0.16)	0.25
6-Methyl purine (0.80)	6-Methylpurine-RP (0.49)	3.09
2-Methylhypoxanthine (0.71)	2-Methyl-IMP (0.40)	4.37
6-Methylaminopurine (0.90)	6-Methylaminopurine-RP (0.53)	3.67
2,6-Diaminopurine (0.85)	2,6-Diaminopurine-RP (0.58)	tr.
	-RPP (0.45)	tr.
	-RPPP (0.33)	2.30
6-Hydroxylaminopurine (0.60)	6-Hydroxyaminopurine-RP (0.37)	0.68
	-RPP (0.20)	1.45
	-RPPP (0.08)	2.77
2-Fluoro adenine (0.87)	2-Fluoro-AMP (0.45)	0.81
	-ADP (0.36)	1.41
	-ATP (0.25)	1.74
2-Methyl adenine (0.88)	2-Methyl-AMP (0.65)	+
	-ADP (0.50)	tr.
	-ATP (0.33)	tr.

† Developing solvent, isobutyric acid:acetic acid:1 N ammonia (10:1:5 v/v).

16.4 PRODUCTION OF SUCCINYL ADENINE DERIVATIVES

Midorikawa *et al.*[23] derived a mutant from an adenine requiring strain of *B. subtilis* by transduction with bacteriophage SP-10 and subsequent *N*-methyl-*N*-nitro-*N*-nitrosoguanidine treatment. The mutant could convert added hypoxanthine or inosine to succinyladenine, succinyladenosine and succinyl-5′-AMP (SAMP) and accumulate these compounds in the culture fluid. The production medium was composed of 10% glucose,

0.6% K_2HPO_4, 0.4% NH_4Cl, 0.5% Na citrate, 0.15% KCl, 0.05% $MgSO_4 \cdot 7H_2O$, 0.015% $CaCl_2 \cdot 2H_2O$ and 1% yeast extract, pH 7.8, in soybean extract, which was prepared by extracting 5% defatted soybean by boiling for 1 h in 0.1% NaOH. The yields of succinyladenine, succinyladenosine and SAMP in the production medium supplemented with 17 mM hypoxanthine were about 4.4 mM, 5.2 mM and 2.2 mM, respectively, after growth for 4 days at 40° C. The formation of succinyl adenine derivatives is thought to be due to a genetic block of adenylosuccinate lyase, which is responsible for the conversion of SAMP to AMP. The conversion of hypoxanthine or inosine to these derivatives might occur as follows: inosine → hypoxanthine → IMP → SAMP → succinyladenosine → succinyladenine.

16.5 PRODUCTION OF NATURAL AND NON-NATURAL PYRIMIDINE NUCLEOTIDES

The conversion of uracil or orotic acid to UMP by cell-free extracts of microorganisms has been demonstrated by Lieberman *et al.* [1,2] and Nagano *et al.*[24] Nara *et al.*[18] also reported that UMP was formed from uracil or cytosine in cultures of *B. insectiphilum*.

Nakayama *et al.*[25] found that *B. ammoniagenes* ATCC 6872 accumulated UMP in the culture broth, when the organism was cultured with the addition of uracil or orotic acid under the same conditions as for non-natural purine nucleotide formation (see section 16.3). As shown in Table 16.3, the yield of UMP reached 2.7–4.7 mg/ml on the 4–5th day of culture,

TABLE 16.3

Accumulation of UMP from Uracil and Orotic Acid by *B. ammoniagenes*[24]

Base added	Addition time (days)	UMP accumulation (mg/ml)			
		2 days	3 days	4 days	5 days
None	—	tr.	tr.	—	—
Uracil (2 mg/ml)	0	0.8	2.7	2.7	—
	1	0.5	3.1	3.8	4.4
	2	—	3.0	3.8	4.7
	3	—	—	3.3	4.8
Orotic acid (2 mg/ml)	0	0.4	2.7	2.8	—
	2	0.7	3.0	3.1	3.5
	3	—	—	3.8	4.3

when 2 mg/ml of uracil or orotic acid was added to the 0–3-day culture. The addition of 6-azauracil (6AU) (0.25–1.0 mg/ml) to the culture led to the accumulation of 2.7–3.6 mg/ml of orotidine 5′-monophosphate (OMP) with inhibition of UMP accumulation. The mechanism of OMP accumulation was thought to be as follows. The formation of UMP from orotic acid had been shown by Kornberg *et al.*[1,2] to involve the two reactions,

orotic acid + PRPP ⟶ OMP + PPi (OMP pyrophosphorylase)
OMP ⟶ UMP + CO_2 (OMP decarboxylase).

Therefore, it seemed reasonable to assume that added 6AU was converted to 6-aza-UMP and then inhibited OMP decarboxylase,[26] resulting in OMP accumulation.

A subsequent study was performed on the ribotidation of 6AU, 5-fluorouracil, 5-hydroxyuracil, 2-thiouracil and 4-thiouracil with *B. ammoniagenes* ATCC 6872 by Tanaka *et al.*[27] When each of the non-natural pyrimidine bases (2 mg/ml) was added to a 3-day culture and cultivation continued until the 4–5th day, the corresponding nucleotide was accumulated, as shown in Table 16.4.

TABLE 16.4

Production of Non-natural Pyrimidine Ribotides[27]

Pyrimidine analog added (*Rf*)†	Ribotide produced	
	Compound (*Rf*)†	Accumulation (mg/*ml*)
6-Azauracil (0.70)	6-Aza-UMP (0.39)	4.58
5-Fluorouracil (0.63)	5-Fluoro-UMP (0.34)	4.62
5-Hydroxyuracil (0.56)	5-Hydroxy-UMP (0.28)	4.24
2-Thiouracil (0.66)	2-Thio-UMP (0.35)	2.04
4-Thiouracil (0.67)	4-Thio-UMP (0.36)	2.75
	-UDP (0.24)	0.51
	-UTP (0.15)	0.14

† Developing solvent, isobutyric acid:acetic acid:1 N ammonia (10:1:5 v/v).

REFERENCES

1. I. Lieberman, A. Kornberg and E. S. Simms, *J. Biol. Chem.*, **215**, 403 (1955).
2. I. Lieberman and A. Kornberg, *J. Am. Chem. Soc.*, **76**, 2844 (1954).
3. A. Kornberg, I. Lieberman and E. S. Simms, *J. Biol. Chem.*, **215**, 417 (1955).
4. A. Kornberg, I. Lieberman and E. S. Simms, *J. Am. Chem. Soc.*, **76**, 2027 (1954).
5. E. D. Korn, C. N. Remy, C. Wasilejko and J. M. Buchanan, *J. Biol. Chem.*, **217**, 875 (1955).
6. C. N. Remy, W. T. Remy and J. M. Buchanan, *ibid.*, **217**, 885 (1955).
7. A. G. Moat and H. Friedman, *Bact. Rev.*, **24**, 309 (1960).

8. J. L. Ott and C. H. Werkman, *Arch. Biochem. Biophys.*. **48**, 483 (1954).
9. T. Hara, Y. Koaze, Y. Yamada and M. Kojima, *Agr. Biol. Chem.*, **26**, 61, 747 (1962).
10. Y. Koaze, Y. Yamada, M. Kojima and T. Hara, *ibid.*, **26**, 655, 758 (1962).
11. Y. Koaze, Y. Yamada, M. Kojima and T. Hara, *ibid.*, **26**, 740 (1962).
12. Y. Koaze, Y. Yamada, M. Kojima, K. Sato and Y. Aoyama, *ibid.*, **26**, 747 (1962).
13. P. Ostern, J. Terszakowec and J. Baranowski, *Hoppe-Seyler's Z. Physiol. Chem.*, **251**, 258 (1938).
14. P. Ostern, J. Baranowski and S. Huble, *ibid.*, **255**, 104 (1938).
15. A. Yamanoi, Y. Hirose, T. Shiro and N. Katsuya, *J. Gen. Appl. Microbiol.*, **11**, 269 (1965).
16. A. Yamanoi, Y. Hirose and T. Shiro, *ibid.*, **12**, 299 (1966).
17. T. Nara, M. Misawa, T. Komuro and S. Kinoshita, *Agr. Biol. Chem.*, **31**, 1224 (1967).
18. T. Nara, I. Kawamoto, M. Misawa and S. Kinoshita, *ibid.*, **32**, 956 (1968).
19. O. Kanamitsu, *ibid.*, **34**, 1424 (1970).
20. A. Yamanoi, Y. Hirose and T. Shiro, *J. Gen. Appl. Microbiol.*, **12**, 299 (1966).
21. H. Tanaka and K. Nakayama, *Agr. Biol. Chem.*, **36**, 464 (1972).
22. H. Tanaka and K. Nakayama, *ibid.*, **36**, 1405 (1972).
23. Y. Midorikawa, T. Akiya, Y. Kato, T. Kiyanagi, A. Kuninaka and H. Yoshino, *ibid.*, **36**, 1523 (1972).
24. Y. Nagano, H. Samejima and S. Kinoshita, *ibid.*, **30**, 83 (1966).
25. K. Nakayama and H. Tanaka, *ibid.*, **35**, 518 (1971).
26. R. E. Hundschumacher, *J. Biol. Chem.*, **235**, 2917 (1960).
27. H. Tanaka and K. Nakayama, *ibid.*, **35**, 989 (1971).

CHAPTER 17

Production of Nucleotide Coenzymes

17.1 ATP

ATP plays a central role in energy metabolism as a carrier of chemical energy from catabolic reactions to anabolic reactions. It has two high-energy phosphate bonds which give a free energy change of 7–12 kcal/mole on hydrolysis. Since ATP was first isolated from a tissue having high glycolysis activity (muscle) by Lohmann,[1)] and Fiske and Subbarow,[2)] the mechanism of its formation has been studied through elucidation of the mechanisms of the glycolytic pathway and respiratory chain.

*1 Yoshiki TANI, Faculty of Agriculture, Kyoto University.
*2 Koichi OGATA, Faculty of Agriculture, Kyoto University.
*3 Ichiro CHIBATA, Tanabe Seiyaku Co. Ltd.
*4 Joji KATO, Tanabe Seiyaku Co. Ltd.

There are various methods for the preparation of ATP which include extraction[3] from rabbit muscle, as originated by Lohmann, inhibiting the activity of ATPase with Mg^{2+}, chemical synthesis[4] using AMP as a starting material, and fermentation.

Regarding ATP formation by the fermentation method, Lutwak-Mann and Mann,[5] and Ostern *et al.*[6] found that added AMP or adenosine was phosphorylated to ATP in the alcohol fermentation of yeast. Recently, the fermentative production of ATP was developed in Japan during the course of the microbial production of nucleic acid, and since then a more effective method has been established.

17.1.1 Cell Method

Tochikura *et al.*[7] found that AMP or adenosine added to a fermentation system with ground or acetone-dried cells of bakers' yeast was effectively phosphorylated to ATP and ADP. In this system, 72% of the AMP was converted to ATP, as shown in Fig 17.1. The phosphorylation of AMP scarcely occurred with air-dried or fresh yeast, while the phosphorylation of adenosine to ATP proceeded only with acetone-dried yeast. The concentration of phosphate buffer in the reaction mixture is a special requirement of this method: the phosphorylation of AMP with acetone-dried yeast

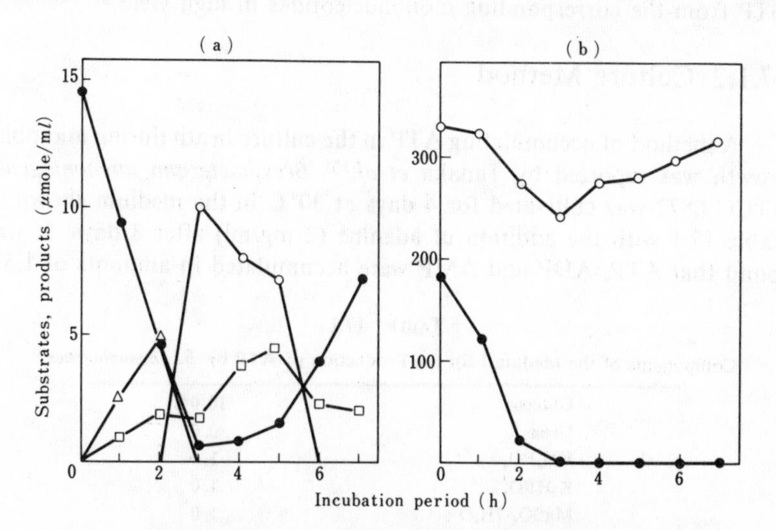

Fig. 17.1. Phosphorylation of AMP by ground bakers' yeast.[7] Reaction mixture (μmoles/ml): AMP 15, glucose 167, potassium phosphate buffer (pH 7.0) 333, and ground bakers' yeast cells (50 mg). Reaction conditions: 37°C, stationary culture.
(a) ●, AMP; ○, ATP; □, ADP; △, adenosine. (b) ○, Pi; ●, glucose.

proceeds at a concentration of 1/3 M but not at 1/9 M or 2/3 M, and that with ground yeast proceeds at 1/4 M. The mechanism of ATP formation is thought to be substrate-level phosphorylation of AMP or adenosine via ADP, using the energy of glycolysis. The essential features of this method are presumably the inhibition of phosphatase by a high concentration of Pi and the release of the inhibited fermentation in high salt concentrations by AMP.

Tanaka and Hironaka[8] modified this method to permit easier cell preparation, and adding higher concentrations of the substrate. They obtained 15.7 g of ATP·Na_2 from 10 g of adenosine in 1.2 liter of the reaction mixture after treatment of the yeast cells with a surfactant, Cation S, and the addition of $MnCl_2$.

Screening to find an enzyme source other than bakers' and *sake* yeasts has been tried by Kawai *et al.*[9] They searched for AMP-phosphorylation activity among 59 strains of 17 genera of yeasts and selected *Debaryomyces nilssonii* as an ATP producer. This yeast could phosphorylate 80–100% of AMP (10 μmoles/ml) to ATP. Watanabe and Takeda[10] found that a hydrocarbon-assimilating yeast, *Candida* sp., possessed the ability to form ATP from AMP in high yield; the activity was markedly affected by the culture conditions for cell production.

This method was also applied to the production of CTP, UTP and GTP from the corresponding mononucleotides in high yield.[11–14]

17.1.2 Culture Method

A method of accumulating ATP in the culture broth during microbial growth was reported by Tanaka *et al.*[15] *Brevibacterium ammoniagenes* ATCC 6872 was cultivated for 4 days at 30° C in the medium shown in Table 17.1 with the addition of adenine (2 mg/ml) after 3 days. It was found that ATP, ADP and AMP were accumulated in amounts of 1.57,

TABLE 17.1

Components of the Medium† for the Production of ATP by *B. ammoniagenes*[15]

Glucose	10.0%
Urea	0.6
KH_2PO_4	1.0
K_2HPO_4	1.0
$MgSO_4 \cdot 7H_2O$	1.0
Yeast extract	1.0
$CaCl_2$	0.01
Biotin	30 μg/l

† pH 7.4.

1.59 and 2.16 mg/ml of the culture filtrate, respectively. GTP, GDP and GMP were similarly accumulated on cultivation with the addition of guanine. The mechanism of phosphorylation is thought to be the further phosphorylation of AMP from adenine, since the accumulation of AMP has been demonstrated with adenine-supplemented cultures of the same organism.

17.2 COENZYME A

Since coenzyme A (CoA) was discovered as a "coenzyme for acetylation" by Lipmann,[17] it has been found that the coenzyme plays an important role in various types of metabolic reactions. Its structural components are 3′-AMP esterified to pantetheine, which is a condensation product of pantothenic acid (PaA) and cysteine (CysSH), through a pyrophosphate bond at the 5′-position of ribose. The biosynthetic pathway

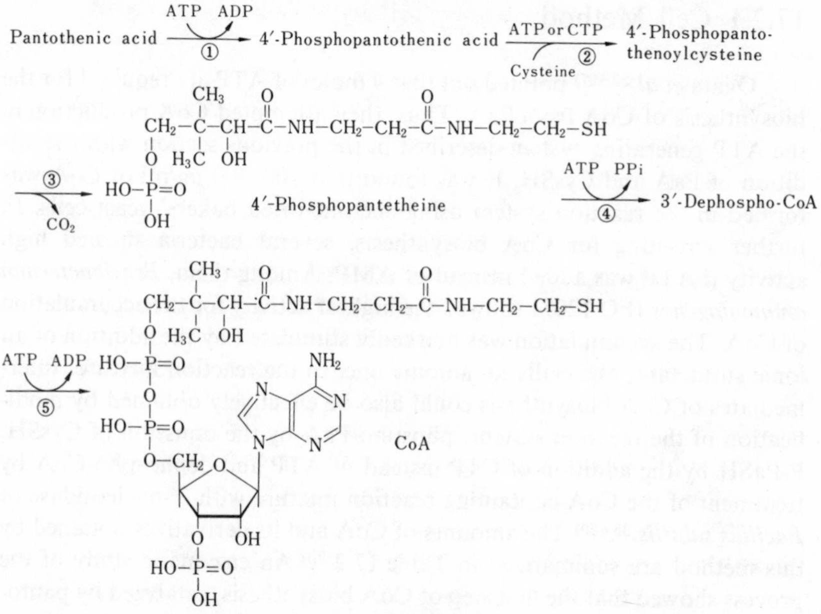

Pantothenic acid —(① ATP → ADP)→ 4′-Phosphopantothenic acid —(② ATP or CTP, Cysteine)→ 4′-Phosphopantothenoylcysteine —(③ −CO_2)→ 4′-Phosphopantetheine —(④ ATP → PPi)→ 3′-Dephospho-CoA —(⑤ ATP → ADP)→ CoA

4′-Phosphopantetheine: $HO-P(=O)(OH)-O-CH_2-C(CH_3)_2-CH(OH)-CO-NH-CH_2-CH_2-CO-NH-CH_2-CH_2-SH$

CoA: $CH_2-C(CH_3)_2-CH(OH)-CO-NH-CH_2-CH_2-CO-NH-CH_2-CH_2-SH$, with the CH_2 linked through $O-P(=O)(OH)-O-P(=O)(OH)-O-CH_2$ to the 5′-position of adenosine (ribose bearing NH_2-adenine), whose 3′-O carries $-P(=O)(OH)_2$ and 2′-OH.

Fig. 17.2. Biosynthetic pathway of CoA.
① Pantothenate kinase (EC 2.7.1.33), ② phosphopantothenoylcysteine synthetase (EC 6.3.2.5), ③ phosphopantothenoylcysteine decarboxylase (EC 4.1.1.36), ④ dephospho-CoA pyrophosphorylase (EC 2.7.7.3), ⑤ dephospho-CoA kinase (EC 2.7.1.24).

was proposed by Brown[18] to be as shown in Fig 17.2, and Abiko *et al.*[19] subsequently confirmed this by a study with rat liver. The recent discovery of acyl carrier protein,[20] which is central to the fatty acid biosynthetic system and contains an intermediate of CoA biosynthesis, 4′-phosphopantetheine (P-PaSH), highlighted the significance of CoA and its derivatives in the biochemical field.

The production methods for CoA can be grouped into chemical and microbial ones. The former, which is represented by the methods of Moffatt and Khorana,[21] and Shimizu *et al.*,[22] involves complex processes and offers rather poor yields. The latter utilizes the CoA-biosynthetic ability of microorganisms to extract CoA from the cells of bakers' or beer yeast. CoA has been produced mainly by the extraction method, which is called the Wisconsin process[23] or the London process.[24] However, these methods gave only 300 mg of CoA of 70% purity from about 25 kg of dried yeast, with a yield of less than 5%.

Recently, a more effective method of preparing CoA using microorganisms has been practically established in Japan.

17.2.1 Cell Method

Ogata *et al.*[25–27] pointed out that 4 moles of ATP are required for the biosynthesis of CoA from PaA. Thus, they attempted CoA production in the ATP-generating system described in the previous section with the addition of PaA and CysSH. It was found that 100–200 μg/ml of CoA was formed in the reaction system using acetone-dried bakers' yeast cells. In further screening for CoA biosynthesis, several bacteria showed high activity if ATP was added instead of AMP. Among them, *Brevibacterium ammoniagenes* IFO 12071 showed the highest activity for the accumulation of CoA. The accumulation was markedly stimulated by the addition of an ionic surfactant, especially an anionic one, to the reaction mixture. Intermediates of CoA biosynthesis could also be effectively obtained by modification of the reaction system: phospho-PaA by the omission of CysSH, P-PaSH by the addition of CTP instead of ATP and dephospho-CoA by treatment of the CoA-containing reaction mixture with 3′-nucleotidase of *Bacillus subtilis*.[28,29] The amounts of CoA and its derivatives obtained by this method are summarized in Table 17.2.[30] An enzymatic study of the process showed that the first step of CoA biosynthesis, catalyzed by pantothenate kinase, was strongly inhibited by the end product, CoA,[31] and that the biosynthetic pathway shown in Fig. 17.2 is distributed in various microorganisms including *B. ammoniagenes*.[32]

The immobilization of *B. ammoniagenes* cells in polyacrylamide gel

TABLE 17.2

Synthesis of CoA and Its Biosynthetic Intermediates by *B. ammoniagenes*[30]

Product	Precursor	Nucleotides	Yield (mg/m*l*)
P-PaA	PaA	ATP	3~4
P-PaSH	PaA+CysSH	ITP+CTP	2~3
P-PaSH	P-PaA+CysSH	CTP	3~4
P-PaSH	PaSH	ITP	2~3
Dephospho-CoA	PaA+CysSH	ATP	1~2
CoA	PaA+CysSH	ATP	2~3

gave a stable cell preparation and yielded 0.5–1.0 mg of CoA per ml of the reaction mixture.[33]

17.2.2 Culture Method

Shimizu *et al.*[34] found that CoA (3–5 mg/ml) could be directly accumulated in the culture filtrate of *B. ammoniagenes* by the addition of PaA, CysSH and a cationic surfactant, CPC, after culture for 3 days in an AMP-containing medium. Fig. 17.3 shows a typical time course, indicating that AMP was initially phosphorylated to ATP and then CoA was formed, accompanied by a decrease of PaA. On the addition of adenine or adeno-

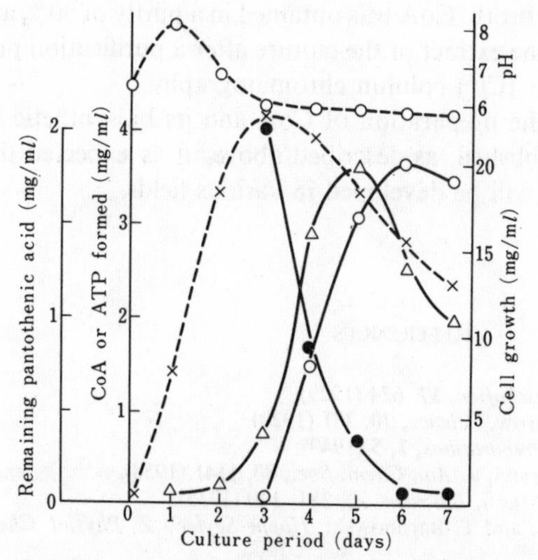

Fig. 17.3. Time course of CoA production by *B. ammoniagenes*.[34] Components of the medium (%): glucose 10, peptone 0.6, yeast extract 0.5, K_2HPO_4 2.0, $MgSO_4 \cdot 7H_2O$ 1.0, biotin 30 μg/ml, AMP 0.2, PaA 0.2, CysSH 0.2, CPC 0.1. —○—, CoA; —●—, PaA; —△—, ATP; ---×---, cell growth; ---○---, pH.

sine instead of AMP, CoA accumulation was also observed. The purification of CoA in the disulfide form, which was formed from CoA in the reaction mixture due to the vigorous shaking during this culture method, was carried out by Duolite S-30, active charcoal and Dowex 1 (Cl^-) column chromatographies.[35]

Nishimura *et al.*[36] screened microorganisms having CoA accumulating activity with the addition of PaA, CysSH and adenine to a 2-day culture and selected *Sarcina lutea* IAM 1099 as the best CoA producer. The strain accumulated 0.6 mg of CoA per ml of culture broth under optimal conditions. Subsequently, a new method for the purification of CoA was devised using affinity chromatography with Sepharose 6B covalently linked to a protein, which was prepared from the cell extract of *S. lutea* and had a specific affinity for CoA.[37] By this method, CoA was obtained in a purity of 92% from 5% pure CoA in 94% yield.

17.2.3 Extraction Method

Kuno *et al.*[38] found that the CoA content of *n*-paraffin assimilating microorganisms was generally superior, and attempted to extract CoA from such cells. *Pseudomonas alkanolytica*, which was selected as a representative strain as regards CoA content and growth rate, formed 80 mg of CoA per liter of culture broth. CoA was obtained in a purity of 90% and at about 13% yield from the extract of the culture after a purification procedure involving Dowex 1 (Cl^-) column chromatography.

Since a method for the preparation of CoA and its biosynthetic intermediates has been established, as described above, it is expected that new applications of CoA will be developed in various fields.

References

1. K. Lohmann, *Naturwissenschaften*, **17**, 624 (1929).
2. C. H. Fiske and Y. Subbarow, *Science*, **70**, 381 (1929).
3. G. A. LePage, *Biochem. Preparations*, **1**, 5 (1949).
4. M. Smith and H. G. Khorana, *J. Am. Chem. Soc.*, **80**, 1141 (1958).
5. C. Lutwak-Mann and T. Mann, *Biochem. Z.*, **281**, 140 (1935).
6. P. Ostern, T. Terszakowec and T. Baronowski, *Hoppe-Seyler's Z. Physiol. Chem.*, **251**, 258 (1938).
7. T. Tochikura, M. Kuwahara, S. Yagi, H. Okamoto, Y. Tominaga, T. Kano and K. Ogata, *J. Ferment. Technol.*, **45**, 511 (1967).
8. A. Tanaka and J. Hironaka, *Agr. Biol. Chem.*, **36**, 867 (1972).
9. H. Kawai, F. Sako and K. Endo, *Amino Acid and Nucleic Acid*, no. 30, 62 (1974).
10. S. Watanabe and I. Takeda, *ibid.*, no. 24, 14 (1971).

11. N. Kitajima, S. Watanabe and I. Takeda, *J. Ferment. Technol.*, **48**, 753 (1970).
12. S. Shirota, S. Watanabe and I. Takeda, *Agr. Biol. Chem.*, **35**, 325 (1971).
13. K. Ogata and K. Kawaguchi, *Hakkokogaku Zasshi* (Japanese), **50**, 46 (1972).
14. T. Tochikura, Y. Kariya, T. Yano, T. Tachiki and A. Kimura, *Amino Acid and Nucleic Acid*, no. 29, 59 (1974).
15. H. Tanaka, Z. Sato, K. Nakayama and S. Kinoshita, *Agr. Biol. Chem.*, **32**, 721 (1968).
16. T. Nara, M. Misawa and S. Kinoshita, *ibid.*, **32**, 561 (1968).
17. F. Lipmann, *J. Biol. Chem.*, **160**, 173 (1945).
18. G. M. Brown, *ibid.*, **234**, 370 (1959).
19. Y. Abiko, T. Suzuki and M. Shimizu, *J. Biochem.* (Tokyo), **61**, 309 (1967).
20. S. J. Wakil, E. L. Pugh and F. Sauer, *Proc. Natl. Acad. Sci. U.S.A.*, **52**, 106 (1964).
21. J. G. Moffatt and H. G. Khorana, *J. Am. Chem. Soc.*, **83**, 663 (1961).
22. M. Shimizu, O. Nagase, S. Okada, H. Hosokawa, H. Tagawa, Y. Abiko and T. Suzuki, *Chem. Pharm. Bull.*, **13**, 655 (1967).
23. H. Beinert, R. W. von Korff, D. E. Green, D. A. Buyske, R. E. Handschumacher, H. Higgins and F. M. Strong, *J. Biol. Chem.*, **200**, 385 (1953).
24. M. C. Reece, M. B. Donald and E. M. Crook, *J. Biochem. Microbiol. Technol. Eng.*, **1**, 217 (1959).
25. K. Ogata, S. Shimizu and Y. Tani, *Agr. Biol. Chem.*, **36**, 84 (1972).
26. K. Ogata, Y. Tani, S. Shimizu and K. Uno, *ibid.*, **36**, 93 (1972).
27. S. Shimizu, T. Tani and K. Ogata, *ibid.*, **36**, 370 (1972).
28. S. Shimizu, S. Satsuma, K. Kubo, Y. Tani and K. Ogata, *ibid.*, **37**, 857 (1973).
29. S. Shimizu, K. Kubo, S. Satsuma, Y. Tani and K. Ogata, *J. Ferment. Technol.*, **52**, 114 (1974).
30. S. Shimizu, Y. Tani and K. Ogata, *Amino Acid and Nucleic Acid*, no. 30, 30 (1974).
31. S. Shimizu, K. Kubo, Y. Tani, and K. Ogata, *Agr. Biol. Chem.*, **37**, 2863 (1973).
32. S. Shimizu, K. Kubo, H. Morioka, Y. Tani and K. Ogata, *ibid.*, **38**, 1015 (1974).
33. S. Shimizu, H. Morioka, Y. Tani and K. Ogata, *J. Ferment. Technol.*, **53**, 77 (1975).
34. S. Shimizu, K. Miyata, Y. Tani and K. Ogata, *Agr. Biol. Chem.*, **37**, 607 (1973).
35. S. Shimizu, K. Miyata, Y. Tani and K. Ogata, *ibid.*, **37**, 615 (1973).
36. N. Nishimura, T. Shibatani, T. Kakimoto and I. Chibata, *Appl. Microbiol.*, **28**, 117 (1974).
37. Y. Matuo, T. Tosa and I. Chibata, *Biochim. Biophys. Acta*, **338**, 520 (1974).
38. M. Kuno, M. Kikuchi, Y. Nakao and S. Yamatodani, *Agr. Biol. Chem.*, **37**, 313 (1973).

17.3 NAD

Since the initial isolation and identification of pyridine nucleotides as coenzymes, great interest has centered around their roles in the oxidation-reduction systems of living cells. Recent discoveries showed that the ADP-ribose moiety of NAD is incorporated into acid-insoluble material by liver nuclear enzyme, producing poly ADP-ribose. It was later shown that an enzyme from *Escherichia coli* catalyzed the NAD-dependent repair of an interrupted strand in a DNA duplex by the formation of phosphodiester bonds.

There have been some reports on the pyridine coenzyme levels of living cells. Among microorganisms, Kitahara and Obayashi[1] found an exceedingly high level of NAD in *Lactobacillus plantarum* and isolated as much as 5 mg of NAD from 1 g of dried cells. Following this discovery, the levels of pyridine coenzymes in various lactic acid bacteria were studied in comparison with those in other microorganisms and most lactic acid bacteria were found to contain a large amount of NAD.[2] However, bakers' yeast has so far been used as the source of pyridine coenzymes since large amounts of the cells are readily available. There have been many reports concerning the extraction and purification of NAD from bakers' yeast. Hayano *et al.*[3] examined a strain which contained the highest level of NAD in yeasts and investigated the changes of NAD level with the growth conditions. They found that the pyridine coenzyme content in yeast cells was 1–2 mg per g of dried cells and was higher in cells grown aerobically than in those grown statically. Using *Candida utilis*, which has a higher pyridine coenzyme level than other yeasts, the level of NAD in cells was found to reach approximately 10 mg per g of dried cells.[4]

Later, an attempt to increase the NAD content of microorganisms was made by adding NAD precursors. Sakai *et al.*[5] found that the NAD content of *Saccharomyces carlsbergensis* reached 30 mg per g of dried cells when adenine and nicotinamide were added to the medium after culture for 24 h. Further, the NAD content was found to reach as much as 42 mg per g of dried cells on adding adenine, nicotinamide and a fatty acid such as sodium myristate. They also found that NAD in cells leaked into the culture broth on adding an anionic surfactant such as sodium lauryl sulfate. Under optimal culture conditions about 0.75 mg/ml of NAD accumulated in the culture broth after culture for 72 h.

On the other hand, Nakayama *et al.*[6] succeeded in the fermentative production of NAD by utilizing the salvage synthesizing ability of bacteria. They found that a large amount of NAD accumulated in the culture broth when *Brevibacterium ammoniagenes* was incubated in a medium containing adenine and nicotinic acid or nicotinamide. When precursors were added to the medium after culture for 48 h, the amount of NAD accumulated reached 2–3 mg/ml at 120 h. This method seems to be very useful for the production of NAD. They discussed the biosynthetic pathway of NAD in *B. ammoniagenes.* A large amount of nicotinate ribonucleotide, together with a small amount of NAD accumulated when nicotinic acid or nicotinamide was added. However, a large amount of NAD accumulated when either nicotinic acid or nicotinamide was added together with adenine. These facts supported the operation of a salvage biosynthetic pathway via nicotinate ribonucleotide in *B. ammoniagenes,* as in *E. coli* (Fig. 17.4).

In parallel with the development of the fermentative production of

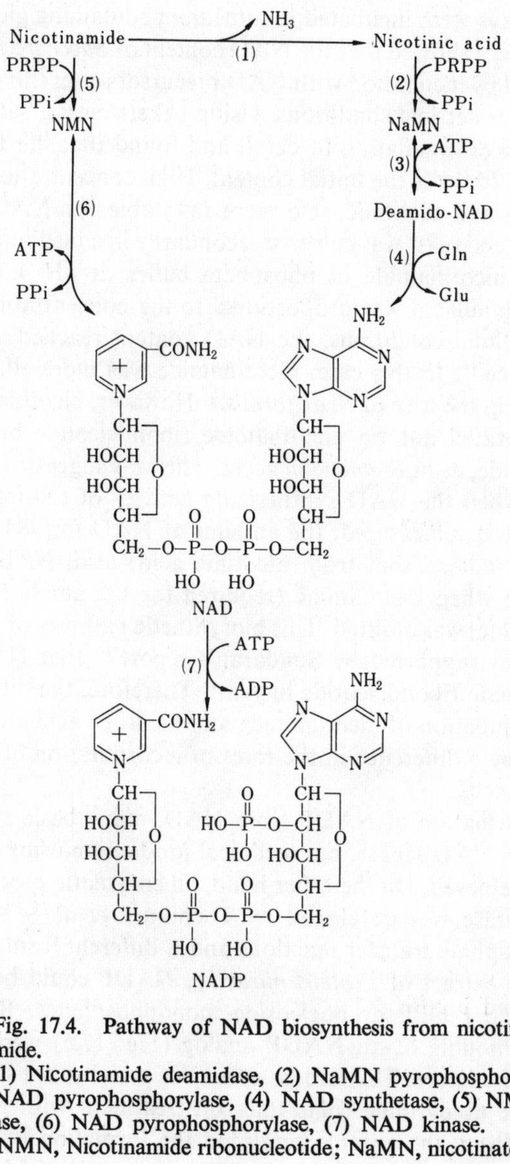

Nicotinamide —(1), NH_3→ Nicotinic acid

Nicotinamide + PRPP —(5)→ NMN + PPi

Nicotinic acid + PRPP —(2)→ NaMN + PPi

NaMN + ATP —(3)→ Deamido-NAD + PPi

Deamido-NAD + Gln —(4)→ NAD + Glu

NMN + ATP —(6)→ NAD + PPi

NAD + ATP —(7)→ NADP + ADP

NAD: NH_2, N, N, N, N, $CONH_2$, +, N, CH, HOCH, HOCH, CH, O, O, O, O, $CH_2-O-P-O-P-O-CH_2$, HO, HO, HOCH, HOCH, CH, CH

NADP: NH_2, N, N, N, N, $CONH_2$, +, N, CH, HOCH, HOCH, CH, O, O, O, O, $HO-P-O-CH$, HO, HOCH, CH, $CH_2-O-P-O-P-O-CH_2$, HO, HO

Fig. 17.4. Pathway of NAD biosynthesis from nicotinic acid or nicotinamide.
(1) Nicotinamide deamidase, (2) NaMN pyrophosphorylase, (3) deamido-NAD pyrophosphorylase, (4) NAD synthetase, (5) NMN pyrophosphorylase, (6) NAD pyrophosphorylase, (7) NAD kinase.
NMN, Nicotinamide ribonucleotide; NaMN, nicotinate ribonucleotide.

NAD, an attempt to increase the NAD content in yeast cells by secondary culture without growth was made. Takakura and Oda[7] found that the NAD content of pressed bakers' yeast increased 4 times when yeast cells stored at 4° C for 15 days were incubated in a medium containing glucose and nicotinamide. Takei[8] showed that the NAD content of *Saccharomyces oviformis* was increased by incubation with NAD precursors after this yeast had been cultured under aerobic conditions. Using bakers' yeast, Sakai *et al.*[9] investigated NAD accumulation in detail and found that the NAD content reached about 20 times the initial content. High concentrations of adenine, nicotinamide and nicotinic acid were favorable for NAD accumulation. When bakers' yeast was cultured secondarily in a medium containing adenine and nicotinamide in phosphate buffer at pH 4.5, the amount of NAD accumulated was proportional to the concentration of precursors. Under optimal conditions, the NAD content reached about 12 mg per g of dried cells. In this case, nicotinamide was more effective than nicotinic acid, as in the case of *S. oviformis*.[8] However, nicotinamide seemed to be incorporated not via nicotinamide ribonucleotide but via nicotinate ribonucleotide, as in *B. ammoniagenes*. This was suggested from the following facts: when the NAD synthesizing activity of cell-free extracts of bakers' yeast was measured, the amount of NAD formed from nicotinamide was the same as that from nicotinic acid; also, NAD formation was negligible when L-glutamine (required for the amidation of nicotinate ribonucleotide) was omitted. This biosynthetic pathway of NAD from nicotinamide was supported by Sundaran's report[10] that NAD is synthesized via nicotinate ribonucleotide in yeast. Therefore, the different effects on NAD accumulation of nicotinamide and nicotinic acid are considered to be caused by a difference in the rates of incorporation of these substances into yeast cells.

Regarding the formation of NADP from NAD, many basic studies have been reported on NAD kinase, but technical production using NAD kinase has not been achieved. On the other hand, an enzymatic procedure using phosphotransferase was developed by Tochikura *et al.*[11,12] This enzyme catalyzes phosphate transfer reactions and is different from NAD kinase. With cell-free extract of *Proteus mirabilis*, NADP could be produced from NAD and *p*-NPP or nucleoside monophosphates. Besides NADP, significant amounts of an NADP analog (Fig. 17.5) were also formed in the reaction mixture. This had no coenzyme activity, but showed the same fluorescence as pyridine coenzymes on treatment with methyl ethyl ketone. In addition, the enzyme catalyzed the formation of NAD diphosphate (Fig. 17.6) when NADP or NADP analog was used instead of NAD.[13] These NAD derivatives showed no coenzymatic or inhibitory activities with most dehydrogenases except for glyceraldehyde 3-phosphate

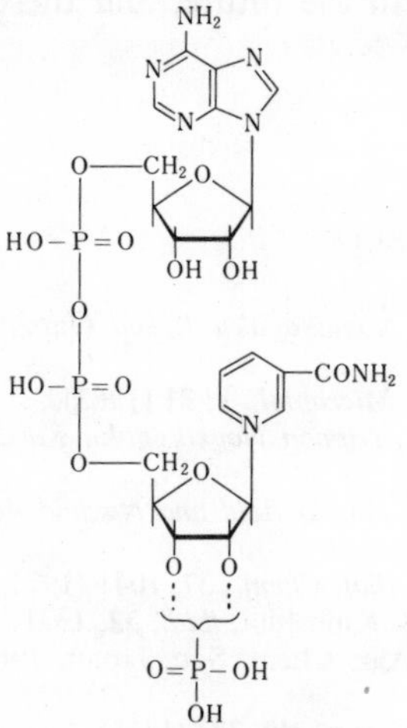

Fig. 17.5. Structure of the NADP analog (see text).

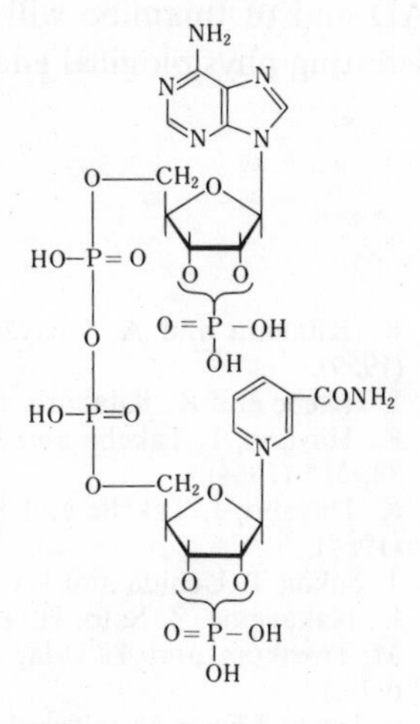

Fig. 17.6. Structure of NAD diphosphate.

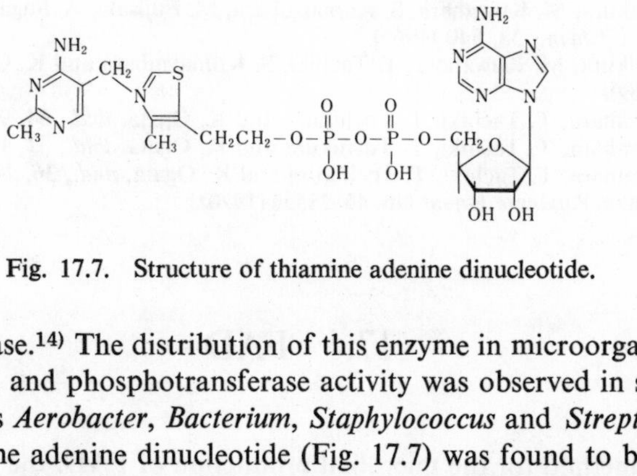

Fig. 17.7. Structure of thiamine adenine dinucleotide.

dehydrogenase.[14] The distribution of this enzyme in microorganisms was investigated, and phosphotransferase activity was observed in several genera such as *Aerobacter*, *Bacterium*, *Staphylococcus* and *Streptomyces*.[15]

Thiamine adenine dinucleotide (Fig. 17.7) was found to be accumulated in the culture broth of *B. ammoniagenes*, an NAD accumulator,[16] though it is different from pyridine nucleotide derivatives. When thiamine and adenine were added to the medium at 72 h after inoculation, 0.92 mg/ml of thiamine adenine dinucleotide was accumulated on incubation for a further 24 h. This compound has no physiological activity.

The development of pyridine nucleotide production using microorganisms has resulted in improved supply of pyridine coenzymes. For the production of compounds having a complicated structure or various isomers, biosynthesis is preferable to organic synthesis, and production using microorganisms is expected to become increasingly important. There is also a possibility that new nucleotide derivatives such as derivatives of

NAD and of thiamine will be found in the future, and these may have interesting physiological effects.

References

1. K. Kitahara and A. Obayashi, *Nippon Nogeikagaku Kaishi* (Japanese), **33**, 497 (1959).
2. I. Takebe and K. Kitahara, *J. Gen. Appl. Microbiol.*, **9**, 31 (1963).
3. K. Hayano, I. Takebe and K. Kitahara, *Nippon Nogeikagaku Kaishi* (Japanese), **38**, 515 (1964).
4. K. Hayano, I. Takebe and K. Kitahara, *Amino Acid and Nucleic Acid*, no. 9, 37 (1964).
5. T. Sakai, T. Uchida and I. Chibata, *Agr. Biol. Chem.*, **37**, 1041 (1973).
6. K. Nakayama, Z. Sato, H. Tanaka and S. Kinoshita, *ibid.*, **32**, 1331 (1968).
7. M. Takakura and T. Oda, Ann. Mtg. Agr. Chem. Soc. Japan, 1966. Abstracts, p 78.
8. S. Takei, *Nippon Nogeikagaku Kaishi* (Japanese), **40**, 283 (1966).
9. T. Sakai, T. Uchida and I. Chibata, *Agr. Biol. Chem.*, **37**, 1049 (1973).
10. T. K. Sundaran, K. V. Rajagopalan, C. V. Pichappa and P. S. Sarma, *Biochem. J.*, **77**, 145 (1960).
11. T. Tochikura, M. Kuwahara, S. Komatsubara, M. Fujisaki, A. Suga and K. Ogata, *Agr. Biol. Chem.*, **33**, 840 (1969).
12. T. Tochikura, M. Kuwahara, T. Tachiki, S. Komatsubara and K. Ogata, *ibid.*, **33**, 848 (1969).
13. M. Kuwahara, T. Tachiki, T. Tochikura and K. Ogata, *ibid.*, **34**, 983 (1970).
14. M. Kuwahara, T. Tachiki, T. Tochikura and K. Ogata, *ibid.*, **32**, 177 (1971).
15. M. Kuwahara, T. Tachiki, T. Tochikura and K. Ogata, *ibid.*, **36**, 745 (1972).
16. M. Tanaka, *Japanese Patent* No. 45–28556 (1970).

17.4 FAD

As a method for the microbial production of FAD, the extraction of yeast cells was first undertaken in 1942, but the purification of FAD was very difficult as the FAD content was very low. In 1954, Kuwada *et al.*[1,2] found culture conditions under which *Eremothecium ashbyii* accumulated a large amount of FAD in the cells and this accelerated the technical production of FAD. Subsequently, the procedure for the extraction and purification of FAD from *E. ashbyii* was improved by Izumiya and Kobayashi,[3] and the production of FAD was industrialized. However, there were many problems; the amounts of FAD accumulated in cells were limited, and the isolation of FAD from the cells was still difficult. Thus, studies were begun on fermentations directly accumulating FAD in the medium.

FAD is biosynthesized as shown in Fig. 17.8, so FAD fermentations adding cheap precursors such as adenine, riboflavin or FMN were investigated. Nakamura *et al.* found that a large amount of FAD was formed in the supernatant of the culture broth when *Brevibacterium ammoniagenes* was incubated in a medium supplemented with adenine.[4] In this fermentation, FAD accumulation was affected by the amount and rate of addition of precursors such as FMN and adenine. Although FAD production was observed on a single addition of FMN itself, simultaneous addition of a suitable amount of adenine increased the accumulation. Under optimum conditions, 1.18 mg/ml of FAD was accumulated. In this case, 86% of the total flavin was FAD. The high conversion rate from FMN to FAD was thought to be a result of the fact that the activity of FMN adenyltransferase in *B. ammoniagenes* is higher than in *E. ashbyii* and *Escherichia coli*.[5]

Sakai *et al.* also attempted to select microorganisms such as bacteria, actinomycetes, yeasts and fungi for the purpose of producing FAD from FMN, adenine or AMP.[6] Of the microorganisms studied, *B. ammoniagenes*, *Sarcina lutea*, and *S. aurantiaca* showed relatively high FAD producing activity. Suitable culture conditions for FAD production were investigated using *S. lutea*, which accumulated the largest amount of FAD. This strain required thiamine, and FAD accumulation increased signifi-

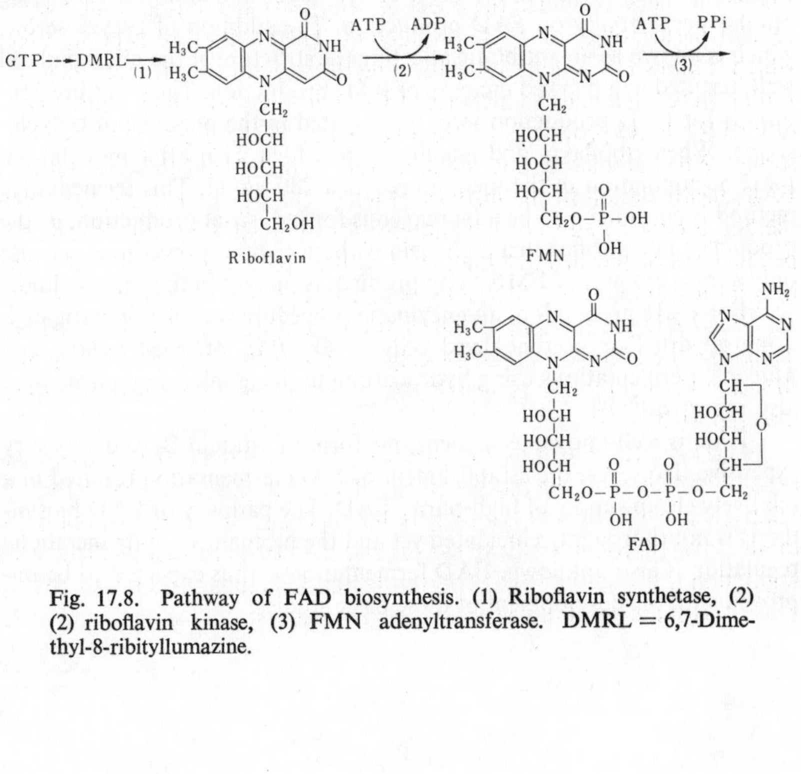

GTP---→DMRL →(1) Riboflavin; ATP → ADP (2) → FMN; ATP → PPi (3) → FAD

Fig. 17.8. Pathway of FAD biosynthesis. (1) Riboflavin synthetase, (2) (2) riboflavin kinase, (3) FMN adenyltransferase. DMRL = 6,7-Dimethyl-8-ribityllumazine.

cantly with sucrose as a carbon source, and peptone, meat extract and yeast extract as nitrogen sources. Ammonium acetate had a marked effect on FAD production; the addition of 2% ammonium acetate increased the amount of FAD about three times. Even under optimal culture conditions, however, the amount of FAD was less than 200 μg/ml. Thus, improvement of the microorganism was attempted to obtain an efficient mutant for the production of FAD.[7] As this microorganism had a high level of adenosine deaminase activity, adenosine formed from adenine was converted to inosine by adenosine deaminase, and ATP was not supplied for FAD production. It was thought that an adenosine deaminase deficient mutant might be suitable for efficient conversion of adenine to ATP. *S. lutea* was treated with *N*-methyl-*N'*-nitro-*N*-nitrosoguanidine and the adenosine deaminase lacking mutant AD-1 was derived from it. The mutant accumulated about 1 mg/ml of FAD (more than three times that produced by the parent strain).

Watanabe *et al.* reported FAD fermentation using added riboflavin instead of FMN as a precursor. Riboflavin is cheaper and easier to separate from FAD than FMN.[8] The amount of FAD accumulated was only about 200 μg/ml when riboflavin was added to the medium instead of FMN. This result was presumed to be due to a smaller permeability of riboflavin. They examined the effect of detergents and antibiotics affecting cellular permeability on FAD production. The addition of D-cycloserine, which is known as an antibiotic affecting the structure of the microbial cell wall, resulted in a marked increase of FAD production. Thus, culture conditions for FAD production were investigated in the presence of D-cycloserine. When riboflavin and adenine were added 24 h after inoculation, FAD accumulation in the medium reached 780 μg/ml. This fermentative method is considered to be advantageous for industrial production, as the product can be purified in a high yield without tedious procedures because only a trace amount of FMN, a by-product, is present in the culture fluid.

For FAD production, an enzymatic procedure was also investigated, using air-dried or acetone-dried cells of *Candida*, *Microbacterium*, and *Mucor*.[9] Fermentations using hydrocarbon utilizing microorganisms were also reported.[10–12]

FAD is well-known as a coenzyme form of vitamin B_2 and was very expensive. However, the establishment of FAD fermentation resulted in a relatively cheap supply of high-purity FAD. The pathway of FAD biosynthesis is not thoroughly elucidated yet and the mechanism of its metabolic regulation is also unknown. FAD fermentation is thus expected to be improved with the development of these basic studies.

References

1. S. Kuwada, T. Masuda, Y. Sawa and M. Asai, Abstracts 73rd Vitamin B Res. Committee Japan: *Bitamin* (Japanese), **7**, 990 (1954).
2. S. Kuwada, *ibid.*, (Japanese), **9**, 453 (1955).
3. M. Izumiya and T. Kobayashi, *ibid.*, **26**, 93 (1962).
4. N. Nakamura, S. Takasawa and M. Tanaka, Abstracts 20th Ann. Mtg. Vitamin Soc. Japan: *ibid.* (Japanese), **37**, 622 (1968), *Japanese Patent* No. 45–22518 (1970).
5. N. Nakamura and M. Tanaka, *Nippon Nogeikagaku Kaishi* (Japanese), **42**, 281 (1968).
6. T. Sakai, T. Watanabe and I. Chibata, *Agr. Biol. Chem.*, **37**, 849 (1973).
7. T. Sakai, T. Watanabe and I. Chibata, *ibid.*, **37**, 2855 (1973).
8. T. Watanabe, T. Uchida, J. Kato and I. Chibata, *Appl. Microbiol.*, **27**, 531 (1974).
9. K. Takeda, S. Watanabe and M. Shirota, *Japanese Patent* No. 44–30113 (1969).
10. Y. Nishimura and H. Iizuka, Ann. Mtg. Agr. Chem. Soc. Japan, 1971. Abstracts, p. 140.
11. Y. Nishimura and H. Iizuka, 4th Int. Fermentation Symp. 1972. Abstracts, p. 199.
12. Y. Nishimura and H. Iizuka, Ann. Mtg. Agr. Chem. Soc. Japan, 1973. Abstracts, p. 22.

CHAPTER **18**

Production of Sugar Nucleotides*

Although mononucleotides have long been known as constituents of nucleic acids, only adenine nucleotides such as ATP, ADP and AMP have been found as free compounds in cells and tissues of organisms. However, in the course of studies on galactose metabolism in *Saccharomyces fragilis*, Leloir and his co-workers (1949–50)[1–3] found a nucleoside diphosphate

* Tatsurokuro Tochikura, Faculty of Agriculture, Kyoto University.

sugar, UDP-glucose (UDPG), which contained UMP and glucose-1-P bound so as to form a pyrophosphate bridge, and which acted as a coenzyme in the transformation of galactose-1-P to glucose-1-P. At the same time (1949–1952), Park and his collaborators[4,5] detected the accumulation of an acid-labile UDP-acetylmuramyl peptide in *Staphylococcus aureus* inhibited with penicillin.

A remarkable variety of nucleoside diphosphate sugars has been found in cells and tissues of various organisms. These include UDPG, UDP-galactose (UDP-Gal), UDP-*N*-acetylglucosamine (UDP-GlcNAc), UDP-*N*-acetylgalactosamine (UDP-GalNAc), UDP-pentoses, GDP-mannose (GDP-Man), GDP-glucose (GDPG), ADP-glucose (ADPG), dTDP-glucose (dTDPG), dTDP-*N*-acetylglucosamine (dTDP-GlcNAc) and others. In general, these nucleoside diphosphate sugar-related compounds are briefly termed sugar nucleotides. Sugar nucleotides play very important roles in cell metabolism, since many of the pathways of sugar metabolism and of the biosyntheses of oligo- and polysaccharides involve various sugar nucleotides as key intermediates. The sugar nucleotides are mainly involved in reactions of the following types:

(1) Isomerization of sugars

$$\text{UDPG} \rightleftharpoons \text{UDP-Gal}$$

$$\text{UDP-GlcNAc} \rightleftharpoons \text{UDP-GalNAc}$$

(2) Oxidation of sugars and decarboxylation of uronic acids (UA)

$$\text{UDPG} \xrightarrow{\text{NAD}} \text{UDP-GlcUA}$$

$$\text{UDP-GlcUA} \longrightarrow \text{UDP-xylose} + CO_2$$

(3) Biosynthesis of oligo- and polysaccharides by glycosyl transfer

$$\text{UDPG} + \text{glucose-6-P} \longrightarrow \text{trehalose phosphate} + \text{UDP}$$

$$(\text{UDPG})_m + \underset{\text{primer}}{(\text{glucose})_n} \longrightarrow \underset{\text{glycogen}}{(\text{Glucose})_{m+n}} + (\text{UDP})_m$$

The sugar nucleotides have been prepared by extraction from cells and tissues of various organisms, by enzymatic synthesis, and by chemical synthesis. However, these methods have been used for only small amounts of the sugar nucleotides. In Japan, purine and pyrimidine nucleoside 5′-monophosphates have been produced on an industrial scale by the microbial degradation of yeast RNA and by fermentation processes.

This chapter deals with the fermentative production and mechanism of formation of some sugar nucleotides, including UDPG (Fig. 18.1),[6–8) UDP-Gal,[9,10] UDP-GlcNAc,[11,12] GDP-Man (Fig. 18.2)[7,8,13–15] and GDPG[16] and their enzymatic conversion reactions such as enzymatic syntheses of UDP-glucuronic acid (UDP-GlcUA)[18] from UDPG and of UDP-GalNAc[22] from UDP-GlcNAc.

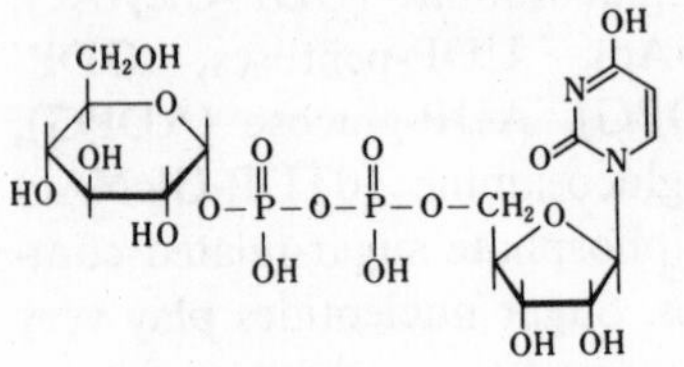

Fig. 18.1. Uridine diphosphate glucose (UDPG).

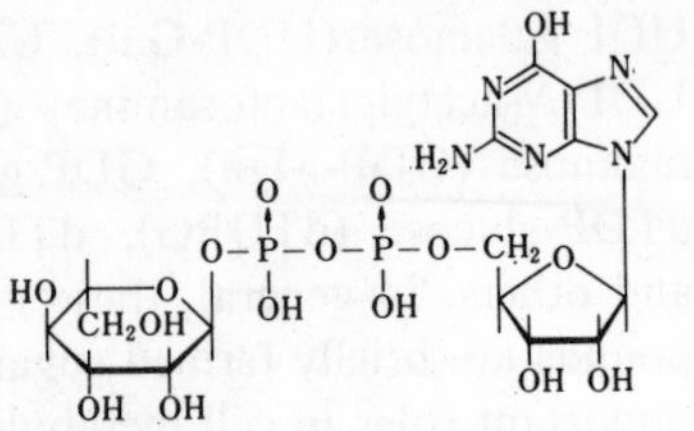

Fig. 18.2. Guanosine diphosphate mannose (GDP-Man).

18.1 Production of Sugar Nucleotides by Fermentation

18.1.1 Distribution of Sugar Nucleotide Producing Activity in Yeasts[17]

A. UDPG producers

The distribution of UDPG producing activity is very wide in yeasts. Potent producers include *Torulopsis candida*, *Candida parapsilosus*, *C. krusei*, bakers' yeast, brewers' yeast and others.

B. UDP-Gal producers

The distribution of UDP-Gal producing activity is very narrow. Potent producers include *T. candida*, *T. sphaerica* and *C. intermedia*.

C. GDP-Man producers

This activity is widely distributed in yeasts. Potent producers include

Hansenula jadinii, *H. saturnus*, *H. suaveolens*, *C. utilis*, *T. xylinus*, *Debaryomyces subglobosus*, bakers' yeast, brewers' yeast and others.

D. UDP-GlcNAc producers

The distribution of activity is relatively narrow. Potent producers include bakers' yeast, *D. subglobosus*, *D. globosus*, *D. cantavellii* and others.

Table 18.1 lists suitable yeast strains and substrates for sugar nucleotide production, together with the fermentation yields.

TABLE 18.1

Sugar Nucleotide Production by Yeasts[17]

Substrates	Sugar nucleotides produced	Yeast†	Yield of sugar nucleotides (mg/ml)
UMP+glucose+Pi	UDPG	bakers' yeast	6.7
		T. candida	10.0
UMP+galactose (lactose) +Pi	UDP-Gal	*T. candida*	51.5
UMP+glucose (fructose) +glucosamine+Pi	UDP-GlcNAc	bakers' yeast	4.1
		Debaryomyces sp.	6.0
GMP+glucose+Pi	GDP-Man	bakers' yeast	5.4
		Hansenula jadinii	10.8

† *Yeasts*: pressed bakers' yeast was purchased from Oriental Yeast Co. Ltd. Some yeasts were grown on a medium containing 50 g of glucose or lactose, 5 g of peptone, 2 g of yeast extract, 2 g of KH_2PO_4, 2 g of $(NH_4)_2HPO_4$ and 1 g of $MgSO_4 \cdot 7H_2O$ in 1000 ml of tap water.

Preparation of dried cells: after incubation at 28°C for 48 h on a reciprocal shaker, the cells were harvested and washed three times with tap water. The washed cells were dried with an electric fan at room temperature for 12–24 h. The air-dried cells were further dried over P_2O_5 *in vacuo*. These cells, containing 4–8% water, were used throughout this work.

18.1.2 Time Course of Fermentation and Effect of Phosphate Concentration

Fig. 18.3 shows the effect of Pi concentration on UDPG production by ground cells of bakers' yeast.[6] UMP and glucose were nearly consumed in the early stage of fermentation with loss of Pi. UTP began to appear with the disappearance of added UMP, and on further incubation after maximal accumulation of UTP, UDPG production took place in parallel with the disappearance of the triphosphate. The maximal yield of UDPG was about 50–60% based on added UMP. The data also show that the rate of UTP degradation and UDPG production were markedly influenced by the Pi concentration. The incubation time required for the maximal accu-

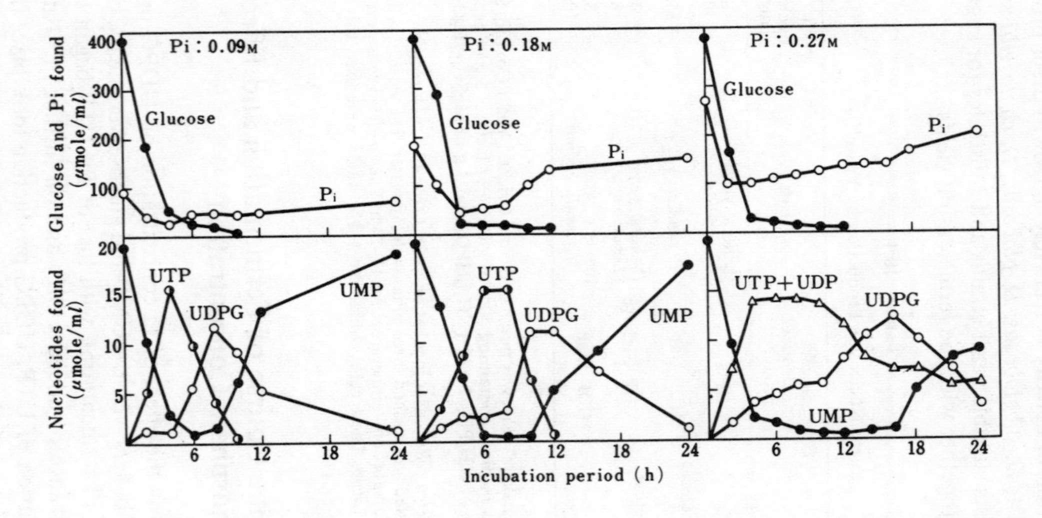

Fig. 18.3. Typical time courses of UDPG production by ground bakers' yeast.[6] The reactions were carried out under the standard conditions except that the Pi concentration was varied as indicated.

mulation of UDPG increased as the Pi concentration increased from 0.09 to 0.27 M, and at very high concentrations of Pi, the consumption of glucose and production of UDPG were completely inhibited. Similar results were obtained in the production of other sugar nucleotides.

18.1.3 Effect of Sugars

Various sugars, including glucose, fructose, sucrose, galactose and lactose, not only supply the energy required for the biosynthesis of sugar nucleotides but also provide the sugar moiety of the molecules. In general, when a high concentration of sugars is used as an energy source, the phosphorylation of nucleotide donors such as UMP and GMP is retarded, resulting in retardation of the overall synthesis of sugar nucleotides. The relationship between kinds of sugar nucleotides produced and substrate sugars can be summarized as follows:

Substrates for UDPG and GDP-Man production	energy source	various sugars
	glycosyl donors (glucose and mannose)	various sugars
Substrates for UDP-Gal production	energy source	various sugars
	glycosyl donors (galactose)	galactose, lactose
Substrates for UDP-GlcNAc production	energy source	various sugars
	glycosyl donor (amino sugar)	glucosamine

18.1.4 Other Factors Affecting Sugar Nucleotide Production

The optimal concentration of dried yeast as an enzyme source varied with the concentrations of sugars as energy sources, glycosyl and nucleotide donors, and Pi; optimal concentrations of cells are usually 100–150 mg/ml. Although Pi was usually used at an initial pH of about 6.5 to 7.0 for the production of GDP-Man and UDP-GlcNAc, good yields were obtained at an initial pH of 7.5–8.5.

Addition of $MgSO_4$ or $MgCl_2$ (10–20 μmole/ml) was effective in promoting sugar nucleotide production. The rate of fermentation with shaking (or agitation) was higher than that without shaking. Aeration was indispensable for sugar nucleotide production by aerobic yeasts such as *T. candida*. Table 18.2 summarizes suitable conditions for sugar nucleotide production.

TABLE 18.2

Fermentation Conditions for Sugar Nucleotide Production

Sugar nucleotides produced	Composition of media and culture conditions
UDPG[6]	UMP 20 μmole/ml, glucose 400 μmole/ml, potassium phosphate (pH 7.0) 180 μmole/ml, $MgSO_4$ 12 μmole/ml, ground bakers' yeast 60 mg/ml; 28℃, 8-12 h with shaking.
UDP-Gal[9,10,17]	UMP 20-100 μmole/ml, galactose 200-400 μmole/ml, potassium phosphate(pH 7.0) 200 μmole/ml, $MgSO_4$ 12 μmole/ml, dried cells of lactose- or galactose-grown *T. candida* 100 mg/ml; 28℃, 8-24 h with shaking.
GDP-Man[13-15]	GMP 20 μmole/ml, glucose 500-800 μmole/ml, potassium phosphate(pH 7.5-8.5) 300-400 μmole/ml, $MgSO_4$ 20 μmole/ml, dried bakers' yeast or *H. jadinii* 100 mg/ml; 28℃, 8-12 h with shaking.
UDP-GlcNAc[11,12]	UMP 20 μmole/ml, glucose 70 μmole/ml, potassium phosphate(pH 7.5-8.5) 170 μmole/ml, $MgSO_4$ 5 μmole/ml, glucosamine 20 μmole/ml, dried bakers' yeast 100 mg/ml; 28℃, 8-12 h with shaking.
GDPG[16]	GTP 20 μmole/ml, glucose-1-P 100 μmole/ml, potassium phosphate(pH 7.0) 270 μmole/ml, $MgSO_4$ 20 μmole/ml, dried cells of *Streptomyces* sp. (AKU 280) 100 mg/ml; 28℃, 8 h with shaking.

18.2 PREPARATION OF SUGAR NUCLEOTIDE-RELATED SUBSTANCES BY ENZYMATIC CONVERSION

It is known that the glycosyl group, including glucose, mannose and *N*-acetylglucosamine, bound to a nucleotide component of a sugar nucleotide can be converted to the corresponding homolog. As typical examples of this enzymatic conversion, the preparations of UDP-GlcUA and UDP-GalNAc will be briefly described here.

18.2.1 Preparation of UDP-Glucuronic Acid

UDP-GlcUA is an important compound which can serve as a detoxifying factor in liver through the formation of alcoholic, phenolic and amine glucuronides. This UDP derivative has been shown to be formed from UDPG by the action of a specific NAD-linked enzyme, UDPG dehydrogease (EC 1.1.1.22), which has been found in some bacteria and yeasts. Recently, Sugimori *et al.*[18] reported that *Bacillus licheniformis* was most useful for the enzymatic preparation of UDP-GlcUA from UDPG, since the organism had very low activity for the decomposition of both UDPG and UDP-GlcUA. A disintegrated cell preparation (sonicate) can easily oxidize UDPG to UDP-GlcUA with high conversion yield. UDPG dehydrogenase of *B. licheniformis* requires NAD for full activity; the optimum pH is 9.5 and the optimum temperature is 40° C. Under aerobic conditions, the NADH oxidizing system of this organism couples with the dehydrogenation of UDPG and facilitates the conversion of UDPG to UDP-GlcUA. In the presence of 0.5 mM NAD, 6 mM UDPG was dehydrogenated to UDP-GlcUA at a conversion rate of 70% in 20 h at 30°C, using a sonicate of a strain of *B. licheniformis*.

18.2.2 Preparation of UDP-*N*-Acetylgalactosamine

Glaser[19)] found an enzyme, UDP-GlcNAc 4-epimerase (EC 5.1.3.7), in *B. subtilis*, which catalyzed the enzymatic conversion of UDP-GlcNAc to UDP-GalNAc. Strominger *et al.*[20)] reported a method for preparing UDP-GalNAc from a naturally occurring mixture of UDP-GlcNAc and UDP-GalNAc contained in hen oviduct, using UDP-GlcNAc pyrophosphorylase (EC 2.7.7.23) from *Staphylococcus aureus*.[21)] Kawai *et al.*[22)] have prepared UDP-GalNAc from UDP-GlcNAc by conversion with a crude epimerase from *B. subtilis*, followed by degradation of residual UDP-GlcNAc with staphylococcal pyrophosphorylase, which is inactive against UDP-GalNAc. In a typical experiment, 0.465 mmole of UDP-GlcNAc was incubated with 5 mmole of Tris buffer (pH 8.0), 0.5 mmole of $MgCl_2$, 0.05 mmole of EDTA and a dialyzed, sonicated extract (770 mg protein) of *B. subtilis* in a total volume of 50 ml. After incubation at 37° C for 3 h, about 35% of UDP-GlcNAc was converted to UDP-GalNAc.

18.3 Mechanism of Sugar Nucleotide Production

18.3.1 Mechanism of Production and Coupled Fermentation with Double or Multiple Energy Transfer

The fermentative synthesis of sugar nucleotides appears to be a relatively complex and refined energy requiring process. The following points are noteworthy. In the first step of biosynthesis, glycosyl donors such as glucose, galactose and glucosamine which are added to the fermentation media are phosphorylated to sugar-6-phosphate (glucose-6-P, mannose-6-P) or sugar-1-phosphate (galactose-1-P) with the consumption of ATP, which is generated by glycolysis in yeasts. Subsequently, sugar-6-phosphates are converted to sugar-1-phosphates. In the production of UDP-

GlcNAc, glucosamine is converted to *N*-acetyl-glucosamine-1-P with consumption of ATP and acetyl-CoA. On the other hand, UMP or GMP, which is used as a nucleotide donor for the synthesis of sugar nucleotides, is phosphorylated to UTP or GTP by coupling with glycolysis. In the next step, the sugar-1-phosphate reacts with UTP or GTP to form sugar nucleotide and pyrophosphate. Finally, the pyrophosphate is broken down by pyrophosphatase, and the biosynthesis of sugar nucleotide proceeds to completion. Accordingly, the fermentative production of sugar nucleotides is accomplished by coupling endergonic reactions in the synthetic process with glycolysis. This type of fermentative production is a practical application of coupled reactions involving double or multiple energy transfer systems. One energy transfer system is an ATP-ADP reaction responsible for the phosphorylation of the glycosyl donors in the synthesis of sugar nucleotides, and the other is an ATP-ADP reaction responsible for the phosphorylation of the nucleotide donors, i.e. for the conversion of UMP or GMP to UDP or GDP and then to UTP or GTP. Thus, this method of microbial production may be termed coupled fermentation with double or multiple energy transfer. The mechanisms of fermentative production of several sugar nucleotides are outlined in Fig.18.4, according to the principle of coupling with energy transfer.

Table 18.3 shows reactions in the biosynthesis of UDPG, which illustrate the pattern involved in the biosynthesis of most sugar nucleotides. For each mole of UDPG formed, three moles of ATP are consumed. One mole of ATP is consumed in the phosphorylation of glucose to glucose-6-P and two moles of ATP in the phosphorylation of UMP to UTP.

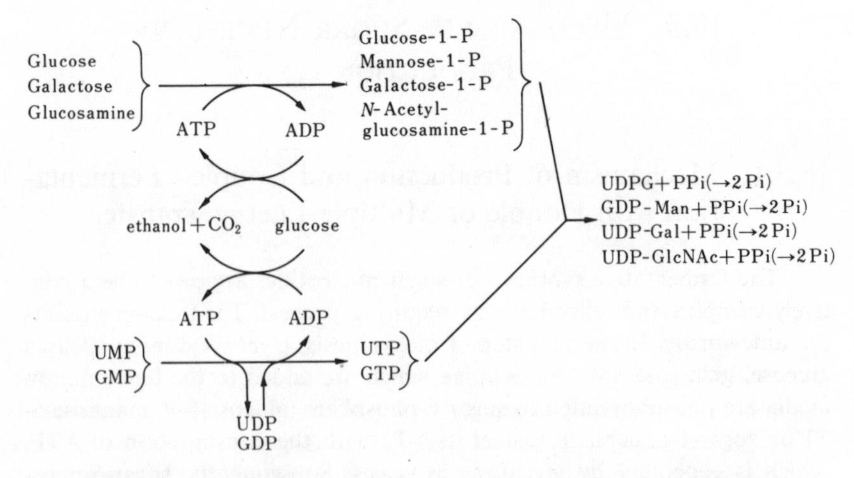

Fig. 18.4. Mechanism of sugar nucleotide production and fermentation coupled with energy transfer.

TABLE 18.3

Reactions Involved in the Biosynthesis of UDPG

ATP generating system:	
	1.5 glucose + 3ADP + 3Pi $\longrightarrow$ $3C_2H_5OH + 3CO_2$ + 3ATP
ATP consuming system:	
	glucose + ATP $\longrightarrow$ glucose-6-P + ADP (hexokinase, EC 2.7.1.1)
	glucose-6-P $\longrightarrow$ glucose-1-P (phosphoglucomutase, EC 2.7.5.1)
	UMP + ATP $\longrightarrow$ UDP + ADP (UMP kinase, EC 2.7.4.4)
	UDP + ATP $\longrightarrow$ UTP + ADP (UDP kinase, EC 2.7.4.6)
	glucose-1-P + UTP $\longrightarrow$ UDPG + PPi (UDPG pyrophosphorylase, EC 2.7.7.9)
	PPi $\longrightarrow$ 2Pi (inorganic pyrophosphatase, EC 3.6.1.1)
Overall:	2.5 glucose + UMP + Pi $\longrightarrow$ UDPG + $3C_2H_2OH + 3CO_2$

18.3.2 Mechanism of UDP-Galactose Production

The biosynthetic route of UDP-Gal is closely related to galactose metabolism in many organisms and can be summarized as follows:

(1) Galactose + ATP $\longrightarrow$ galactose-1-P + ADP
(galactokinase, EC 2.7.1.6)

(2) Galactose-1-P + UDPG $\rightleftharpoons$ UDP-Gal + glucose-1-P
(Gal-1-P uridylyltransferase, EC 2.7.7.12)

(3) UDP-Gal $\rightleftharpoons$ UDPG
(UDP-Gal epimerase, EC 5.1.3.2)

A catalytic amount of UDPG, required for the initiation of reaction (2), is supplied by reaction (4).

(4) Glucose-1-P + UTP $\rightleftharpoons$ UDPG + PPi
(UDPG pyrophosphorylase EC 2.7.7.9)

Another pathway leading to the direct formation of UDP-Gal is:

(5) Galactose-1-P + UTP $\rightleftharpoons$ UDP-Gal + PPi
(UDP-Gal pyrophosphorylase)

The authors[10,24)] investigated enzyme activities involved in galactose metabolism with cell-free extract and ammonium sulfate fractions from *T. candida* grown on a galactose or lactose medium. Remarkable activities of galactokinase, galactose-1-P uridylyltransferase, UDPG pyrophosphorylase and UDP-Gal 4-epimerase were demonstrated, whereas UDP-Gal pyrophosphorylase activity was weak. It is of interest to confirm whether UDP-Gal production proceeds through a coupling of reaction (2) with reaction (4), where UDPG or glucose-1-P acts as a catalyst. Either UDPG or glucose-1-P was added to a cell-free system containing excess galactose-1-P or UTP. It was found that the addition of catalytic amounts of UDPG or glucose-1-P stimulated UDP-Gal production greatly. The effect of added glucose-1-P is shown in Fig. 18.5. At the early stage of incubation, relatively larger amounts of UDPG accumulated with the addition of glucose-1-P prior to the formation of UDP-Gal. When, however, the concentration of UDPG reached a certain level, it began to decrease due to consumption for UDP-Gal synthesis. Moreover, the cell-free extract was demonstrated to contain a strong inorganic pyrophosphatase activity.

(6) $$PPi + H_2O \longrightarrow 2Pi$$
(inorganic pyrophosphatase, EC 3.6.1.1)

This hydrolysis of inorganic pyrophosphate proceeds with a large negative free energy change. Therefore, the pyrophosphatase plays a very significant role in UDP-Gal synthesis, i.e. the enzyme operates as a chemical device in cells to ensure that UDP-Gal synthesis proceeds essentially to completion. From all the results obtained, it appears that UDP-Gal pro-

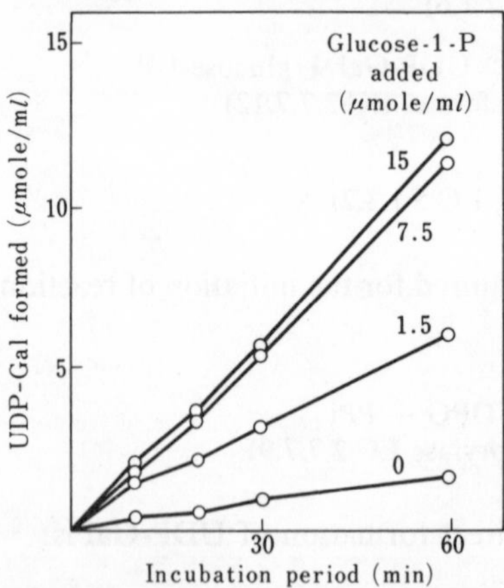

Fig. 18.5. Effect of glucose-1-P on UDP-Gal production from galactose-1-P and UTP by a cell-free extract of *T. candida*.[24] The reaction mixture contained 20 μmole of glactose-1-P, 30 μmole of UTP, 300 μmole of potassium phosphate buffer (pH 7.2), 10 μmole of $MgCl_2$, 10.7 mg of protein and the indicated amounts of glucose-1-P in a total volume of 1.5 ml; incubation at 30°C.

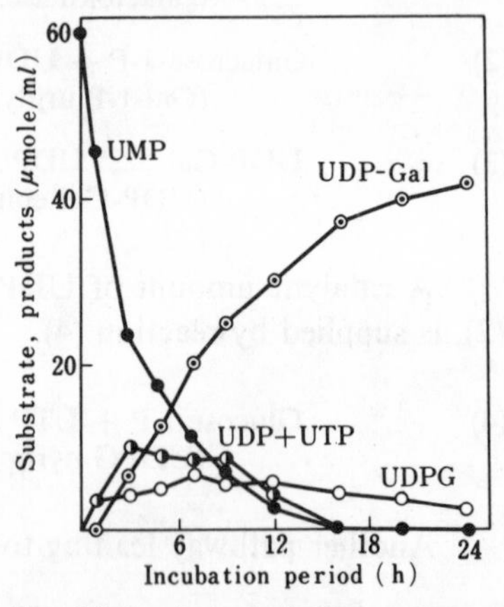

Fig. 18.6. A typical time course of UDP-Gal production by *T. candida*.[17] The fermentation was carried out under the standard conditions.

duction in *T. candida* proceeds through a coupled reaction catalyzed by UDPG pyrophosphorylase and galactose-1-P uridylyltransferase, where the hydrolysis of pyrophosphate produced causes the coupled reaction to shift overwhelmingly toward UDP-Gal synthesis.

Another striking characteristic of UDP-Gal production by *T. candida* is that UDP-Gal, once accumulated, is very stable under fermentative conditions. Fig. 18.6 shows a typical time course of UDP-Gal production by dried cells of *T. candida*. UDP-Gal was produced in 70% yield (42 μmole/ml) with respect to UMP. Furthermore, as shown in Table 18.4, it was possible to raise the UMP concentration to 140 μmole/ml and to convert 80% of the nucleotide to UDP-hexoses (UDP-Gal 51.5 mg/ml and UDPG 11.3 mg/ml). No marked degradation of UDP-Gal was observed even on prolonged incubation. The amount of UDPG present as a contaminant in UDP-Gal was reduced by long incubation. If UDP-Gal 4-epimerase is actually operating under fermentation conditions, the amount of UDPG produced should exceed that of UDP-Gal after equilibrium is reached,

TABLE 18.4

Effect of UMP Concentration on UDP-hexose Production by Lactose-grown *T. candida*[10]†

UMP added (μmole/ml)	Reaction time (h)	Total UDP-hexose produced (μmole/ml)	UDPG produced (μmole/ml)	UDP-Gal produced (μmole/ml)
	8	17.2	—	—
20	16	17.5	—	—
	24	17.1	3.3	13.7
	8	35.4	—	—
60	16	54.4	—	—
	24	51.0	7.8	43.0
	8	46.5	—	—
100	16	86.8	—	—
	24	84.7	10.4	74.2
	8	28.5	—	—
140	16	51.6	—	—
	24	112.1	20.2	91.7

† The reaction was carried out under the standard conditions except that the UMP concentration was varied as indicated; incubation at 28°C with shaking.

since the epimerase reaction lies far toward UDPG; the equilibrium corresponds to 75% UDPG and 25% UDP-Gal in *Saccharomyces fragilis*. As this is not the case for UDP-Gal production by dried cells of *T. candida*, it is thought that the UDP-Gal 4-epimerase activity of the yeast may be inhibited for some reason under fermentation conditions. It has recently been reported that UDP-Gal 4-epimerase obtained from *S. fragilis* is strongly inactivated by UMP, provided that some specific sugars such as D-galactose, D-fucose or L-arabinose are present.[25,26] The author *et al.*[24] therefore investigated whether the epimerase activity of *T. candida* was inhibited in the presence of UMP and galactose. As shown in Fig. 18.7, about 65% of UDP-Gal is rapidly converted to UDPG by dried cells in the control system, which indicates the presence of a high UDP-Gal 4-epimerase activity in the yeast. However, the conversion of UDP-Gal to UDPG was strongly inhibited in the presence of both UMP and galactose. This inhibition was also seen with a partially purified preparation of the enzyme and

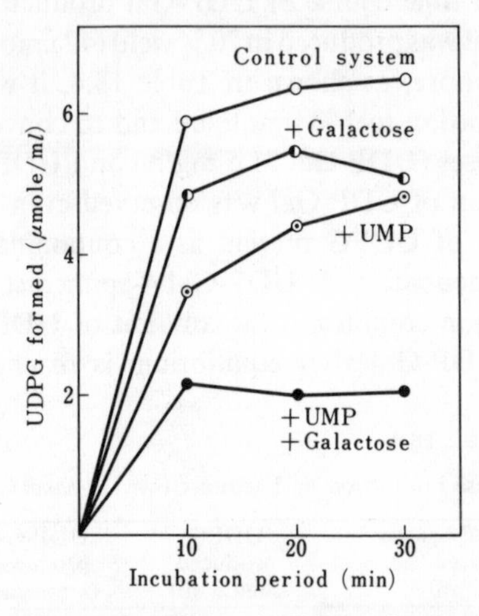

Fig. 18.7. Effect of UMP and galactose on the conversion of UDP-Gal to UDPG by dried cells of *T. candida*.[24] The control system contained 9.7 μmole of UDP-Gal, 200 μmole of potassium phosphate buffer (pH 7.0) and 100 mg of dried cells in a total volume of 1 ml. To the control was added either UMP (20 μmole/ml) or galactose (200 μmole/ml), or a combination of the two; incubation at 30°C with shaking.

the concentration of UMP added was more critical than that of galactose for inhibition; 5×10^{-3} M UMP was enough to cause 75% inhibition, provided that 1×10^{-1} M galactose was present.[27] Therefore, it appears that the epimerase of *T. candida* may be unable to operate under conditions where both UMP and galactose are present. This means that metabolic stability of certain fermentation products in cultures may be brought about by the presence of substrates and, in general, this may give interesting pointers for screening factors to increase fermentation yields.

18.3.3 Effect of the Water Content of Dried Yeast Cells on Sugar Nucleotide Production[10]

One of the most important factors affecting the fermentative production of sugar nucleotides is the water content of the dried yeast cells used as enzyme sources. For example, in UDP-Gal production by lactose-grown cells of *T. candida*,[10] the amount of UDP-Gal produced increased on using dried cells containing less than 20% water, a maximal yield being attained with dried cells containing 5–8% water (Fig.18.8). No UDP-Gal was produced with dried cells containing more than 20% water. Similar results were obtained in UDPG production by incubation of lactose-grown cells with glucose as a glycosyl donor. On the other hand, in UDPG production by glucose-grown cells, remarkable yields of UDPG were obtained even with dried cells containing about 20% water. The above results suggest that a difference in the carbon source in the growth media for the yeast might affect the structure of the cell membrane, changing its permeability to reaction substrates or the leakage level of enzymes responsible for sugar nucleotide synthesis, and that these effects are enhanced by decreasing the

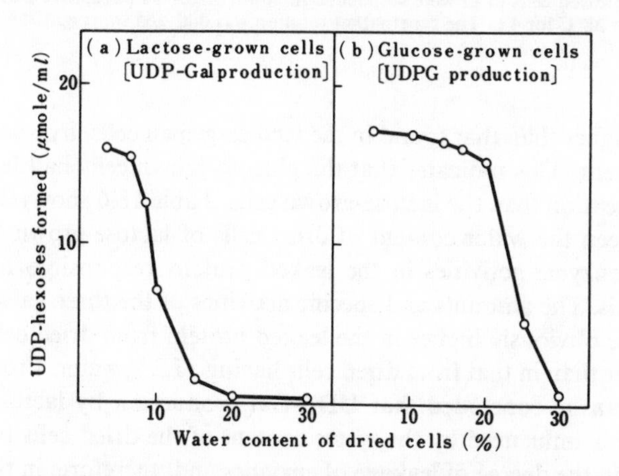

Fig. 18.8. Effect of water content of glucose- or lactose-grown dried cells on UDP-hexose production by *T. candida*.[10] Reaction mixtures (per ml) contained 20 μmole of UMP, 12 μmole of $MgSO_4$, 200 μmole of Pi (pH 7.0) and the following additions: in UDP-Gal production, 200 μmole of galactose and 100 mg of the lactose-grown dried cells; in UDPG production, 200 μmole of glucose and 100 mg of the glucose-grown dried cells. Incubation at 28°C for 8 h with shaking.

water content of the yeast cells. As shown in Table 18.5, the amounts of protein which leaked from glucose-grown and lactose-grown cells with different water contents parallel the sugar nucleotide production activities of the corresponding dried cells. A small amount of protein leaked from lactose-grown cells containing 18.1% water, and the cells did not produce UDP-Gal; however, marked leakage of protein took place from glucose-grown cells containing 17.1% water, and the cells produced a large amount of UDPG. Moreover, the percentage of dead cells found in the glucose-

TABLE 18.5

UDP-hexose Producing Activities, Amounts of Leaked Protein and Percentages of Dead Cells in Dried Cell Preparations of *T. candida*[10]

Cell preparations and their water content (%)	UDP-hexose formed[†1] (μmole/ml/8h)		Leaked protein[†2] (mg/ml)	Dead cells found (%)
	UDPG	UDP-Gal		
Glucose-grown cells (6.0)	18.5	—	3.73	88
Glucose-grown cells (17.1)	17.5	—	2.75	84
Lactose-grown cells (7.8)	—	18.5	2.55	55
Lactose-grown cells (18.1)	—	0	0.46	32

†1 The reactions were carried out under the standard conditions (Table 18.2).
†2 Leaked protein: dried cells (5 g) were suspended in 50 ml of 0.1 M phosphate buffer (pH 7.0) and shaken at 28°C for 4 h. The supernatant solution was dialyzed overnight at 4°C a d assayed.

grown cells was higher than that found in the lactose-grown cells, irrespective of water content. This indicates that the glucose-grown cells had less resistance to desiccation than the lactose-grown cells. Table 18.6 shows the relationship between the water content of dried cells of lactose-grown *T. candida* and the enzyme activities in the leaked protein responsible for UDP-Gal synthesis. The amounts and specific activities of the three galactose enzymes were obviously higher in the leaked protein from dried cells having 5.9% water than in that from dried cells having 17.2% water. From these results, it can be concluded that UDP-Gal production by lactose-grown *T. candida* is influenced by the water content of the dried cells because of changes in the degree of leakage of enzymes and, therefore, in the degree of permeability of substrates, as a result of the degree of drying of the yeast.

TABLE 18.6

Effect of the Water Content of Dried Cells of Lactose-grown *T. candida* on Enzyme Activities[10]

Water content (%)	UDPG pyrophosphorylase		Galactose-1-P uridylyltransferase		Galactokinase	
	TU[†1]	SA[†2]	TU[†1]	SA[†2]	TU[†1]	SA[†2]
5.9	1528	10.4	948	6.5	846	5.8
17.2	270	6.1	113	2.5	197	4.4

†1 TU = total units. †2 SA = specific activity (units/mg protein). The cell-free extract (leaked protein) was prepared by the method described in Table 18.5.

18.4 FUTURE PROSPECTS

In many instances, the immediate precursors in syntheses of homo- and heteropolysaccharides are sugar nucleotides (nucleoside diphosphate sugars), though a nucleoside monophosphate sugar, CMP-*N*-acetylneuraminic acid, is involved in the biosynthesis of colominic acid. However, the mechanisms involved in biosyntheses of many complex heteropolysaccharides and the participating enzymes are still little known. Accordingly, it is important to prepare large amounts of specific sugar nucleotides which are key substrates and to study the mechanisms of enzymatic synthesis, the sites of synthesis in the cells and the physiological functions of complex polysaccharides.

In the application of isotopic tracers to studies on carbohydrate metabolism, great difficulty arises in the organic syntheses of sugar nucleotides labeled with isotopes; in general, organic syntheses are complicated and not suitable for the preparation of tracer samples.

On the other hand, isotopically labelled compounds can be synthesized without great difficulty by using the microbial processes described in this section. Furthermore, one might expect that sugar nucleotides could be useful for chemical modification or transformation by the introduction of their glycosyl groups into a variety of useful molecules.

REFERENCES

1. R. Caputto, L. F. Leloir, R. E. Trucco, C. E. Cardini and A. C. Paladini, *J. Biol. Chem.*, **179**, 497 (1949).
2. C. E. Cardini, A. C. Paladini, R. Caputto and L. F. Leloir, *Nature*, **165**, 191 (1950).
3. R. Caputto, L. F. Leloir, C. E. Cardini and A. C. Paladini, *J. Biol. Chem.*, **184**, 333 (1950).
4. J. T. Park and M. J. Johnson, *ibid.*, **179**, 585 (1949).
5. J. T. Park, *ibid.*, **194**, 877, 885, 897 (1952).
6. T. Tochikura, H. Kawai, S. Tobe, K. Kawaguchi, M. Osugi and K. Ogata, *J. Ferment. Technol.*, **46**, 957 (1968).
7. S. Shirota, S. Watanabe and I. Takeda, *Agr. Biol. Chem.*, **35**, 325 (1971).
8. S. Watanabe and I. Takeda, *ibid.*, **36**, 2265 (1972).
9. T. Tochikura, K. Kawaguchi, H. Kawai, Y. Mugibayashi and K. Ogata, *J. Ferment. Technol.*, **46**, 970 (1968).
10. T. Tochikura, Y. Mugibayashi, H. Kawai, K. Kawaguchi and K. Ogata, *Amino Acid and Nucleic Acid*, no. 22, 144 (1970).
11. T. Tochikura, H. Kawai and T. Gotan, *Agr. Biol. Chem.*, **35**, 163 (1971).

12. T. Tochikura, H. Kawai and T. Gotan, *Amino Acid and Nucleic Acid*, no. 22, 38, 46 (1970).
13. T. Tochikura, K. Kawaguchi, T. Kano and K. Ogata, *J. Ferment. Technol.*, **47**, 564 (1969).
14. K. Kawaguchi, K. Ogata and T. Tochikura, *Agr. Biol. Chem.*, **34**, 908 (1970).
15. K. Kawaguchi, S. Tanida, Y. Mugibayashi, Y. Tani and K. Ogata, *J. Ferment. Technol.*, **49**, 195 (1971).
16. K. Kawaguchi, S. Tanida, K. Matsuda, Y. Tani and K. Ogata, *Agr. Biol. Chem.*, **37**, 75 (1973).
17. T. Tochikura, H. Kawai, K. Kawaguchi, Y. Mugibayashi and K. Ogata, Proc. 4th Int. Fermentation Symp.: *Fermentation Technology Today*, p. 463, 1972.
18. Y. Tazuke, Y. Tsukada and T. Sugimori, 4th Int. Fermentation Symp., 1972. Abstracts, p. 193.
19. L. Glaser, *J. Biol. Chem.*, **234**, 2801 (1959).
20. J. L. Strominger and M. S. Smith, *ibid.*, **234**, 1828 (1959).
21. J. L. Strominger and M. S. Smith, *ibid.*, **234**, 1822 (1959).
22. H. Kawai, K. Yamamoto, A. Kimura and T. Tochikura, *Agr. Biol. Chem.*, **37**, 1741 (1973).
23. T. Tochikura, Y. Kariya, T. Yano, T. Tachiki and A. Kimura, *Amino Acid and Nucleic Acid*, no. 29, 59 (1974).
24. H. Kawai and T. Tochikura, *Agr. Biol. Chem.*, **35**, 1578 (1971).
25. A. U. Bertland and H. M. Kalckar, *Proc. Natl. Acad. Sci. U.S.A.*, **61**, 629 (1968).
26. H. M. Kalckar, A. U. Bertland and B. Bugge, *ibid.*, **65**, 1113 (1970).
27. H. Kawai and T. Tochikura, *Agr. Biol. Chem.*, **35**, 1587 (1971).

CHAPTER **19**

Production of Other Nucleoside Derivatives

19.1 POLYNUCLEOTIDES

19.1.1 Historical Background

It has been reported that polynucleotides can be synthesized by the following enzymes:

*1 Masahiko YONEDA, Takeda Chemical Industries Ltd.
*2 Sawao MURAO, Faculty of Agriculture, Osaka Metropolitan University.
*3 Mitsuru SHIBUKAWA, Asahi Chemical Industry Co. Ltd.

(1) Polynucleotide phosphorylase (PNPase)[1)]
(2) RNA polymerase[2)]
(3) DNA polymerase[3)]
(4) Terminal nucleotidyl transferase[4)]
(5) Others

These enzymes are widely distributed in animals, plants and microorganisms and are related to the synthesis and degradation of nucleic acids.

In 1955, Grunburg-Manago *et al.*[5)] succeeded in the synthesis of polynucleotides from nucleoside diphosphates by the action of PNPase obtained from extracts of *Azotobacter agilis*. Subsequently, many laboratories have studied the distribution of the enzyme, and its applications. Many reports have been presented on the enzymes from *E. scherichia coli*, *Micrococcus lysodeikticus* and *Azotobacter vinelandii*. Microorganisms which produced PNPase are listed in Table 19.1.

TABLE 19.1

Distribution of PNPase in Microorganisms

Microorganisms	Ref.
Azotobacter agilis	5)
Escherichia coli	6)
Micrococcus lysodeikticus	7)
Alkaligenes faecalis	8)
Bacillus cereus	9)
Eubacterium sarcoginosenum	10)
Clostridium perfringens	11)
Streptococcus faecalis	12)
Pseudomonas aeruginosa	13)
P. aeruginosa R-399	14)
Agrobacterium tumefaciens	15)
Aerobacter aerogenes	14)
Bacillus subtilis	14)
Proteus vulgaris	14)
Serratia marcescens	14)
Xanthomonas begoniae	19)
Achromobacter delmarvae	19)

RNA and DNA polymerases are also widely distributed in microorganisms and have been studied by many laboratories. However, terminal nucleotidyl transferase was found only in animal tissues. This section will deal with PNPase, which is suitable for the synthesis of polynucleotides.

19.1.2 Nature and Measurement of the PNPase Reaction

The reaction may be written as follows:

$$(NMP)^*_n + n\text{Pi} \xrightleftharpoons{Mg^{2+}} n\text{NDP}^*$$

PNPase catalyzes the formation of polynucleotides from NDP and *vice versa* in the presence of inorganic phosphate and Mg^{2+}. This is an equilibrium reaction, and is generally measured by following the formation of ADP^{32} in the presence of ^{32}Pi and poly(A).[16] The activity can also be measured by the addition of a uranyl reagent to the reaction mixture on completion of the reaction, while the viscosity of the reaction mixture has also been used as a measure of activity. For instance, the viscosity increases about 10-fold when poly(U) with a molecular weight of 58,000 to 70,000 is newly synthesized.[17]

PNPase exhibits two types, one requiring and the other not requiring precursors for its action. The former is obtained from *E. coli* and the latter from *M. lysodeikticus*.[18] PNPase generally does not synthesize poly(G) but the *E. coli* enzyme does so. This reaction requires the presence of Mg^{2+}. Recently, Rokugawa *et al.* found that the enzyme obtained from *Achromobacter* sp. requires Mn^{2+} rather than Mg^{2+} in the reaction mixture.[19]

The distribution of this enzyme in cells is not yet clear. However, it was reported that more than 40% of the total PNPase activity is present in the ribosome fraction of *E. coli* and *Pseudomonas aeruginosa*,[20,21] and a major part is bound to the cell membrane in *S. faecalis*.[12]

19.1.3 Purification of PNPase

When using bacterial cells or extracts for polynucleotide synthesis, substrates and polynucleotides are sometimes decomposed by the action of coexisting ribonuclease, phosphodiesterase, phosphatase, etc. Thus, a highly purified enzyme preparation is desirable for the synthesis of polynucleotides of high molecular weight. As an example, the purification method of Singer,[16] using *E. coli*, will be described.

E. coli was grown in 1 liter of the culture medium shown in Table 19.2 for 16 h. The culture was then inoculated into fresh medium (about 2%). When the absorption had increased to 2.2 at 420 nm, the cells were harvested by centrifugation. The cells were triturated with alumina and extracted with 0.02 M Tris buffer (pH 8.0). The extracts thus obtained were fractionated by the addition of ammonium sulfate. Further purification was carried out by protamine treatment and chromatography on DEAE-Sephadex, DEAE-cellulose and Sephadex G-200. Finally, a highly purified enzyme fraction was obtained by sucrose density gradient centrifugation.

* NMP, Nucleoside monophosphate; NDP, nucleoside diphosphate.

TABLE 19.2
Culture Medium for *E. coli*

$(NH_4)_2SO_4$	2 g
Na-K-phosphate buffer(pH 7.4)	100 μ mole
$MgSO_4 \cdot 7H_2O$	0.1 g
Glucose (sterilized separately)	2 g
$FeSO_4 \cdot 7H_2O$	5 mg
Peptone	20 g
Total	1 liter

TABLE 19.3
Summary of Purification Procedure for PNPase

	Protein (mg/ml)	Specific activity	Total units
Crude extract	67,800	0.16	10,865
$(NH_4)_2SO_4$ fraction I	38,800	0.28	10,865
Protamine treatment	12,560	0.37	4,648
$(NH_4)_2SO_4$ fraction II	3,674	1.7	6,338
DEAE-Sephadex	345	16	5,575
DEAE-Cellulose	160	16	2,550
Sephadex G-200	15.4	132	2,038
Sucrose density gradient centrifugation	2.5	199	498

These results are summarized in Table 19.3. This purified enzyme was stable for at least 1 week at pH 8.1 to 9.7, and at 4° C. Generally, the enzyme was stored at −20° C.

The PNPases from *M. lysodeikticus* and *A. vinelandii* were also purified by the same procedure. The purified enzymes were free from ribonuclease, phosphatase, ribonucleoside kinase, etc.

19.1.4 Synthesis of Polynucleotides

The reaction mixture consisted of 150 mM Tris-HCl buffer (pH 8.1), 20 mM NDP, 2 mM $MgCl_2$ and 3 to 7 units of purified enzyme in 1 ml. The mixture was incubated for 3 to 5 h at 37° C. However, 150 mM glycine-NaOH buffer (pH 9.2) with incubation at 50° C was used for the synthesis of polynucleotides containing guanine. In the case of copolynucleotide synthesis, the corresponding NDP's were used at a constant ratio. The base ratio in the polynucleotides was similar to that of NDP's at the beginning of the reaction, but did not remain constant. This problem has not yet been resolved.

After the reaction, the same volume of phenol and 1/10 volume of 5%

Na laurylsulfate were added to the mixture, which was shaken vigorously. The water layer thus obtained was separated and polynucleotides were precipitated by the addition of 2 volumes of ethanol and an equal volume of 1 M Na acetate. The precipitate was dissolved in a small amount of water and dialyzed against distilled water. Purified polynucleotides were stored in a lyophilized state.

Many polynucleotides have been synthesized by these methods: poly (A), poly(G), poly(I), poly(C), poly(U) and poly(X) as homopolynucleotides, and poly(IA), poly(IG), (poly(IU), poly(IC), poly(AC), poly(XC), poly(GC), poly(CU), poly(AU), poly(AX), poly(GU), poly(GA), poly (UX), etc. However, the molecular weights of these polynucleotides and the base ratios in copolymers were not constant.

19.1.5 Physiological Properties of Polynucleotides

Many nucleic acid-related substances, from bases to high polymers, show physiological activity. However, no clear physiological role for polynucleotides has been reported. In 1967, Lampson *et al.*[23] reported that double-stranded RNA obtained from *Penicillium funiculosum* acted as an interferon inducer. Furthermore, double-stranded polynucleotides, poly (I) and poly(C), and poly(I_2G) and poly(C), also appeared to show the same activity.[22] On the other hand, poly(A) yielded a biphasic depressor response.[24] The mechanisms of action still remain unknown.

19.1.6 Concluding Remarks

PNPases are widely distributed in microorganisms. Although the enzyme was investigated for industrial application, production has not been started, probably because the only major demand for polynucleotides is as reagents for research. In addition, this enzyme reaction gives products with differing molecular weights and base ratios, which is not suitable for industrial use. However, it is well known that polynucleotides and nucleic acid play key roles in living cells, and applications may well be developed. Many problems remain in this field, while the progress of related studies is very rapid.

REFERENCES

1. J. N. Davidson and W. E. Cohn, *Progress in Nucleic Acid Research*, **1**, 93 (1963).
2. J. N. Davidson and W. E. Cohn, *ibid.*, **1**, 1 (1963).

3. J. N. Davidson and W. E. Cohn, *ibid.*, **1**, 27 (1963).
4. M. Yoneda and F. J. Bollum, *J. Biol. Chem.*, **240**, 3385 (1965).
5. M. Grunberg-Manago and S. Ochoa, *J. Am. Chem. Soc.*, **77**, 3165 (1955).
6. U. Z. Littauer and A. Kornberg, *J. Biol. Chem.*, **226**, 1077 (1957).
7. R. F. Beers, Jr., *Nature*, **177** (1956).
8. D. O. Brummond, M. Staehelin and S. Ochoa, *J. Biol. Chem.*, **225**, 835 (1957).
9. J. Skoda, J. Kara and Z. Sormova, Proc. 5th Int. Congr. Biochem., p. 81, 1961.
10. J. T. Tildon and J. Szulmajster, *Biochim. Biophys. Acta*, **47**, 199 (1961).
11. M. I. Dolin, *Biochem. Biophys. Res. Comm.*, **6**, 11 (1961).
12. P. McNamara and A. Abrams, *Fed. Proc.*,**20**, 362 (1961).
13. G. A. Strasdine, L. A. Hogg and J. J. Campbell, *Biochim. Biophys. Acta* , **55**, 231 (1962).
14. Y. Katoh, A. Kuninaka and H. Yoshino, *Agr. Biol. Chem.*, **37**, 1537 (1973).
15. A. Vardanis and R. M. Hochster, *Can. J. Biochem. Physiol.*, **39**, 1695 (1961).
16. G. L. Cantoni and D. R. Davies, *Procedures in Nucleic Acid Research*, p. 246, 1964.
17. M. M. Cherepak and R. W. Hansen, *Biotechnol. Bioeng.*, **10**, 350 (1968).
18. R. A. Harvey, C. S. McLaughlin and M. Grunberg-Manago, *Eur. J. Biochem.*, **9**, 50 (1969).
19. K. Rokugawa, Y. Katoh, A. Kuninaka and H. Yoshino, *Agr. Biol. Chem.*, **39**, 1455 (1975).
20. H. E. Wade and S. Lovett, *Biochem. J.*, **81**, 319 (1961).
21. G. A. Strasdine, L. A. Hogg and J. J. Campbell, *Biochim.Biophys. Acta* , **47**, 199 (1962).
22. S. Matsuda, M. Kida, H. Shirafuji, M. Yoneda and H. Yaoi, *Arch. Virusforsch.*, **34**, 105 (1971).
23. G. P. Lampson, A. A. Tytell, A. K. Field, M. M. Nemes and M. R. Hilleman, *Proc. Natl. Acad. Sci. U.S.A.*, **58**, 782 (1967).
24. T. Kokubu, E. Ueda, Y. Yamamura, A. Nagaoka, N. Fukuda, K. Kawazoe, K. Kikuchi, M. Yoneda and M. Kida, *J. Takeda Res. Lab.*,**30**, 580 (1971).

19.2 Higher Polyphosphates of Nucleosides

19.2.1 Historical Background

In 1973 ATP:nucleotide pyrophosphotransferase, a new enzyme which catalyzed pyrophosphoryl transfer to nucleotides, was found by Murao and Nishino[1,2] in a culture broth of *Streptomyces adephospholyticus* A-4668 (newly isolated). This enzyme was purified and isolated as a homogenous preparation, and its enzymatic and physicochemical characteristics were elucidated.[3–7] This enzyme catalyzes pyrophosphoryl transfer from pyrophosphate at the β,γ-position of ATP to C-3′ of 5′-purine nucleotides to synthesize new nucleotides; adenosine, guanosine or inosine 5′-(mono-, di- or tri-) phosphate 3′-diphosphates; pppApp (adenosine 5′-triphosphate 3′-diphosphate), ppApp, pApp, pppGpp (guanosine 5′-triphosphate 3′-diphosphate), ppGpp, pGpp, pppIpp, ppIpp and pIpp.[6,7]

ppGpp and pppGpp were named Magic spot I (MS I) and Magic

spot II (MS II), respectively, and were found in *Escherichia coli* by Cashel *et al.*[8,9] Haseltine *et al.* found the stringent factor (protein), a factor which was essential for the synthesis of ppGpp and pppGpp, in 0.5 M NH_4Cl washings of ribosomes derived from stringent strains of *E. coli* and showed that ppGpp and pppGpp were synthesized from GDP or GTP and ATP in a reaction system containing ribosomes and this factor.[10] The factor was further studied, and some ppGpp and pppGpp was found to be synthesized even in the absence of ribosomes; GMP or ITP was shown to be an acceptor of pyrophosphate.[11–13] The factor was found to be similar to ATP:nucleotide pyrophosphotransferase in that both enzymes catalyzed pyrophosphoryl transfer, but the amount of this factor produced in cells and its specific activity were both very low.

ppGpp can also be synthesized chemically from pGp in a yield of 10–20%.[14,15]

ppGpp and pppGpp were found not only in *E. coli* but also in *Bacillus subtilis*,[16] amoeba[17] and mammals.[18,19]

19.2.2 Production of Nucleoside 5′-(Mono-, Di- or Tri-) phosphate 3′-Diphosphates

As mentioned above, an enzymatic method using ATP:nucleotide pyrophosphotransferase is considered to be most effective for the production of new nucleotides from ATP as a donor and purine nucleotides as acceptors. The production of this enzyme, the enzymatic reaction and isolation of the nucleotides will be described below.

A. Production of the enzyme

S. adephospholyticus was inoculated into a medium containing 2% glycerol, 4% polypepton and salts, and was cultured aerobically at 30° C for 35 h. The enzyme activity reached 4000 units per liter of culture filtrate.[2] One unit of enzyme activity was defined as the amount of enzyme which synthesized 1 μmole of pppApp in one min at 37° C. Purified enzyme or culture broth could be used as an enzyme source.

B. Enzyme reaction

This ATP:nucleotide pyrophosphotransferase catalyzes the following reaction:

$$2\ ATP \rightarrow AMP + pppApp,$$

where ATP is used as both donor and acceptor, and the β,γ-pyrophosphate of ATP is transferred to C-3′ of the ribose moiety of the 5′-nucleotide (ATP) to yield pppApp.[4)]

As shown in Table 19.4, which gives the acceptor specificity of this enzyme reaction, the enzyme synthesizes new, highly phosphorylated nucleotides from all purine 5′-nucleotides as acceptors, but the amounts of nucleotide synthesized vary depending on the kind of nucleotide.[4,5)] Various phosphate compounds were examined as pyrophosphate donors, and ATP is the best substrate (100), followed by dATP (25) and pppApp (25).[4)] In this reaction a divalent cation, usually Mg^{2+}, is essential.[4,5)]

C. Isolation of new nucleotides

For the production of new nucleotides, the reaction system shown in Table 19.4 was used on a large scale (the incubation time and amount of enzyme were adjusted appropriately). The reaction mixture was neutralized with 1.0 N HCl and applied to a DEAE-Sephadex A-25 column previously equilibrated with 0.05 M Tris-HCl buffer (pH 7.5) containing 0.1 M LiCl. The new nucleotide was eluted with LiCl in a linear gradient system and isolated as a precipitate by adding ethanol.[6,7)]

TABLE 19.4

Acceptor Specificity of ATP: Nucleotide Pyrophosphotransferase

Acceptor	Amount of product formed (μ mole/10 min/unit) A		B	Product specificity[†1] A/(A+B) (%)
ATP	10.0 pppApp		—	100
ADP	1.8 ppApp	+	4.8 pppApp	27
AMP	pApp[†2]	+	pppApp	—
GTP	15.0 pppGpp			100
GDP	15.0 ppGpp			100
GMP	6.9 pGpp	+	2.0 pppApp	78
ITP	4.9 pppIpp	+	3.2 pppApp	70
IDP	5.9 ppIpp	+	2.2 pppApp	75
IMP	1.8 pIpp	+	5.4 pppApp	25

The enzymatic reaction was carried out at 37°C for 10 min in an incubation mixture containing 100 mM glycine-NaOH buffer (pH 10.0), 5 mM donor ATP, 5 mM of each acceptor nucleotide, 10 mM $MgCl_2$ and 0.065 unit of enzyme in a total volume of 1.0 ml.

†1 Average values.
†2 This nucleotide was produced with a higher concentration of enzyme or on prolonged incubation.

19.2.3 Concluding Remarks

ppGpp and pppGpp are now known to play an important role in metabolic control, while pppApp, ppApp, etc. have some effects on sporulation and germination in sporogenic bacteria.[20,21] The physiological role of these new nucleotides which have pyrophosphate at C-3′ of the ribose moiety will be of great interest.

For the production of these new nucleotides, nucleotide acceptors and donors are still expensive. Fermentative production is expected to be developed, and in fact fermentative production of ppGpp and pppGpp using *Brevibacterium ammoniagenes* was recently reported (Sato and Furuya, 50th Annual Meeting of the Agricultural Chemical Society of Japan, July 1975. Abstracts, p. 38).

REFERENCES

1. S. Murao and T. Nishino, *Agr. Biol. Chem.*, **37**, 2929 (1973).
2. S. Murao and T. Nishino, *ibid.*, **38**, 2483 (1974).
3. T. Nishino and S. Murao, *ibid.*, **38**, 2491 (1974).
4. T. Nishino and S. Murao, *ibid.*, **39**, 1007 (1975).
5. T. Nishino and S. Murao, *ibid.*, **39**, 1827 (1975).
6. S. Murao, T. Nishino and Y. Hamagishi, *ibid.*, **38**, 887 (1974).
7. Y. Hamagishi, T. Nishino and S. Murao, *ibid.*, **39**, 1015 (1975).
8. M. Cashel and J. Gallant, *Nature*, **221**, 838 (1969).
9. R. A. Lazzarini, M. Cashel and J. Gallant, *J. Biol. Chem.*, **246**, 4381 (1971).
10. W. A. Haseltine, R. Block, W. Gilbert and K. Weber, *Nature*, **238**, 381 (1972).
11. J. Sy, Y. Ogawa and F. Lipmann, *Proc. Natl. Acad. Sci. U.S.A.*, **70**, 2145 (1973).
12. J. W. Cochran and R. Byrne, *J. Biol. Chem.*, **249**, 353 (1974).
13. R. Block and W. A. Haseltine, *ibid.*, **250**, 1212 (1975).
14. A. Simoncsits and J. Tomasz, *Biochim. Biophys. Acta*, **340**, 509 (1974).
15. J. W. Kazarich, A. Craig and S. M. Hecht, *Biochemistry*, **14**, 981 (1975).
16. P. Fortnagal and R. Bergmann, *Biochem. Biophys. Res. Commun.*, **56**, 264 (1974).
17. C. Klein, *FEBS Letters*, **38**, 149 (1974).
18. J. D. Irr, M. S. Kaulenas and B. R. Unswarth, *Cell*, **3**, 249 (1974).
19. H. J. Rhaese, *FEBS Letters*, **53**, 113 (1975).
20. H. J. Rhaese and R. Groscurth, *ibid.*, **44**, 87 (1974).
21. S. Murao, Y. Hamagishi and T. Nishino, *Agr. Biol. Chem.*, **39**, 1893 (1975).

19.3 CYTIDINE DERIVATIVES

The phospholipids are basic components of mitochondria and cell membranes, where many important biochemical reactions take place in living cells. The roles played by phospholipids have long attracted the attention of many workers since their distribution is especially high in the brain. The biosynthetic mechanism of phosphatidyl choline (lecithin) was studied by Kornberg *et al.*[1] and in 1956 CDP-choline was discovered by Kennedy *et al.*[2] to be an important intermediate in phospholipid biosynthesis.

Geiger *et al.*[3,4] carried out perfusion experiments in 1957 using cats and found that prolonged perfusion lowered the cerebral activity and, at the same time, reduced the content of phospholipids in the brain. Further, Geiger *et al.* showed that the addition of liver extract not only brought about an increase of phospholipids but also an improvement in symptoms. Such effects could also be obtained by the use of cytidine.

In 1961, Hayaishi *et al.*[5] experimentarily prepared animals carrying cerebral edemas, and showed that lowering of the cerebral activity corresponded to a drop in the content of either phosphatidyl choline or phosphatidyl ethanolamine (cephalin) and that the injection of either CMP or CDP-choline (1) into the carotid artery of a cat suffering from cerebral edema served to improve its electroencephalogram and to relieve its drowsiness pattern. Preclinical and clinical research by Kondo *et al.*[6] and Araki *et al.*[7] showed that CDP-choline is effective in overcoming the clouding of consciousness caused by brain injury.

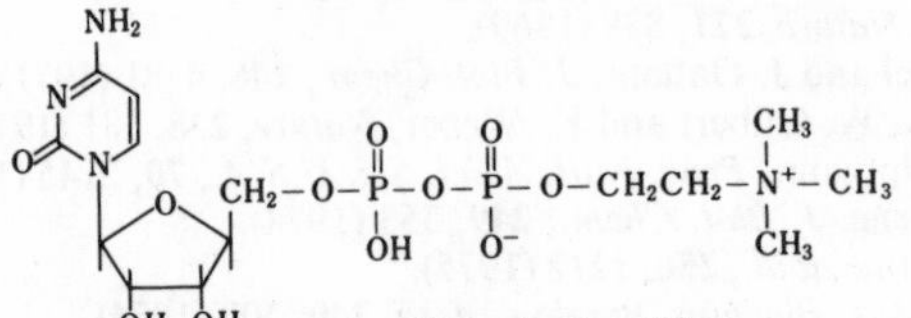

CDP-choline (1)

Regarding methods of preparing these intermediates of phospholipid biosynthesis, studies have been carried out on both chemical synthesis and biosynthesis. The chemical synthesis of CDP-choline was first attempted by condensation between CMP and phosphoryl choline (P-choline) in the presence of *N,N'*-dicyclohexylcarbodiimide by Kennedy *et al.*[8] The next approach was the alkylation[9] of CDP-ethanolamine in the presence of hydrochloric acid; other methods included reaction[10] between the amides of CMP and P-choline, reaction[11] between CMP and P-cholines (Honjo *et al.*), and condensation[12] between esters of CMP and the phosphate ester of choline in the presence of trichloroacetonitrile (Susai *et al.*). Chemical synthetic methods have now become less expensive and more convenient.

As for the biosynthesis of CDP-choline using microorganisms, Kennedy *et al.*[2] described this in 1956 in yeasts. However, clear-cut isolation and identification from microorganisms was first achieved in the same year by Lieberman *et al.*[13] Research relating to so-called "fermentative production" of CDP-choline was started around 1970 by Tochikura *et al.*, Nakayama *et al.*, and Takeda *et al.*

Concerning methods for the production of CDP-ethanolamine, papers by Tochikura *et al.*[14,15] and Takeda *et al.*[16] have appeared. Their methods, however, were not basically different from the production techniques for CDP-choline in terms of the strains used, reaction conditions and isolation method, with the exception that phosphoethanolamine (P-ethanolamine) is replaced by P-choline in the reaction. The production method will therefore be described below with special emphasis on the biosynthesis of CDP-choline.

Cytidine derivative having biological activity have also been studied, particularly in the search for anti-cancer agents. Cytosine arabinoside[17] and its N^4-acyl derivative,[18] as well as the 2,2′-anhydro compound[19] have proved interesting, though details will not be given here.

19.3.1 Production Method

A. Strains, cultivation method and treatment

CDP-choline accumulating ability has been found in a vast range of microorganisms such as yeasts,[20–23] bacteria,[23,24] molds and actinomycetes.[24,25] The following yeasts and bacteria are believed to be suitable for commercial production.

Yeasts: *Saccharomyces cerevisiae*, *S. carlsbergensis*, *S. rouxii*, *Torulopsis candida*, *T. sphaeria*, *Rhodotorula mucilaginosa*, *Brettanomyces petrophilum*, *Hansenula jadinii* and *Pichia miso*.

Bacteria: *Brevibacterium ammoniagenes*, *Escherichia coli* and *Bacillus cereus*.

Regarding the cultivation of these yeasts or bacteria, however, there has been no report of any specific factor being effective. The organisms can be cultivated in either glucose-peptone-inorganic salts or molasses-inorganic salts for 24–48 h. The cells thus cultivated are centrifugally collected and then either air-dried or freeze-dried. According to Tochikura *et al.*[20] in the case of yeast, the transformation ratio in the reaction falls with increase of the water content in the cells and no reaction occurs if the water content is 70% or more. On the other hand, Miyauchi *et al.*[23] found that, in the case of *Rhodotorula*, even intact cells can react although they show lower activity than freeze-dried cells. Further, it has been demonstrated that, in the case of bacteria, intact cells have higher activity than dried cells. Accordingly, the critical water content is believed to vary with the nature of the microorganisms.

As for the accumulation of CDP-choline by growing cells, it is known that some bacteria accumulate CDP-choline from CTP in good yield,[23] but no yeast has been so far reported to have such activity.

B. Substrates, reaction conditions and yield

A large amount of CDP-choline accumulates in reaction mixtures having the following starting composition, after reaction for 8 h at about 30° C with shaking:

CMP (salt)	20–30 μmole/ml
P-choline (salt)	40–100
Glucose	200–800
Potassium phosphate buffer (pH 7.0)	200
$MgSO_4 \cdot 7H_2O$	10–15
Cells (as dry weight)	60–100 mg/ml

In addition, a small amount (0.02–0.1 ml/ml) of toluene may sometimes be added to prevent deamination of the cytosine moiety. It is also possible to use CDP, CTP or cytosine in place of CMP. In the case of CMP, 20 μmole/ml is regarded as the optimum level. At concentrations higher than this value, with a view to improving tank efficiency, it is desirable to follow the method of feeding CMP during a suitable period after the start of the reaction.[26] The concentration of P-choline has a marked effect on the reaction. It is necessary to add this in a large excess with respect to CMP; about 5 times as much is believed to be adequate. The use of choline[21,29] and its derivatives, such as choline chloride, in place of P-choline is also possible. Further, this being an energy requiring reaction, it is essential to add either glucose or some other sugar; less than 200 μmole/ml will markedly reduce the formation of CDP-choline. Moreover, the presence of $MgSO_4$ at not less than 10 μmole/ml is also indispensable to obtain CDP-choline in good yield, especially when bacteria are used as the source of enzymes. The optimum pH of the reaction lies in the neighborhood of 7. In the case of yeasts, the productivity is comparatively stable, but with bacteria the productivity falls off suddenly on either side of pH 7. Further, this being a phosphorylation reaction, the concentration of phosphate is an especially important factor. The optimum concentration lies in the neighborhood of 200 μmole/ml.

It has been reported that, by a suitable combination of the factors mentioned above, the yield of CDP-choline from CMP can reach 80–100% on a molar basis.[26]

C. Isolation and assay

The basic process for isolation is described below.[21,23,27] After completion of the reaction, the reaction mixture is heated and the pH adjusted to either the acidic or alkaline side of neutral. Protein is removed centrifugally. Next, CDP-choline in the supernatant is adsorbed on active carbon, eluted with ammonia-ethanol-water ($EtOH:NH_4OH:H_2O = 50:2\text{–}5$: remainder % v/v) and subjected to vacuum concentration. The concentrate (pH 8) is adsorbed on Dowex 1×2, and eluted with hydrochloric acid or formic acid. The resulting CDP-choline fraction is adsorbed on active carbon again, eluted with ammonia-alcohol-water, then subjected to vacuum concentration, and, if necessary, subjected to Dowex 50 treatment. Decolorization, if necessary, is carried out only after isolation and concentration. Crystallization is effected by adding alcohol to the final condensate, yielding pure crystals in the form of white plates.

The standard assay method is as follows. The supernatant obtained by centrifugally removing protein after the reaction, followed by heating and pH adjustment, is subjected to paper chromatography. The UV absorbing spot corresponding to CDP-choline is cut out, eluted with 0.01 N HCl and subjected to spectrophotometry. Development is carried out with 1M ammonium acetate-95% EtOH (1:2–2:5 v/v).

19.3.2 Metabolism of CDP-choline

CDP-choline was discovered by Kennedy *et al.*[2] to be an important intermediate on the route to lecithin, as mentioned previously. Kennedy[28] discussed the situation on the basis of data relating mainly to the higher animals.

Lecithin biosynthesis:

(1) CMP + ATP $\rightleftharpoons$ CTP + AMP (CMP pyrophosphokinase)
(2) Choline + ATP $\rightleftharpoons$ P-choline + ADP (choline kinase)
(3) P-Choline + CTP $\rightleftharpoons$ CDP-choline + pyrophosphate (P-choline pyrophosphorylase = P-choline cytidyl transferase)
(4) CDP-choline + D-α,β-diglyceride $\rightleftharpoons$ lecithin + CMP (P-choline-glyceride transferase)

Subsequently, many researchers showed that the path indicated in the above scheme operates also in microorganisms. Tochikura *et al.*[27,29] elucidated the following points:

(1) The formation of CDP-choline from choline and CMP by yeasts is very much restricted at pH 6 or less.

(2) Even under the above conditions, however, the phosphorylation of CMP is not greatly restricted. Accordingly, the energy supply system, including glycolysis, is active even at low pH. The low pH is therefore believed to affect the choline kinase side.

(3) When choline is used, competition between reactions (1) and (2) in the above scheme is assumed to occur, since CTP formed by reaction (1) inhibits reaction (2). The effects of CTP on the activities of both enzymes have been determined, confirming this hypothesis.

REFERENCES

1. A. Kornberg and W. E. Pricer, *J. Biol. Chem.*, **204**, 345 (1953).
2. E. P. Kennedy and S. B. Weiss, *ibid.*, **222**, 195 (1956).
3. A. Geiger and S. Yamazaki, *J. Neurochem.*, **1**, 95 (1957).
4. A. Geiger, *Metabolism of the Nervous System*, p. 245, Pergamon Press, 1957.
5. O. Hayaishi, K. Ozawa, C. Araki, S. Ishii and H. Kondo *Nisshin Igaku* (Japanese), **48**, 519 (1961).
6. H. Kondo, *Nippon Geka Hokan* (Japanese), **32**, 487 (1963).
7. C. Araki, S. Ishii, O. Hayaishi, S. Kuno, K. Ozawa and H. Tsuji, *Shinkei Shimpo* (Japanese), **4**, 820 (1963).
8. E. P. Kennedy, *J. Biol. Chem.*, **222**, 185 (1956).
9. K. Tanaka and Y. Mino, *Japanese Patent Publication* No. 39–6541 (1964).
10. M. Honjo and S. Furukawa, *Japanese Patent Publication* No. 42–1384 (1967).
11. M. Honjo and S. Furukawa, *Japanese Patent Publication* No. 45–4747 (1970).
12. M. Susai and T. Takamatsu *Japanese Patent Publication* No. 45–4505 (1970).
13. I. Lieberman, L. Berger and W. T. Giminez, *Science*, **124**, 81 (1956).
14. T. Tochikura, A. Kimura, H. Kawai and T. Gotan, *J. Ferment. Technol.*, **49**, 1005 (1971).
15. T. Tochikura, A. Kimura, H. Kawai and T. Gotan, *ibid.*, **50**, 178 (1972).
16. I. Takeda, S. Watanabe and S. Shirota, *Japanese Patent Publication* No. 45–9872 (1970).
17. J. S. Evans, E.A. Musser, G. D. Mengel, K. R. Forsblad and H. Hunter, *Proc. Soc. Exptl. Biol. Med.*, **106**, 350 (1961).
18. S. Tsukagoshi, M. Aoshima, Y. Sakurai and T. Ishida, 95th Ann. Mtg. Pharm. Soc. Japan, 1975. Abstracts, p. 44.
19. A. Hoshi, F. Kanazawa, K. Kuretani, M. Saneyoshi and Y. Arai, *Gann*, **62**, 145 (1971).
20. T. Tochikura, A. Kimura, H. Kawai, T. Tachiki and T. Gotan, *J. Ferment. Technol.*, **48**, 763 (1970).
21. S. Shirota, S. Watanabe and I. Takeda, *Agr. Biol. Chem.* **35**, 325 (1971).
22. K. Nakayama and H. Hagino, *Japanese Patent Publication* No. 48–40757 (1973).
23. K. Miyauchi, K. Uchida and H. Yoshino, *Amino Acid and Nucleic Acid*, no. 25, 47 (1972).

24. I. Takeda, S. Watanabe and S. Shirota, *British Patent* No. 1294769.
25. K. Nakayama and H. Hagino, *Japanese Patent Publication* No. 48–4155 (1973).
26. A. Kimura, M. Morita and T. Tochikura, *Agr. Biol. Chem.*, **35**, 1955 (1971).
27. T. Tochikura, Y. Kariya and A. Kimura, *Hakkokogaku Zasshi* (Japanese), **52**, 637 (1971).
28. E. P. Kennedy *The Enzymes*, 2nd ed., p. 63, Academic, 1960.
29. Y. Kariya, K. Aisaka, A. Kimura and T. Tochikura, *Amino Acid and Nucleic Acid*, no. 29, 75 (1974).

PART V

UTILIZATION OF NUCLEIC ACID-RELATED SUBSTANCES

CHAPTER **20**

Utilization in Foods*

20.1 TASTE COMPONENTS OF NATURAL FOODS

It is water-soluble components of foods that cause chemical sensations of taste, and these components may well be considered to constitute

* Yasushi KOMATA, Ajinomoto Co. Inc.

Animal food

free Amino acids (Gly·Ala·L-Pro···) →
Peptides (Carnosine·Anserine·Balenine···) →
Organic acides (Lacticacid·Succinic acid···) →
IMP
GMP
+
L-Glutamic acid
← Organic bases (Betaine·*Tri*-Methyl-Amine·→Oxide·Creatine···)
← Sugars (Ribose·Glucose·Glycogen···)

Vegetable food

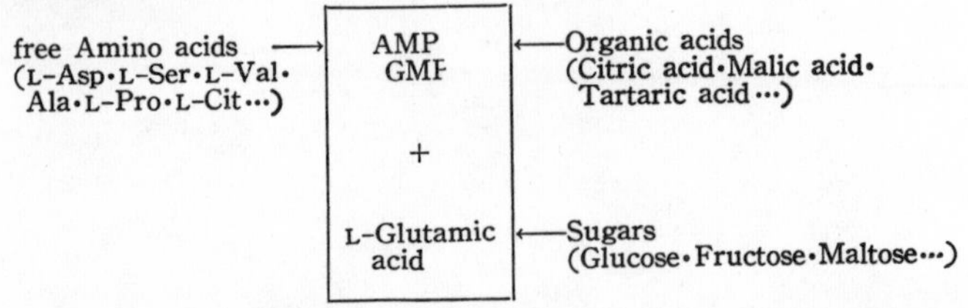

Fig. 20.1. Contributions of various components to the taste of natural foods.

the basis of the food flavors. Extractible components of natural foods include nucleic acid-related substances such as nucleotides,* nucleosides, etc., amino acids, peptides, organic acids, sugars, organic bases such as betaines, creatine, creatinine, etc., and inorganic ions.

There are five fundamental tastes; sourness, saltiness, sweetness, bitterness and savory taste, but most of the above substances change their taste quality according to concentration, pH and complicated interactions with other substances. Consequently, it can be said that the tastes of natural foods are produced by complex combinations of extractible components at specific concentrations.

There are some experimental data on the relations between taste and extractible components based on chemical analyses and taste tests.[1–8] From these studies it appears that, as shown in Fig. 20.1, major taste components in both animal foods and vegetable foods include 5′-ribonucleotides such as IMP, AMP, GMP, etc., and free amino acids such as L-glutamic acid, L-aspartic acid, etc.

20.2 Savory Compounds

In addition to the four taste sensations of sweetness, sourness, salti-

---

* Including ribonucleotides and deoxyribonucleotides.

ness and bitterness, a savory taste is also regarded as a fundamental taste in food chemistry. Savory compounds in natural foods include IMP, GMP, L-glutamic acid, L-aspartic acid, trichoromic acid,[9] ibotenic acid,[9] succinic acid,[10,11] theanine,[12] etc. The relations between chemical structure and savory taste have been well studied for nucleotide-related substances and L-glutamic acid-related substances. Some of the L-glutamic acid-related substances are illustrated in Fig. 20.2.

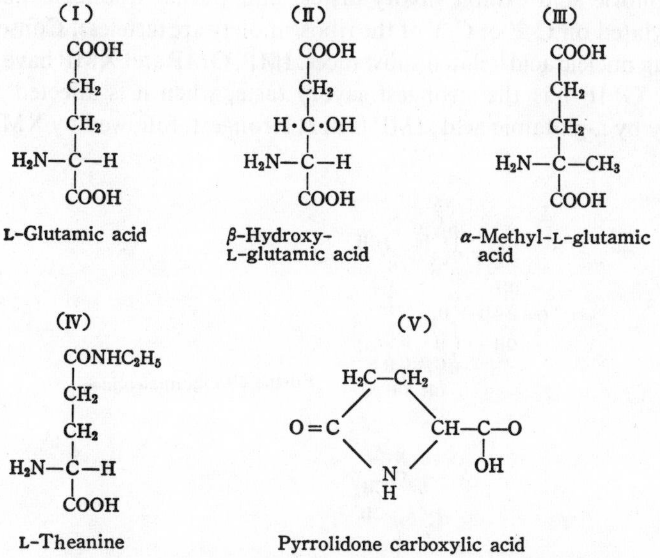

(I)
COOH
|
CH_2
|
CH_2
|
$H_2N—C—H$
|
COOH

L-Glutamic acid

(II)
COOH
|
CH_2
|
H·C·OH
|
$H_2N—C—H$
|
COOH

β-Hydroxy-L-glutamic acid

(III)
COOH
|
CH_2
|
CH_2
|
$H_2N—C—CH_3$
|
COOH

α-Methyl-L-glutamic acid

(IV)
$CONHC_2H_5$
|
CH_2
|
CH_2
|
$H_2N—C—H$
|
COOH

L-Theanine

(V)
$H_2C—CH_2$
O=C　　CH—C—O
　N　　　OH
　H

Pyrrolidone carboxylic acid

Fig. 20.2. L-Glutamic acid-related substances.

Among these substances, β-hydroxy-L-glutamic acid (II) and L-theanine (IV) have a savory taste similar to that of L-glutamic acid, but α-methyl-L-glutamic acid (III), in which the α-H atom is replaced by CH_3, and pyrrolidone carboxylic acid (V), which is formed by the loss of water from α-NH_2 and γ-COOH of L-glutamic acid, are tasteless. Although the taste intensities are lower, L-α-aminoadipic acid, which has one more carbon than L-glutamic acid, and L-aspartic acid, which has one less carbon, also have a savory taste, like L-glutamic acid. Moreover, several synthetic derivatives of L-glutamic acid are known to have a savory taste.[13–15]

20.2.1 Relation between Taste and Chemical Structure for Nucleic Acid-related Substances

Kuninaka[16] systematically studied the relation between the savory taste and chemical structure of nucleic acid-related substances. He concluded that, as shown in Fig. 20.3, purine ribonucleotides which have 6-OH on the purine ring and 5′-OH on the ribose moiety esterified with phosphoric acid exhibit savory tastes, and purine ribonucleotides phosphorylated on C-2′ or C-3′ of the ribose moiety are tasteless. Consequently, among nucleic acid-related substances, IMP, GMP and XMP have a savory taste. GMP has the strongest savory taste, when it is affected synergistically by L-glutamic acid; IMP is next strongest, followed by XMP.

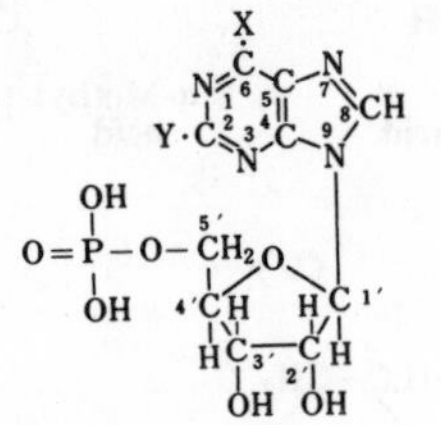

Purine 5′-ribonucleotides

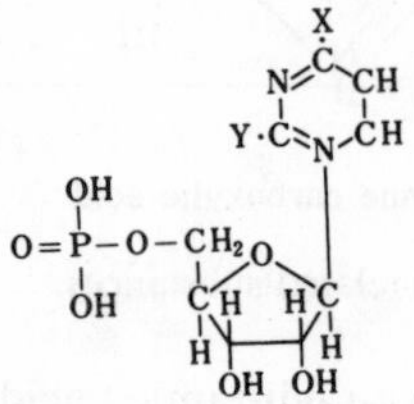

Pyrimidine 5′-ribonucleotides

Fig. 20.3. Chemical structures of 5′-nucleotides.

Nakao *et al.*[17] reported that although 5′-deoxyribonucleotides have a savory taste, the taste intensities are weaker than those of 5′-ribonucleotides. Moreover, Honjo *et al.*[18] reported that a phosphate ester linkage at C-5′ of the ribose moiety is necessary for a savory taste, and sulfate ester loses the savory taste. In addition, the C-5′ phosphate residue must have both primary and secondary dissociation of OH to exhibit a savory taste; if the phosphate residue is replaced by a phosphate monoester or a phosphoric acid amide, the compounds lose savory taste. On the other hand, among synthetic derivatives of 5′-ribonucleotides,[19] 2-methyl-IMP, 2-ethyl-IMP, 2-*N*-methyl-GMP, 2-*N*, *N*-dimethyl-GMP, 2-methyl-thio-

IMP, 2-ethyl-thio-IMP, etc. are known to have a strong savory taste when they are affected synergistically by L-glutamic acid.[20]

20.2.2 Taste Synergism among Savory Substances

Taste synergism among L-glutamic acid, IMP, GMP and succinic acid has been investigated by Sakaguchi,[21] Kuninaka,[16] Toi *et al.*[22] and Nakao *et al.*[17] The results of these studies can be summarized as follows: (1) Strong synergistic effects are observed between monosodium L-glutamate (MSG) and sodium salts of purine 5′-ribonucleotides: $IMP \cdot Na_2$, $GMP \cdot Na_2$ and $XMP \cdot Na_2$. Similar synergistic effects are observed between MSG and sodium salts of purine 5′-deoxyribonucleotides. (2) The synergistic effects mainly operate to enhance the intensity of the savory taste, but some difference of taste quality exists between a simple solution of MSG and a mixture of MSG and 5′-rinonucleotides prepared to have equal intensity of savory taste. (3) There are no marked synergistic effects between sodium succinate and MSG, between sodium succinate and sodium salts of 5′-ribonucleotides, and among 5′-ribonucleotides. On the other hand, L-aspartic acid and L-α-aminoadipic acid, which are in the same group as L-glutamic acid, manifest synergism with savory-tasting 5′-ribonucleotides; and ibotenic acid and trichoromic acid, which were isolated by Takemoto as derivatives of L-glutamic acid, also have strong synergism with savory-tasting 5′-ribonucleotides. Other nucleic acid-related substances which show a certain degree of taste synergism with MSG include $AMP \cdot Na_2$, the sodium salts of ADP, ATP, ITP, GTP, etc.

20.2.3 Distribution of Savory-tasting Substances in Natural Foods

L-Glutamic acid is widely distributed in almost all foods of animal or vegetable origin (Tables 20.1, 20.2).[23–43] On the other hand, 5′-ribonucleotides show particular distributions (Tables 20.3, 20.4).[44–49] Poultry, cattle and fish abound in IMP but contain little of the other 5′-ribonucleotides. In crustacea and mollusca, AMP is predominant. Almost all kinds of vegetables also contain AMP. Since the latter manifests some degree of taste synergism with L-glutamic acid, the basic components of the savory taste of these foods are AMP and L-glutamic acid. Moreover, the GMP content of mushrooms is particularly high, especially in "*Shiitake*" species; GMP produces a rich taste, having strong synergism with L-glutamic acid. Many reports have been published on the distribution of free 5′-ribonucleotides; those by Hashida[50] and Shimazono[51,52] are excellent.

TABLE 20.1

Composition (mg %) of L-Glutamic Acid in Animal Foods

Material	L-Glutamic acid	Literature
	mg%	
Beef (tenderloin)	33	24)
Beef (shank)	11	〃
Pork (tenderloin)	23	〃
Chicken meat	44	〃
Chicken (bone)	40	〃
Duck meat (breast)	50	〃
Mutton meat	6	37)
Yellowfin tuna ("Kiwada Maguro")	4～8.9	23)
Albacore ("Binnaga Maguro")	5.3	24)
Frigate mackerel ("Soda-gatsuo")	37～59	25)
Yellow tail ("Buri")	18	23)
Spring herring ("Nishin")	7	27)
Pilchard ("Ma-iwashi")	280	26)
Mackerel pike ("Sanma")	36	〃
Horse mackerel ("Aji")	19	28)
Mackerel ("Saba")	20	〃
Black sea bream ("Chinu")	19	〃
Red sea bream ("Ma-dai")	9.3	25)
White croaker ("Guchi")	13	26)
Cod ("Ma-dara")	9	〃
Trunk-fish ("Fugu")	6.8	24)
Arrow toothed halibut ("Abura-Garei")	9.9	30)
Bastard halibut ("Hirame")	9.8	〃
Fishing frog ("Ankoo")	51	26)

material	L-Glutamic acid	Literature
	mg%	
Carp	7.3～17.6	31)
prussian Carp ("Funa")	15.5	〃
Loach ("Dojoo")	22	26)
Eel ("Unagi")	10	24)
Prawn ("Kuruma-ebi")	51	〃
Spiny lobster ("Ise-ebi")	7	32)
Common scallop ("Hotate-gai")	150.5	33)
Hard clam ("Hamaguri")	249	34)
Surf clam ("Baka-gai")	151	24)
Corbicula ("Shijimi")	23	〃
Little neck clam ("Asari")	233	34)
Oyster ("Kaki")	264	33)
Common abalone ("Awabi")	109	35)
Squid ("Surume-ika")	41.5	36)
Squid ("Koo-ika")	44	28)
Southern squid ("Kensaki-ika")	14.5	36)
Squid ("Aori-ika")	3.1	〃
Octopus ("Ii-dako")	29	35)
Horse-dung sea urchin ("Bafun-Uni")	300～400	2)
Purple sea urchin ("Murasaki-Uni")	352	2)

TABLE 20.2

Composition (mg %) of L-Glutamic Acid in Vegetable Foods

Material	L-Glutamic acid	Literature	material	L-Glutamic acid	Literature
	mg%			mg%	
Japanese radish ("Daikon")	1.9	38)	Citrus orange ("Unshu-Mikan")	11.6	40)
Carrot ("Ninjin")	3.02	〃	Kumguat ("Kinkan")	15.2	〃
Edible burdock ("Gobo")	0.86	〃	Citrus ("Hassaku")	9.7	〃
Onion ("Tama-Negi")	0.69	〃	Citrus ("Ponkan")	10.4	〃
Garlic ("Ninniku")	1.29	〃	Sour Orange ("Daidai")	28.6	〃
Lotus root ("Renkon")	1.00	〃	Sour orange ("Yuzu")	18.8	〃
Ginger ("Shooga")	0.92	〃	Lemon	7.3	〃
Tomato ("Tomato")	3.99		Summer orange ("Natsu-mikan")	18.8	〃
Egg plant ("Nasu")	0.84		Japanese Pear	15.6	〃
Cucumber ("Kyuri")	0.65		Grape	34.3	〃
Oriental pickling melon ("Shirouri")	0.28		Japanese persimmon ("Fuyu-gaki")	trace	〃
Pumpkin & Squash ("Kabocha")	3.03		Apple ("Kokko-Ringo")	3.6	〃
Red-pepper ("Toogarashi")	1.21		Green tea (raw leaf)	208～504	41)
Spinach ("Hoorenso")	3.85	39)	Dried tangle	1780～4226	42)
			Purple laver (fresh leaf)	640	43)

TABLE 20.3

Composition (mg %) of Free 5′-Ribonucleotides in Animal Meats[44]

Material	AMP	IMP	GMP	UMP	CMP
Beef	6.6	106.9	2.2	1.6	1.0
Pork	7.6	122.2	2.5	1.6	1.9
Chicken meat	11.5	75.6	1.5	1.3	2.6
Whale meat	2.1	214.5	3.6	1.9	±

TABLE 20.4

Composition (mg %) of Free 5′-Ribonucleotides in the Muscle of Marine Animals[44–46, 49]

Material	AMP	IMP	material	AMP	IMP
	mg%	mg%		mg%	mg%
Horse mackerel ("Ma-aji")	6.4	212.6	Squid ("Surume-ika")	163.2	0
Horse mackerel ("Shima-aji")	7.2	285.4	Common octopus ("Ma-dako")	23.3	0

TABLE 20.4—*Continued*

Material	AMP	IMP	material	AMP	IMP
	mg%	mg%		mg%	mg%
Sweet fish ("Ayu")	7.2	189.0	Squid ("Surume-ika")	163.2	0
Common sea bass ("Suzuki")	8.4	124.9	Common octopus ("Ma-dako")	23.3	0
Pilchard ("Ma-iwashi")	0.7	188.7	Spiny lobster ("Ise-ebi")	72.5	0
Black sea bream ("Kuro-dai")	11.0	277.1	Hairy Crab ("Ke-gani")	10.1	0
Pike mackerel ("Sanma")	6.7	149.5	Squilla ("Shako")	32.6	17.4
Mackerel ("Saba")	5.7	188.1	Common abalone ("Awabi")	71.7	0
Keta salmon ("Sake")	6.9	154.5	Round clam ("Baka-gai")	86.5	0
Tuna ("Maguro")	5.2	188.0	Common scallop ("Hotate-gai")	102.3	0
Globefish ("Fugu")	5.6	188.7	Short-neck clam ("Asari")	10.8	0
Eel ("Unagi")	17.6	108.6	Dried bonito ("Katsuobushi")		416~862

20.3 TASTE CHARACTERISTICS OF 5′-RIBONUCLEOTIDES

20.3.1 Differences of Taste Characteristics between Disodium 5′-Ribonucleotides and MSG

5′-Ribonucleotides which are authorized for use as food additives and which are industrially produced in Japan are IMP·Na_2, GMP·Na_2, and mixtures of sodium or calcium salts of these two 5′-ribonucleotides.[53] The threshold values, i.e. the minimal concentrations which can be tasted, of IMP·Na_2, GMP·Na_2 and MSG in aqueous solutions are 0.025, 0.0125 and 0.03 g/dl, respectively. These values are quite similar, but the relations between the concentrations and intensities of savory taste at higher ranges of concentration are very different. As shown in Figs. 20.4 and 20.5, the taste intensity of MSG increases linearly with the logarithm of concentration, whereas for 5′-ribonucleotides the increase of taste intensity with concentration is slight.[54] Accordingly, while MSG can exhibit a marked savory taste by itself, disodium 5′-ribonucleotides cannot do so in the absence of amino acid-related savory substances such as L-glutamic acid or L-aspartic acid.

As for qualitative differences of taste between a simple solution of MSG and mixtures of disodium 5′-ribonucleotides with MSG at equal intensities of savory taste, the two can be discriminated in that a certain degree of "fullness", like meat, is felt with the mixed solutions even if the level of disodium 5′-ribonucleotide is relatively low.

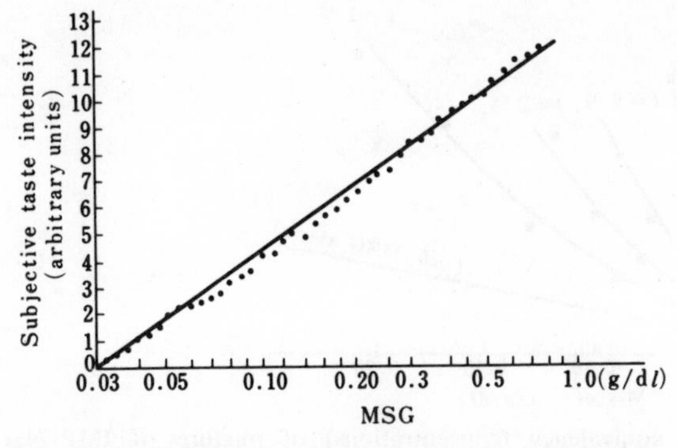

Fig. 20.4. Relation between concentration and taste intensity for MSG.

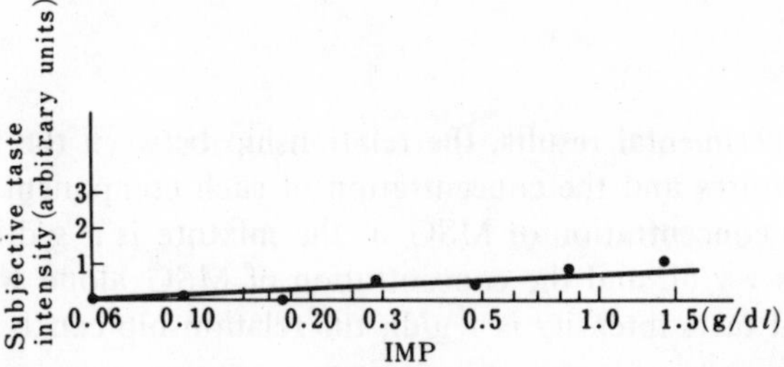

Fig. 20.5. Relation between concentration and taste intensity for IMP·Na_2.

20.3.2 Synergism of Savory Taste between Disodium 5′-Ribonucleotides and MSG

As described previously, there is remarkable synergism of savory taste between nucleic acid-related savory-tasting substances and amino acid-related savory-tasting substances. Yamaguchi[54] tried to measure this synergism in careful taste tests. She prepared 20 samples composed of four different ratios of IMP·Na_2 and MSG at five levels of concentration, and compared these concentrations at each ratio with five levels of concentration of MSG alone in order to determine equivalent intensities of savory taste of the mixtures and of MSG alone. Taste tests were carried out 50 times for each mixture. The results were analyzed by the method of Probit and the equivalent concentrations are shown in Fig. 20.6. As can be seen, the taste intensities of mixtures of IMP·Na_2 with MSG increase exponentially, while MSG alone shows a linear increase of taste intensity. The synergy between IMP·Na_2 and MSG depends upon the concentrations of the two components in the mixture.

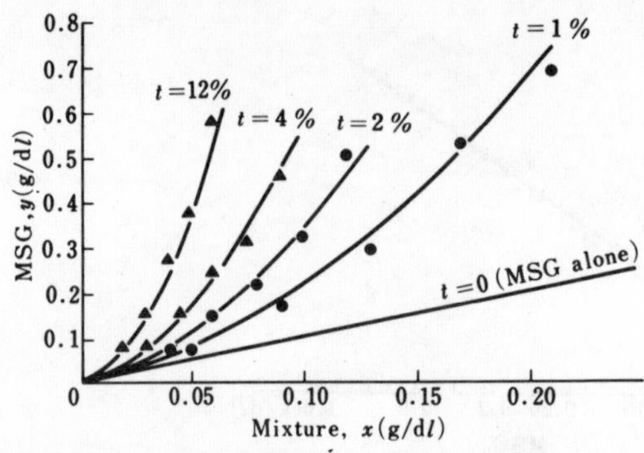

Fig. 20.6. Taste equivalency (concentrations) of mixtures of $IMP \cdot Na_2$ with MSG, and MSG alone. t represents the fraction of $IMP \cdot Na_2$ in the mixtures.

From these experimental results, the relationship between the taste intensity of the mixtures and the concentration of each component was obtained; when the concentration of MSG in the mixture is u g/dl and that of $IMP \cdot Na_2$ is v g/dl, and the concentration of MSG alone which shows an equivalent taste intensity is y g/dl, the relationship can be expressed as

$$y = u + 1218\, uv \tag{20.1}$$

Similarly, in the case of mixtures of $GMP \cdot Na_2$ and MSG, replacing v in Eq. 20.1 by v' g/dl to represent the concentration of $GMP \cdot Na_2$ and retaining the other variables, the relationship can be expressed as

$$y = u + 2800\, uv' \tag{20.2}$$

Comparing this with Eq. 20.1, the only difference is the coefficient, which is 2.3 times greater. This indicates that the synergistic effect of $GMP \cdot Na_2$ is 2.3 times stronger than that of $IMP \cdot Na_2$.

For mixtures of $IMP \cdot Na_2$ and $GMP \cdot Na_2$ with MSG, if the concentration of 5′-ribonucleotides is v''g/dl and the ratio of $IMP \cdot Na_2$ to $GMP \cdot Na_2$ is $p{:}q$, the relationship becomes

$$y = u + \frac{(p + 2.3q)}{p + q} \times 1218uv'' \tag{20.3}$$

Yamaguchi[55] determined the variations of taste intensity for mixtures of $IMP \cdot Na_2$ with MSG at a fixed concentration, 0.05 g/dl, of the mixture but with the components present in various ratios from 0:100 to 100:0, using the method of Gulliksen.[56] The results are shown in Fig. 20.7. The

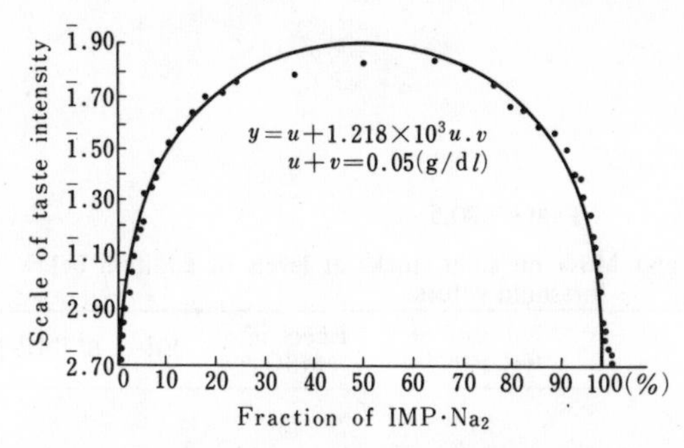

Fig. 20.7. Consistency between determined taste intensities (by Gulliksen's method) and values calculated from Eq. 20.1.

theoretical results calculated from Eq. 20.1 are shown by a solid line, and are in good agreement with the experimental points.

20.4 Seasoning Effect of Disodium 5′-Ribonucleotides

Disodium 5′-ribonucleotides are clearly potentially valuable as seasoning agents in view of their flavor synergism with amino acid-related savory substances. Since L-glutamic acid is almost universally distributed, the palatability of foods is improved by the addition of 5′-ribonucleotides. In particular, when the contents of amino acid-related savory compounds are lower than their threshold values, the savory tastes of these compounds will not be detectable, but the taste is enhanced above the threshold value by the synergistic action of small amounts of added 5′-ribonucleotides. Table 20.5 shows the seasoning effects of IMP·Na_2 and MSG on soups made from popular foods at addition levels below their respective threshold values. As shown in the table, IMP·Na_2 exhibited a remarkable seasoning effect in almost all the soups, except for those which already contained relatively high amounts of 5′-ribonucleotides. The seasoning effect is not only due to the enhancement of savory taste, but also to the sensation of "fullness", which may be regarded as the formation of a full-bodied taste.

Table 20.6 summarizes the effects of IMP·Na_2 on the individual flavor notes of foods as determined by Kurtzman and Sjöström[57] for various foods through taste analysis by the flavor profile method. Thus, IMP·Na_2 not only enhances the effect of MSG, but also increases such sensations

TABLE 20.5

Seasoning Effect of IMP·Na_2 and MSG on soup stocks at levels of addition below the threshold values.

Kind of soup stock	Number of test panelists	Effect of MSG	Effect of IMP·Na_2
Short-neck clam ("Asari")	25	◎†	
Common abalone ("Awabi")	28	◎	
Dried shrimp ("Sakura-ebi")	28	◎	◎
Beef (shank)	28	◎	△
Chicken bone	29	○	◎
White fish	28	◎	◎
Dried bonito ("Katsuobushi")	55	◎	○
Dried scallop ligament ("Hotate-kai-bashira")	29	◎	◎
Dried "Shiitake" mushrooms	25	◎	◎
Dried tangle ("Konbu")	26	×	◎
Potato	30	○	◎
Sweet corn	25	◎	◎
Green peas	25	◎	◎
Asparagus	25	◎	◎
Tomato	25	×	◎
"Shimeji" mushrooms	25	◎	◎

† *Symbols*: ◎ = effective at addition level III
○ = effective at addition level II
△ = effective at addition level I
× = not effective

Addition levels of IMP·Na_2 and MSG.

I	II	III
0.025%	0.0125%	0.00625%

TABLE 20.6

Effects of IMP·Na_2 on Individual Flavor Characteristics of Foods

Characteristic of flavor complex	Aroma†	Flavor†
Sweet		0
Salty		0
Sour	0	0
Bitter		0
Meaty	+	+
Brothy	0	+
HVP (hydrolyzed vegetable protein	−	−
Fatty (hard fat and oil	0 −	0 −
Buttery	0 +	0 +
Sulfury	−	−
Burnt	0	0 −
Starch	0 −	0 −
Herb spice	0	0 −
MSG		+
Impression of greater viscosity		+
Drying		+
Fullness		+
Astringent		+

† 0 = No change; + = increased; − = decreased.
(Source: ref. 57. Reproduced by kind permission of the Institute of Food Technologists, U.S.A.)

as meatiness, viscosity, and fullness of flavor. In addition, it decreases such flavors as sulfury and HVP*. GMP·Na_2 exhibits the same seasoning effects as IMP·Na_2 at 1/2.3 times the concentration of IMP·Na_2.

20.5 UTILIZATION OF DISODIUM 5′-RIBONUCLEOTIDES IN FOODS

5′-Ribonucleotides which are presently used as chemical seasoning agents are IMP·Na_2, GMP·Na_2 and mixtures of the two substances (referred to as disodium 5′-ribonucleotides). Disodium 5′-ribonucleotides are mostly used alone in food processing and used in combination with MSG for home or restaurant cooking.

* Hydrolyzed vegetable protein.

20.5.1 Use in Cooking

Mixtures of disodium 5′-ribonucleotides with MSG, which are called compound chemical seasonings, are on the market for home and restaurant use. Compound chemical seasonings show remarkable flavor enhancing effects for almost all dishes. For general use, compound chemical seasonings composed of 92% MSG and 8% disodium 5′-ribonucleotides are used in amounts corresponding to 5% of the salt content in the case of weak-tasting dishes and 10% of the salt content in the case of strong-tasting dishes. The standard level of compound chemical seasoning for cooked foods is about 0.1–0.2 g per single serving; this level is a half to one-third of the amount of MSG alone which would be necessary. The compound chemical seasoning agents also permit a saving of the natural materials of soup stocks; in the case of Japanese soup stock made from dried bonito or dried tangle, compound chemical seasoning can reduce the amount of natural materials required to half of the original level without loss of quality, and in the case of stock for Japanese noodle soup it can reduce the amount of dried bonito required to a half to two-thirds, or that of dried tangle to one-third. With beef or chicken consommé soups, it can reduce the requirement for beef shank meat or chicken bone to 80% of the original level, and in the case of stock for Chinese noodle soup it can reduce the pork or chicken bone requirement to 60–70% of the original level. Table 20.7 shows the usage of compound chemical seasoning in beef consommé.

TABLE 20.7
Utilization of Compound Chemical Seasoning for Beef Consommé

Ingredients	Beef consommé		
	Customary method	Chemical seasoning method	
Bouillon	10 l	8 l	10 l
Water	—	2 l	2.5 l
NaCl	100 g	100 g	125 g
MSG	28 g	—	—
Chemical seasoning†	—	10∼15 g	13∼19 g
Total	10 l	10 l	12 l

† Combination of MSG (92%) and disodium 5′-ribonucleotide (8%).

Moreover, compound chemical seasoning is more effective in softening vinegary sourness than MSG alone, both as regards stronger effect and higher solubility, since the solubility of MSG in vinegar is less than 0.8%. Standard levels of use of compound chemical seasoning in cooked foods are summarized in Table 20.8.

TABLE 20.8
Standard Levels of Use of Compound Chemical Seasoning for Cooked Foods

Kinds of prepared foods	Use levels† (g)	Kinds of prepared foods	Use levels† g
Japanese clear soup	4~6	Chicken consommé	15~20
Clear soup with soybean paste	8~10	Boiled vegetables in seasoned liquids	4~5
Seasoned soy sauce for 'tempura'	6~7.5	Thick soup	4~6
Seasoned soy sauce for vermicelli	4~6	Curried rice	10~20
Vegetable soup	6~7.5	Crab meat in vinegared dressing	8~10
Chinese soup	7.5~10	'Shao-mai'	4~6
Chinese vermicelli soup	15~20	Hamburgers	4~6

† 50 servings.

20.5.2 Use in Food Processing

In food processing, IMP•Na_2, GMP•Na_2 and 1:1 mixtures of the two are generally used. They are also used in combination with MSG in some processed foods.

Although the levels of use of disodium 5′-ribonucleotides vary, the standard level for a 1:1 mixture of IMP•Na_2 and GMP•Na_2 is 2–5% of added MSG.

At present, the largest consumption of disodium 5′-ribonucleotides is for instant foods, especially for instant soups and instant noodle soups, and they are indispensable to produce a satisfactory savory quality in the products. Table 20.9 shows standard levels of use of disodium 5′-ribonucleotides in various processed foods.

A. Use for instant foods

In seasoned processed foods, disodium 5′-ribonucleotides and MSG are as indispensable as salt and sugar to produce a satisfactory savory taste. The remarkable growth in sales of instant foods nowadays has been largely based on improvements of seasoning technique using disodium 5′-ribonucleotides and MSG. Disodium 5′-ribonucleotides are mostly used in instant Chinese noodle soups, instant Japanese soups, instant sauce mixes, instant curry, etc.

B. Use in seasoned sauces

Seasoned sauces for roast meats and fish are generally made of natural soup stocks, soy sauce or hydrolyzed vegetable protein, sugar, vinegar, spices and chemical seasonings (disodium 5′-ribonucleotides and MSG). Disodium 5′-ribonucleotides are indispensable to produce a rich savory taste in various sauces for roast meat, roast chicken, and cereal crackers.

C. Use in fermented soy sauce (*shoyu*)

Protection against enzymes which degrade disodium 5′-ribonucleotides is a serious problem in using them in food processing. While disodium 5′-ribonucleotides are scarcely affected by heat in food processing, they are seriously affected by degrading enzymes, such as phosphatase, which exist widely in raw food materials. Phosphatase readily decomposes disodium 5′-ribonucleotides into tasteless compounds. However, phosphatase is unstable to heat and can be easily inactivated by heating at 70–80° C. Heating foods before or immediately after adding disodium 5′-ribonucleotides is thus effective in protecting their activity. Since soy sauce contains a large amount of L-glutamic acid, even slight addition of disodium 5′-ribonucleotides causes marked synergism of savory taste, producing an improved quality of savory and full-bodied taste.

Disodium 5′-ribonucleotides should be added to fresh soy sauce after heating to inactivate phosphatase. Heating at 85° C for 5 min or at 80° C for 30 min (slightly more severe conditions than normal for fresh soy sauce) is necessary to retain the activity of disodium 5′-ribonucleotides. Soy sauce with added disodium 5′-ribonucleotides which is processed under these conditions shows no change of taste or flavor, and retains 90% of added disodium 5′-ribonucleotides after storage for 100 days at room temperature.

D. Use for fermented soy paste (*miso*)

Disodium 5′-ribonucleotides have the effect of reducing the saltiness of foods containing large amounts of salt, and they are especially effective for fermented soy paste (*miso*). Soy paste also contains phosphatase, and disodium 5′-ribonucleotides are decomposed very rapidly in unprocessed material. However, heating is effective in retaining the taste activity of disodium 5′-ribonucleotides, and therefore a heating machine was developed to inactivate the degrading enzymes without any loss of texture or flavor quality of the product. Thus, disodium 5′-ribonucleotides could be used in this application.

TABLE 20.9

Standard Levels of Use of Either IMP·Na_2 Alone or Mixture of IMP·Na_2 and GMP·Na_2 (1:1) with MSG (*continued on p. 316*)

Kinds of processed foods	Use levels[†1]			Remarks[†3]
	MSG	IMP·Na_2	IMP·Na_2+ GMP·Na_2[†2]	
	g	g	g	
"Kamaboko" fish cakes "Yakichikuwa" baked fish cakes	60~150	3~8	2~5	
"Agekamaboko" Fried fish cakes	50~100	2~4	1.5~3	
Fish meat ham, Sausage	30~60	1.5~3	1~2	For fish meat
Delicacies (Cuttlefish, Octopus, globefish)	75~150	3~8	2~5	
Preserved marine food boiled in soy sauce (laver, tangle)	15~40	2~5	1~3	
"Ajitsuke nori" (Seasoned and roasted laver)	1500~4000	130~300	80~250	Per 18 *l* of seasoned liquid
"Shiofuki kombu" (processed tangle)	600~1000	17~50	10~30	
Prepared frozen foods	100~150	8~15	5~10	Per 10 kg of bread crumbs
Canned foods (Seasoned)	20~30	1~1.5	0.6~1	
Consommé soup (powder)	800~1300	25~40	15~24	
Potage soup (powder)	200~500	15~25	10~15	
Instant chinese vermicelli, Instant vermicelli	1000~1400	30~60	20~35	For powdered soup
Instant spaghetti sauce	400~500	30~50	20~30	
Condensed buckwheat noodle soup	150~300	16.5~33	10~20	Per 18 *l*
Instant curry	100~180	8~13	5~8	

E. Use in fish cakes (*kamaboko*)

Various kinds of shoreline fish were previously used as raw materials for fish cakes (*kamaboko*), but the catch has been decreasing and at present frozen Alaskan pollack is the main raw material. Since fish cakes made of frozen pollack are less savory than those made from fresh shoreline fish meat, the use of disodium 5′-ribonucleotides and MSG was introduced in the manufacture of *kamaboko*.

In the production of *kamaboko*, disodium 5′-ribonucleotides are exposed to attack by phosphatase, and the following measures must be taken:

(1) The raw meat is washed and bleached to reduce the enzyme activity.

(2) The fish meat is kept as cool as possible in all processes until the final steaming.

TABLE 20.9—*Continued*

Kinds of processed foods	Use levels[†1]			Remarks[†3]
	MSG	IMP·Na_2	IMP·Na_2+ GMP·Na_2[†2]	
	g	g	g	
Bottled sauce for rice	300~400	15~20	10~15	
"Miso" (soy paste)	80~150	5~13	3~8	
Soy sauce	—	3.5~10	2~6	Per 18 *l*
"Tare" seasoned soy sauce for meat	100~200	8~16	5~10	Per 10 *l*
Sauce	50~100	10~15	6~10	Per 18 *l*
Ketchup	20~30	3~5	2~3	
Meat sausage, Pressed ham	30~60	2~3	1.5~2	
Corned beef	15~40	0.8~2.5	0.5~1.5	
Pickles (seasoned with soy sauce)	350~500	15~40	10~30	Per 10 *l* of seasoned soy sauce
Pickled radishes (in airtight containers)	80~200	15~40	10~30	Per 10 *l* of seasoned liquid
Rice candies (seasoned with soy sauce)	100~150	3~5	2~3	Per 10 *l* of seasoned soy sauce
Rice candies (seasoned with oil and seasoning)	750~1000	25~35	15~20	Per 10 kg of covering powder
Instant japanese clear soup (powder)	1000~1400	30~60	20~35	
Instant miso soup (powder)	400~600	25~35	15~20	

†1 Per 10 kg of final products unless otherwise stated in the remarks.
†2 Combination of IMP·Na_2 (50%) and GMP·Na_2 (50%).

(3) The period of the mixing process is shortened and the product is steamed immediately after shaping.

In some local types of *kamaboko*, shaped meat is held at 40–50° C for about 3 h before steaming in order to increase the elasticity of the product. In this case, added disodium 5′-ribonucleotides are almost completely enzymically decomposed, and these flavor enhancers cannot be used.

F. Use in fish meat ham and sausage

The main raw materials for fish meat ham and sausage were previously albacore, whale, shark and horse mackerel, but these materials were replaced by frozen Alaskan pollack about 10 years ago because of the shortage of albacore. As in the case of *kamaboko*, frozen pollack meat is of inferior quality as regards savory taste, and thus for fish meat ham and sausage it is necessary to use disodium 5′-ribonucleotides and MSG to improve the flavor.

The loss of disodium 5′-ribonucleotides by enzymatic decomposition in the manufacturing process of fish meat ham and sausage is about 10%, which is less significant than in the case of *kamaboko*. Since the product is heated at 90° C for 1 h, which completely inactivates phosphatase, disodium 5′-ribonucleotides are not lost on storage.

G. Use for cattle meat ham and sausage

The taste of cattle meat originates from its free IMP content and free amino acids, including L-glutamic acid. The average content of free IMP in cattle meat is about 0.1 %, but the content varies considerably with the type of meat. In addition, the present shortage of raw meat for ham and sausage has led to a reduction of quality. In order to improve the quality of the raw material, disodium 5′-ribonucleotides can be used, since they are the principal components producing a meaty taste. Fortunately, the addition of disodium 5′-ribonucleotides to cattle meat ham and sausage is feasible, even though the steaming conditions of these products are milder than those used for fish products. The decomposition of disodium 5′-ribonucleotides is slight except in some cases of especially large products or with certain additives. There is almost no loss of flavor components in small products such as Vienna sausage.

H. Use for pickled vegetables

Pickled vegetables have a high salt content, and disodium 5′-ribonucleotides and MSG are very effective in softening the saltiness of these products. Since the phosphatase activity of pickled vegetables is strong, disodium 5′-ribonucleotides are decomposed completely in a short period. Recently, a machine for heating salted vegetables without loss of texture or flavor has been developed, so that disodium 5′-ribonucleotides can be added.

20.6 Prospects for the Utilization of Disodium 5′-Ribonucleotides in Foods

Extractible components, which form the aroma and taste of cooked or processed foods, originate mainly from food and soup stock materials, and also from several kinds of seasonings which can be added to produce a well-balanced, complex characteristic and full-bodied aroma and flavor.

Although preferences in foods are becoming more varied, quantitative shortages of natural raw materials of good quality, as well as price increases, mean that it is essential to develop unutilized resources and to improve the utilization of food raw materials. Thus, the use of seasonings, which have advantages as regards flavor quality, storage life, convenience and cost, is very desirable during both cooking and processing of foods.

A new type of seasoning, which is called flavor seasoning and which has aroma and taste characteristics similar to those of soup stocks, has already been introduced for sale. The basic ingredients of flavor seasoning are disodium 5′-ribonucleotides, MSG, sugars and salts. Materials or extracts of soup stocks such as dried bonito, dried tangle, shellfish, chicken meat, etc. can be combined in powder or granule form. In food processing, various kinds of seasonings, natural extract or amino acids have been developed and utilized. These seasonings are mostly combinations of cheap animal meat or vegetable protein hydrolyzate and yeast extract with natural materials or extracts of soup stocks. Disodium 5′-ribonucleotides and MSG are the most important ingredients of these seasonings; they enhance the aroma and taste of other constituents and harmonize the overall seasoning activity.

A future application of disodium 5′-ribonucleotides will be to enhance the taste characteristics and aroma of natural or synthetic food materials, especially of newly developed food resources.

References

1. Y. Hashimoto, F.A.O. Symposium on the Significance of Fundamental Research in the Utilization of Fish, 1964.
2. Y. Komata, N. Kosugi and Y. Ito, *Bull. Jap. Soc. Sci. Fisheries*, **28**, 623 (1962).
3. Y. Komata, and H. Eguchi, *ibid.*, **28**, 631 (1962).
4. Y. Komata, A. Mukai and Y. Okada, *ibid.*, **28**, 747 (1962).
5. Y. Komata, *ibid.*, **30**, 749 (1964).
6. T. Fujita and Y. Hashimoto, *ibid.*, **26**, 907 (1960).
7. A. E. Bender, T. Wood and J. A. Palgrave, *J. Sci. Food Agr.*, **9**, 812 (1958).
8. T. Wood and A. E. Bender, *Biochem. J.*, **67**, 366 (1957).
9. T. Takemoto and T. Nakashima, 4th Mtg. Tohoku Branch Pharm. Soc. Japan, 1964.
10. T. Takahashi, *J. Soc. Brewering Japan*, **7**, (12), 7 (1912).
11. K. Aoki, *Nippon Nogeikagaku Kaishi* (Japanese), **8**, 867 (1932).
12. Y. Sakato, *ibid.*, **23**, 262 (1949).
13. S. Akabori and T. Kaneko, *J. Chem. Soc. Japan*, **59**, 433 (1938).
14. T. Kaneko, R. Yoshida and H. Katsura, *ibid.*, **80**, 316 (1959).
15. T. Kaneko, R. Yoshida and I. Takano, 14th Mtg. Chem. Soc. Japan, 1961. Abstracts, p. 305.
16. A. Kuninaka, *Nippon Nogeikagaku Kaishi* (Japanese), **34**, 489 (1960).
17. Y. Nakao and K. Ogata, 2nd Kanto Branch Congr. Agr. Chem. Soc. Japan, 1960. Abstracts, p. 9.

18. M. Honjo, K. Imai, S. Furusawa, H. Moriyama, T. Imada and K. Yasumatsu, Ann. Congr. Agr. Chem. Soc. Japan, 1963. Abstracts, p. 40.
19. A. Yamazaki, I. Kumashiro and T. Takenishi, *Chem. Pharm. Bull.*, **16**, 338 (1968).
20. S. Yamaguchi, T. Yoshikawa, S. Ikeda and T. Ninomiya, *Agr. Biol. Chem.*, **32**, 797 (1968).
21. K. Sakaguchi, *Shokuhin Kogyo* (Japanese), **2** (12), 29 (1959).
22. B. Toi, S. Maeda, S. Ikeda and H. Furukawa, 2nd Kanto Branch Congr. Agr. Chem. Soc. Japan, 1960. Abstracts, p. 2.
23. W. Shimizu, *Suisan Neriseihin* (Japanese), p. 86, Korin Shoin, 1966.
24. S. Maeda, S. Eguchi and H. Sasaki, *Nippon Kaseigaku Zasshi* (Japanese), **9** (4), 163 (1958).
25. K. Sugimura, *Nippon Yakurigaku Kaishi* (Japanese), **56**, 300 (1958).
26. S. Maeda, S. Eguchi and H. Sasaki, *Nippon Kaseigaku Zasshi* (Japanese), **12** (2), 5 (1961).
27. R. B. Hughes, *J. Sci. Food Agr.*, **10**, 558 (1959).
28. K. Ito, *Bull. Jap. Soc. Sci. Fisheries*, **23**, 497 (1957).
29. K. Makino, *J. Chem. Soc. Japan*, **73**, 737 (1952).
30. K. Oishi, *Mem. Fac. Fish. Hokkaido Univ.*, **10**, 558 (1959).
31. G. Duchâteau and M. Florkin, *Compt. Rend. Soc. Biol.*, **148**, 1287 (1954).
32. M. Fujita and W. Shimizu, Ann. Congr. Japan. Soc. Sci. Fisheries, 1959.
33. Y. Tsuchiya, Autumn Congr. Japanese Soc. Sci. Fisheries, Symposium Text, p. 44, 1957.
34. K. Ito, *Bull. Jap. Soc. Sci. Fisheries*, **25**, 658 (1959).
35. S. Konosu and Y. Maeda, *ibid.*, **27**, 251 (1960).
36. K. Endo, M. Fujita and W. Shimizu, *ibid.*, **28**, 833 (1962).
37. R. L. Macy, H. D. Naumann and M.E. Bailey, *J. Food Sci.*, **29**, 142 (1964).
38. A. Matsushita and A. Yamada, *Nippon Nogeikagaku Kaishi* (Japanese), **31**, 578 (1957).
39. A. Matsushita, *Eiyo to Shokuryo* (Japanese), **10**, 138 (1957).
40. A. Matsushita, *Nippon Nogeikagaku Kaishi* (Japanese), **31**, 211 (1957).
41. Z. Nakashima and C. Nakagawa, *ibid.*, **31**, 169 (1957).
42. K. Oishi, *Bull. Jap. Soc. Sci. Fisheries*, **27**, 601 (1961).
43. Y. Tsuchiya and Y. Suzuki, *ibid*, **20**, 1092 (1955).
44. N. Nakashima, K. Ichikawa, M. Kamata and E. Fujita, *Nippon Nogeikagaku Kaishi* (Japanese), **35**, 797 (1961).
45. M. Ohara, S. Maeda, Y. Komata and T. Matsuno, 5th Kanto Branch Congr. Agr. Chem. Soc. Japan, 1964.
46. T. Saito, *Bull. Jap. Soc. Sci. Fisheries*, **27**, 461 (1961).
47. D. Hashida, Ann. Mtg. Soc. Fermentation Technol. Japan, 1964.
48. D. Hashida, *Hakko Kogyo Zasshi* (Japanese), **41**, 420 (1963).
49. T. Fujita and Y. Hashimoto, *Bull. Jap. Soc. Sci. Fisheries*, **25**, 312 (1959).
50. D. Hashida, *Chori Kagaku* (Japanese), **1**, 1 (1968).
51. H. Shimazono, *Amino Acid and Nucleic Acid*, no. 10, 179 (1964).
52. H. Shimazono, *Food Technol.*, **18** (3), 36 (1964).
53. *The Japanese Standards of Food Additives*, 3rd ed., 1974.
54. S. Yamaguchi, *J. Food Sci.*, **32**, 473 (1967).
55. S. Ikeda, H. Furukawa and S. Yamaguchi, *Hinshitsu Kanri* (Japanese), **13**, 768 (1962).
56. S. Gulliksen, *Psychometrica*, **21**, 125 (1907).
57. C. H. Kurtzman and L. B. Sjöström, *Food Technol.*, **18**, 1467 (1964).

CHAPTER **21**

Utilization for Medical Treatment*

Natural nucleic acid analogs have been employed as drugs since the 19th century, before the term "nucleic acid" had been introduced.[1] Today they are widely used in medical treatment and their utilization is expected to increase in the future.

Caffeine was probably the first nucleic acid analog to be used as a drug. Since then, xanthine derivatives have become available for medical use. As pharmacological knowledge on nucleotide vitamins and coenzymes increased, these substances were introduced for therapeutic use in the 1930's. It was not until the 1940's that the concept of a metabolic antagonist was introduced and nucleic acid analogs came into practical use as drugs. Antitumor agents were then developed, offering most interesting objects for pharmaceutical research and development, and are still being extensively studied. It was also discovered in the 1960's that certain nucleic acids and synthetic polynucleotides produce interferon (IF), and their possible use in the medical field has attracted much attention.

This chapter briefly describes the application of nucleic acid analogs

* Masao NARUSE, Kyowa Hakko Kogyo Co. Ltd.

to medical treatment, including antitumor agents now in practical use. The latter are described in more detail in Chapter 22.

21.1 AGENTS FOR THE CENTRAL NERVOUS SYSTEM

Natural xanthine analogs stimulate the cerebrum. Caffeine,[2] theobromine[3] and theophylline,[4] which are typical of these, have coronary dilating, hypertensive, diuretic and striated muscle contracting effects in addition to causing cerebral stimulation, and are used in the treatment of poisoning by central nervous system depressants, cardiac diseases, angina pectoris, etc. The order of decreasing effect of the three agents on the central nervous system is caffeine > theobromine > theophylline, while their diuretic effect is in the reverse order. No xanthine analogs are used practically as central nervous system depressants. However, some C^8-substituted derivatives of theophylline[5,6] are anticonvulsive and some N^6-substituted adenine[7] and tetrazolepyridine derivatives[8] are hypnotic. As cerebral lecithin levels decrease with depressed cerebral function, CDP-choline[9,10] is used in the treatment of disturbance of consciousness caused by injuries to the head or following cerebral operations.

21.2 AGENTS FOR THE CIRCULATORY AND URINARY ORGANS

Agents for the circulatory organs include adenosine, oxyetophylline, proxyphylline, diprophylline, TH-24 and ST-7. Theobromine[3] and theophylline[4] are diuretics. Adenosine,[11] which dilates the coronary vessels and promotes muscular metabolism, is administered to patients with cardiac insufficiency or angina pectoris. Among xanthines that dilate the coronary vessels, increase the renal blood flow and promote glomerular filtration, oxyetophylline,[12] proxyphylline,[13] diprophylline,[14] TH-24[15] and ST-7[16] are prescribed for bronchial asthma, coronary disorders and cardiac edema, while theobromine and theophylline are used to treat cardiac or renal edema and hypertension. Theophylline is superior to other xanthine analogs so far synthesized. Adenosine and its derivatives[17–19] inhibit platelet agglutination caused by ADP. They may possibly be used as anticoagulants, but are not yet used clinically due to the complex mechanism of blood coagulation.

21.3 Metabolic Agents

Nucleotide vitamins and coenzymes used in medical treatment include NAD, FAD, ATP, CoA, vitamin L_2 and vitamin B_{12} coenzyme. In an organism, vitamin B_2 is activated to FAD and thus becomes pharmacologically effective. Hence FAD is used in the treatment of vitamin B_2 deficiency, arabinosis and hepatic or intestinal diseases. It also constitutes a supplemental aid in stomatology, dermatology and ophthalmology.[20,21) Pernicious anemia is a specific indication for vitamin B_{12} coenzyme.[22) This agent promotes regeneration of nervous tissue and is nonspecifically used in the treatment of nervous diseases including diabetic nervous disorder.[22,23) Vitamin L_2 is a lactagog.[24) The activating effect of ATP on muscular metabolism has been applied to the treatment of muscular diseases such as muscular atrophy and coronary or cerebrovascular disorders.[25)

Metabolic regulators other than vitamins and coenzymes include inosine and adenine. Inosine is prescribed for coronary insufficiency, in which it promotes sugar metabolism under an enzymoprivic environment due to ATP deficiency,[26) for hepatic diseases in which it activates pyruvic acid oxidase,[27) for leukopenia during chemotherapy and radiotherapy of cancer, and for digitalism.[28) Adenine is also applied in the treatment of leukopenia of various origins.[29)

Nucleic acid analogs are rarely used as antiinflammatory drugs in practice, except for allopurinol, which is used for the treatment of chronic gout.[30) Allopurinol was synthesized during the search for antitumor drugs. Though its antitumor activity was slight, it proved to inhibit xanthine oxidase and was then indicated for gout.[31) Though some N^6-substituted derivatives of adenine are effective against edemas, they are not used clinically.[32)

cAMP is of clinicopharmacological interest, since it has been reported to have such pharmacological properties as transmission of hormones, inhibition of HeLa cells,[33) prolongation of survival of rats with cerebral tumors,[34) improvement of cerebral function[35) and elevation of blood sugar.[36)

21.4 Antitumor Agents

Since it was confirmed that the antibacterial action of sulfonamides against infection originates from its inhibition of folic acid synthesis by antagonism between sulfonamides and *p*-aminobenzoic acid, this concept has been applied to antitumor agents. A number of nucleic acid analogs have been prepared and tested for antitumor activity since the 1940's. Of purine analogues, 8-azaguanine (8AG) was first developed and used against solid cancer.[37] 6-Mercaptopurine (6MP) was developed next and is one of the most effective drugs against acute leukemia.[38] Azathioprine is the only agent that is used clinically among *S*-substituted derivatives of 6MP. Azathioprine, which has long-lasting effectiveness, is biologically transformed into 6MP in the organism and plays a role as an immunosuppressant suppressing the rejection of transplanted tissues.[39] 6MP riboside has recently been confirmed to have clinical usefulness.[38] 8AG was indicated for solid cancer, and thioguanine[40] for acute leukemia: these are analogous to guanine and competitively antagonize it. Arabinosyl-6-MP is under clinical trial.[41]

Pyrimidine analogs include 5-fluorouracil, 5-fluoro-2′-deoxyuridine, FT-207, 6-azauridine, cytosine arabinoside and cyclocytidine. The most interesting of these is 5-fluorouracil[42] and thousands of case reports have been published. It controls solid cancer by antagonistically inhibiting thymidine synthetase. FT-207, an N^1-substituted derivative of 5-fluorouracil, is transformed into 5-fluorouracil in organisms, and its high absorbability in the intestine makes it a good drug for oral use.[43] Cytosine arabinoside controls the growth of leukemic cells L-1210 and thus prolongs the survival of mice transplanted with such cells into the brain. The mechanism of action of cytosine arabinoside, involving inhibition of cytidine diphosphate reductase, is useful in the treatment of acute leukemia.[44] Cytosine arabinoside, which is inactivated by cytidine deaminase, has been replaced by cyclocytidine. The latter is gradually transformed into cytosine arabinoside in organisms and thus becomes effective. Cyclocytidine is six times less active but 20 times less toxic compared with cytosine arabinoside.[45] It is now under clinical evaluation.

21.5 ANTIVIRAL AGENTS AND INTERFERON INDUCERS (IFI)

There are a number of metabolic antagonists possessing antibacterial or antiviral action, but they are of little practical value because of weak activity or severe side effects. 5-Iodo-2′-deoxyuridine has been synthesized during attempts to find antitumor agents.[46] Its effectiveness against herpes

TABLE 21.1
Major Nucleic Acid-related drugs

Agents	Pharmacological properties	Indications
For the central nervous system:		
Caffeine 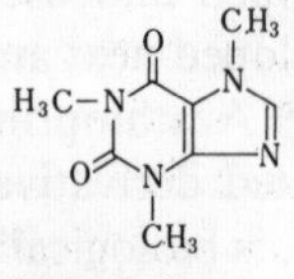	Cerebral stimulation Myocardiac stimulation Vasodilation Coronary vasodilation Striated muscle contraction Diuresis	Cardiac insufficiency Poisoning by central nervous system depressants Pyrexia
CDP-choline 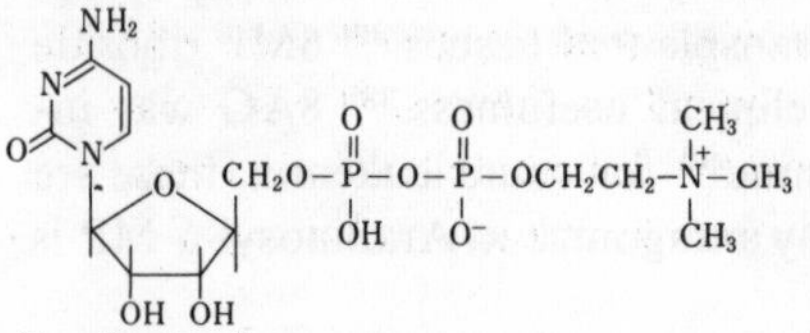	Improvement of cerebral circulation and metabolism	Disturbance of consciousness caused by injuries to the head or following cerebral operations
For the circulatory organs:		
Oxyetophylline 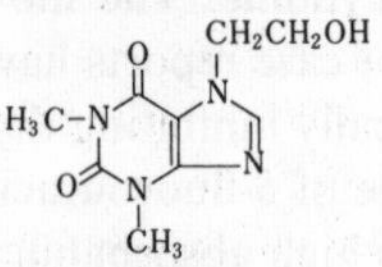	Vasodilation Increase of stroke volume Smooth muscle relaxation Diuresis	Bronchial asthma Coronary disorders Cardiac edema
Proxyphylline 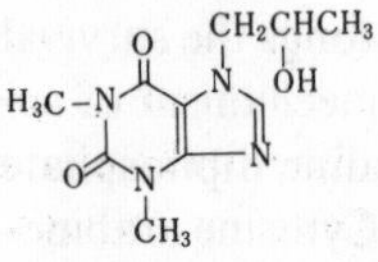	"	"
Diprophylline 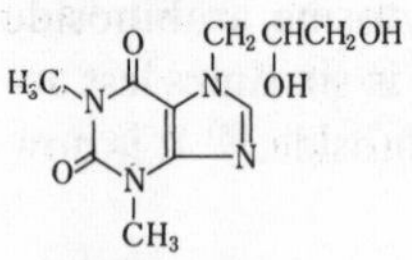		
TH-24 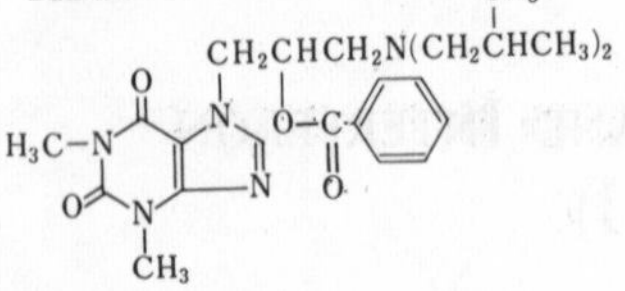	"	"

TABLE 21.1—*Continued*

Agents	Pharmacological properties	Indications
ST-7 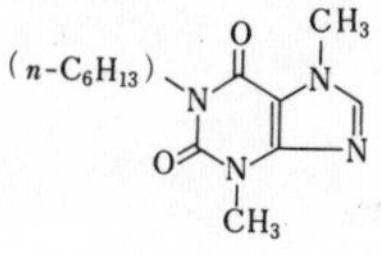 $(n\text{-}C_6H_{13})$, CH_3, CH_3 (xanthine derivative)	Vasodilation Increase of stroke volume Smooth muscle relaxation Diuresis	Bronchial asthma Coronary disorders Cardiac edema
Adenosine 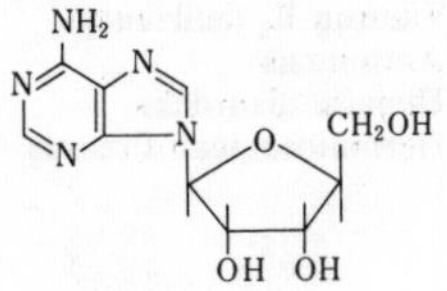 (NH_2, CH_2OH, OH OH)	Increase of coronary blood flow Vasodilation	Coronary disorders Angina pectoris Arteriosclerosis Hypertension
Lentinacin† 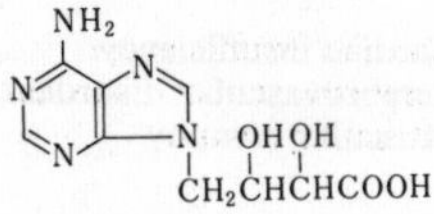 (NH_2; $CH_2CHCHCOOH$ with OH OH)	Decrease of cholesterol	Arteriosclerosis
For the urinary organs:		
Theobromine 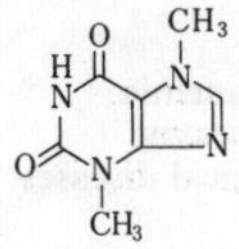 (CH_3, CH_3)	Cerebral stimulation Myocardiac stimulation Vasodilation Coronary vasodilation Striated muscle contraction Diuresis	Arteriosclerosis Cardiac edema Hypertension Coronary disorders
Theophylline 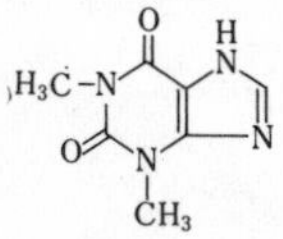 (H_3C, CH_3)	"	"

TABLE 21.1—*Continued*

Agent	Pharmacological properties	Indications
Metabolic agents:		
NAD 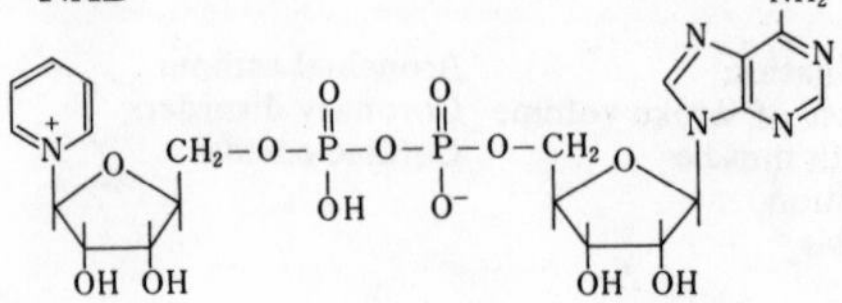		Pellagra
FAD 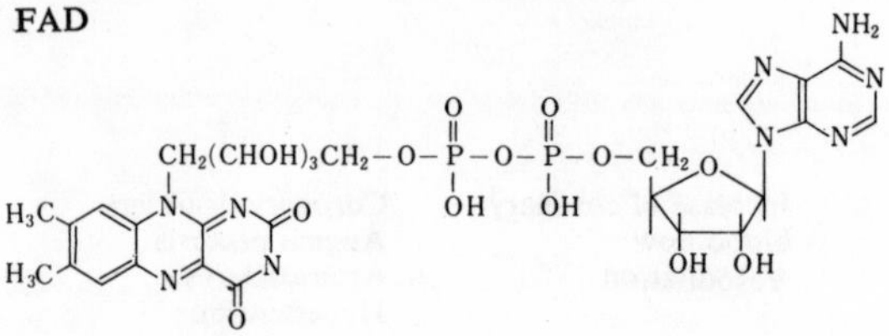		Vitamin B_2 deficiency Arabinosis Hepatic disorders Dermatological diseases
ATP 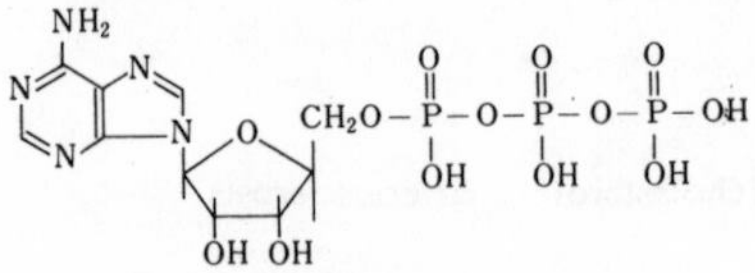		Cardiac insufficiency Cerebrovascular disorders Muscular atrophy
Vitamin B_{12} coenzyme 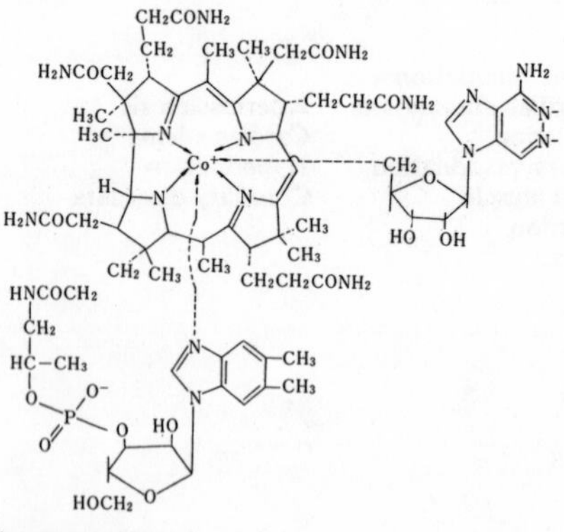		Pernicious anemia Nervous diseases Dermatological diseases
Coenzyme A 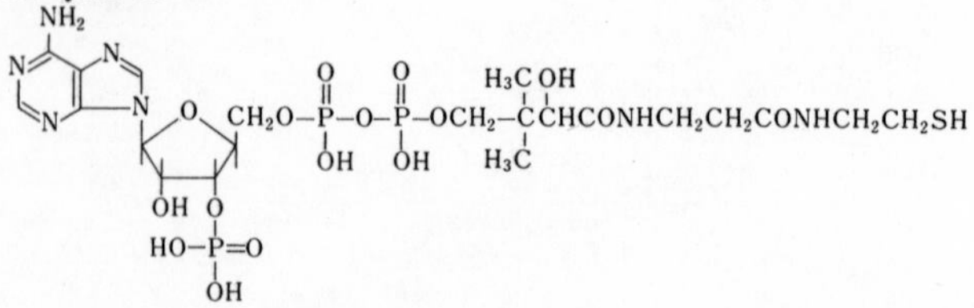		″

TABLE 21.1—*Continued*

Agent	Pharmacological properties	Indications
Vitamin L_2 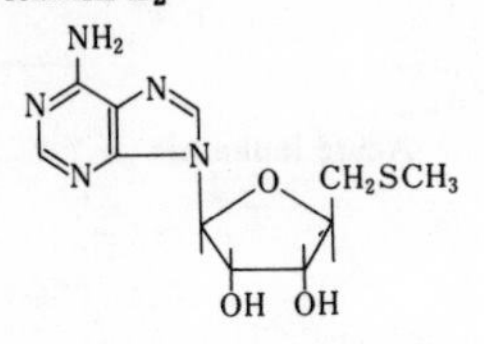		Used as a lactagog
Allopurinol	Inhibition of uric acid biosynthesis	Hyperuricacidemia
Inosine	Reactivation of cells	Coronary insufficiency Hepatic diseases Angina pectoris Leukopenia Digitalism
Adenine		Leukopenia
cAMP	Transmission of hormones	Not used clinically
Antitumor agents: 8-Azaguanine	Antagonism with guanine	Solid cancer

Table 21.1—*Continued*

Agents	Pharmacological properties	Indications
6-Mercaptopurine 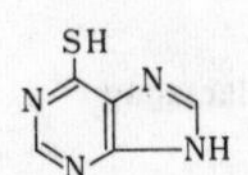	Inhibition of the transformation of IMP into AMP or GMP	Acute leukemia
6-Mercaptopurine riboside	″	″
6-Methylthiopurine riboside	″	″
Thioguanine	Antagonism with guanine	″
Arabinosyl-6-mercaptopurine†		″
β-2′-Deoxythioguanine†		″

TABLE 21.1—*Continued*

Agents	Pharmacological properties	Indications
5-Fluorouracil	Inhibition of thymidine synthetase	Solid cancer
5-Fluoro-2′-deoxyuridine	″	″
FT-207	″	″
6-Azauridine		″
Cytosine arabinoside	Inhibition of cytidine diphosphate reductase	Acute leukemia
Cyclocytidine†	″	″

TABLE 21.1—*Continued*

Agent	Pharmacological properties	Indications
Immunosuppressant:		
Azathioprine 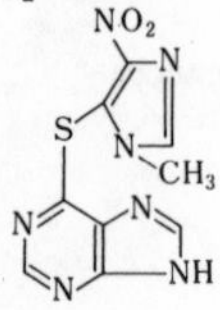		Inhibition of the rejection of transplanted tissues
Antivirotic:		
Idoxuridine 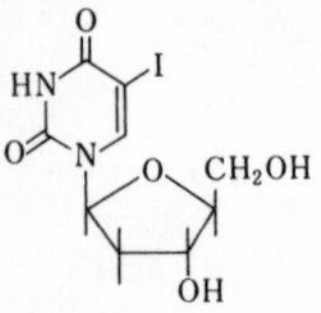	Antagonism with thymidylic acid	Corneal lesions

† Under clinical study.

virus has led to its use in the treatment of corneal lesions.[47] This agent is said to exert antiviral action by antagonizing thymidylic acid and inhibiting DNA synthesis.

Most double-stranded RNA's are highly active in inducing IF. The discovery of IFI goes back to the 1960's when natural statolon[48] and helenine[49] were found to produce IF. Since then, similar RNA's have been extracted from reovirus,[50] mycophage,[51] bacteriophage,[52] fungi,[53] etc. Synthetic poly(I:C)[54] is also an IF inducer. IFI are effective preventives against experimental viral infections and viral tumors. In spite of their reputation as most promising drugs, their usefulness in treatment has not yet been confirmed. This is probably because IFI cause various side effects[55] and induce tolerance. However, appropriate administration of IFI does give encouraging results and their use in medical treatment may therefore be worthwhile.[56,57]

Table 21.1 summarizes the major nucleic acid-related drugs.

REFERENCES

1. Y. Asahina, *Dai-Roku Kaisei Nippon Yakkyokuho Chukai* (Japanese), p. 383, Nankodo, 1956.
2. M. Ishidate, *Dai-Hachi Kaisei Nippon Yakkyokuho Dai-Ichibu Kaisetsusho* (Japanese), p. C–553, Hirokawa Shoten, 1970.
3. L. S. Goodman and A. Gilman, *The Pharmacological Basis of Therapeutics*, 4th ed., p. 358, Collier-Macmillan, 1970.
4. M. Ishidate, *Dai-Hachi Kaisei Nippon Yakkyokuho Dai-ichibu Kaisetsusho* (Japanese), p. C-1131, Hirokawa Shoten, 1971.
5. E. B. Goodsell, H. H. Stein and K. Juhasz, *Pharmacologists*, **12**, 291, Abstr. 499 (1970).
6. W. Szirma, *Praxis*, **58**, 412 (1969).
7. R. B. Moffett, A. Robert and L. L. Skaletzky, *J. Med. Chem.*, **14**, 963 (1971).
8. D. Todorov and D. Paskov, *Agressologie*, **12**, 113 (1971).
9. E. P. Kennedy, *Fed. Proc. Balt.*, **16**, 847 (1957).
10. O. Hayaishi, T. Araki, S. Ishii and Y. Kondo, *Nisshinigaku* (Japanese), **48**, 519 (1961).
11. *New Drugs in Japan* (Japanese), **15**, 85, Yakuji-Nippo-Sha, 1964.
12. M. Ishidate, *Dai-Hachi Kaisei Nippon Yakkyokuho Dai-Ichibu Kaisetsusho* (Japanese), p. C-519, Hirokawa Shoten, 1971.
13. *Ibid.*, p. C-1455, 1971.
14. M. Ishidate, *Dai-Hachi Kaisei Nippon Yakkyokuho Dai-Nibu Kaisetsusho*, p. 320, Hirokawa Shoten, 1970.
15. N. Yago, M. Tsunoho and S. Ohashi, *Folia Pharmacol. Jap.* (Japanese), **56**, 180§ (1960).
16. *New Drugs in Japan* (Japanese), **15**, 172, Yakuji-Nippo-Sha, 1962.
17. G. V. R. Born, Nature, **202**, 95 (1964).
18. K. Kikugawa, K. Iizuka, Y. Higuchi, H. Hirayama and M. Ichino, *J. Med. Chem.*, **15**, 387 (1972).
19. K. Kikugawa, H. Suehiro and M. Ichino, *ibid.*, **16**, 1381 (1973).
20. A. Shigekawa, *J. New Remedies and Clinics* (Japanese), **16**, 1551 (1967).
21. S. Ogawa, M. Abe, Y. Nishio, S. Kihara, M. Nishibashi, M. Hamada, H. Yasunari and N. Kanekawa, *Clinical Reports* (Japanese), **2**, 580 (1968).
22. N. Uchino, *The Pharmaceuticals Monthly* (Japanese), **12**, 165 (1970).
23. S. Kito, T. Muro, M. Asamaga and K. Kuboshima, *Shinryo* (Japanese), **23**, 1948 (1970).
24. T. Takebe, *Saishin-Yakubutsugaku* (Japanese), p. 554, Hirokawa Shoten, 1954.
25. *ATP No Kiso To Rinsho* (Japanese), ATP Kenkyukai, 1961.
26. I. Wada, *Med. Consultation and New Remedies* (Japanese), **4**, 1513 (1967).
27. H. Saito, T. Niimi, K. Uemura and K. Hirano, *J. New Remedies and Clinics* (Japanese), **15**, 1331 (1961).
28. K. Ashikawa, U. Samaru and T. Inoue, *Med. Consultation and New Remedies* (Japanese), **5**, 1303 (1968).
29. *New Drugs in Japan* (Japanese), **12**, 81, Yakuji-Nippo-Sha, 1961.
30. T. Yoshimura, *Rheumati* (Japanese), **13**, 168 (1972).
31. H. E. Skipper, R. K. Robin, J. R. Thomson, C. C. Cheng, R. W. Brockman and F. M. Schabel, *Cancer Res.*, **17**, 579 (1957).
32. A. S. Bhargva and, T. Diamantstein: Arzneimitelforsch., **24**, 6 (1974).
33. M. Nagai, *Igaku No Ayumi* (Japanese), **79**, 357 (1971).

34. T. Kamino, T. Nakazawa and U. Kato, *Clinical Neurology* (Japanese), **13,** 726 (1973).
35. Y. Marukado and M. Kodama, *Saigaiigaku* (Japanese), **16,** 909 (1973).
36. R. A. Levine and S. Lewis, *Biochem. Pharmacol.*, **18,** 15 (1969).
37. A. Kidder, *Proc. Nat. Acad. Sci. U.S.A.*, **34,** 566 (1948).
38. J. H. Burchenal, M. P. Sykes, T. C. Tan, L. A. Leone, L. F. Craver, M. L. Murphy, R. R. Ellison, D. A. Karnofsky, H. W. Darglon and C. P. Rhoads, *Blood*, **8,** 965 (1953).
39. T. Markinodan, G. W. Santos and R. P. Quinn, *Pharmacol. Rev.*, **22,** 189 (1970).
40. G. H. Hitchings and C. P. Rhoads, *Ann. N.Y. Acad. Sci.*, **60,** 183 (1954).
41. A. P. Kimball, G. A. Lepage, B. Bowman, P. S. Buch and J. Herriot, *Cancer Res.*, **26,** 1337 (1966).
42. *New Drugs in Japan* (Japanese), **24,** 93, Yakuji-Nippo-Sha, 1973.
43. N. Itomi, *Jap. J. Cancer Clinics* (Japanese), **17,** 731 (1971).
44. J. S. Evans, E. A. Misser, G. D. Mengel, K. R. Forsblad and J. H. Hunter, *Proc. Soc. Exptl. Biol. Med.*, **106,** 350 (1961).
45. K. Kuretani, *Igaku No Ayumi* (Japanese), **89,** 795 (1972).
46. W. H. Prusoff, *Biochim. Biophys. Acta*, **32,** 295 (1959).
47. S. Uchida, *Acta Soc. Ophthalmol. Jap.* (Japanese), **66,** 7 (1962).
48. H. M. Powell, C. G. Culbertson, J. M. Mcguire, M. Hoehn and L. A. Baker, *Antibiot. Chemother.*, **2,** 432 (1952).
49. R. E. Shope, *J. Exptl. Med.*, **97**, 601 (1953).
50. A. A. Tyfell, G. P. Lampson, A. K. Field and M. R. Hillman, *Proc. Nat. Acad. Sci. U.S.A.*, **58,** 1719 (1967).
51. G. P. Lampson, A. A. Tyfell, A. K. Field, M. M. Names and M. R. Hillman, *ibid.*, **58,** 782 (1967).
52. A. K. Field, G. P. Lampson, A. A. Tyfell, M. M. Nemes and M. R. Hillman, *ibid.*, **58,** 2102 (1967).
53. A. Tsunoda and N. Ishida, *Ann. N. Y. Acad. Sci.*, **173,** 719 (1970).
54. M. R. Hillman, A. K. Field, A. A. Tytell and G. P. Lampson, *Proc. Nat. Acad. Sci. U.S.A.*, **61,** 340 (1968).
55. R. H. Adamson and S. Fabro, *Nature*, **223,** 718 (1968).
56. D. A. Hill, S. Baron, C. Perkins, M. Worthington, J. E. Vankirk, J. Mills, A. Z. Kapikian and R. M. Chenock, *J. Am. Med. Assoc.*, **219,** 1179 (1972).
57. R. Guerra, R. Frezzotti, R. Bonanni, F. Dianzani and G. Rita, *Ann. N.Y. Acad. Sci.*, **173,** 823 (1970).

CHAPTER **22**

Nucleoside-related Antibiotics

More than half the known nucleoside-related antibiotics were discovered in Japan. Most of them are nucleoside analogs and many of them have one or two amino acids or amino sugars attached. Thirty-five antibiotics are listed in Table 22.1. Since some of them contain several components, the total number of antibiotics in this group exceeds 50. It is worth noting that aminoacylnucleosides (blasticidin S and the polyoxins) have been used practically as agricultural fungicides in Japan. However, almost no outstanding medicinal applications of nucleoside antibiotics in this group have been achieved. The mechanism of action of most of this group of antibiotics is related to the inhibition of nucleic acid and protein synthesis and often lacks the selective toxicity necessary for medicinal applications. On the other hand, many of these compounds have proved to be powerful tools for biochemical studies. Puromycin, tubercidin, formycin, and cordycepin are outstanding in this respect.

On the basis of chemical structure, the nucleoside-related antibiotics

* Kiyoshi ISONO, Institute of Physical and Chemical Research, Wako-shi.

TABLE 22.1

Classification of Nucleoside-related Antibiotics

Class	Subclass	Antibiotic
1. Base analogs		8-Azaguanine (I)
		Bacimethrin (II)
		Emimycin (III)
2. *N*-Nucleoside	Sugara analogs	3′-Amino-3′-deoxyadenosine (IV)
		3′-Deoxyadenosine (Cordycepin) (V)
		Arabinosyladenine (Ara A) (VI)
		Psicofuranine (Angustmycin C) (VII)
		Decoyinine (Angstmycin A) (VIII)
		Aristeromycin (IX)
		Nucleocidin (X)
		2′-Amino-2′-deoxyguanosine (XI)
	Base analogs	Tubercidin (XII)
		Toyocamycin (XIII)
		Sangivamycin (XIV)
		5-Azacytidine (XV)
		Neburaline (XVI)
		Bredinin (XVII)
3. C-Nucleosides		Formycin (XVIII)
		Formycin B (Laurusin) (XIX)
		Pyrazomycin (XX)
		Showdomycin (XXI)
		Minimycin (Oxazinomycin) (XXII)
4. Aminoacylnucleosides		Homocitrullylaminoadenosine (XXIII)
		Lysylaminoadenosine (XXIV)
		Puromycin (XXV)
		Gougerotin (XXVI)
		Blasticidin S (XXVII)
		Polyoxins (XXVIII)
5. Aminoglycosylnucleosides		Amicetin (XXIX)
		Bamicetin (XXX)
		Plicacetin (XXXI)
		Oxamicetin (XXXII)
		Hikizimycin (XXXIII)
6. Anhydrouronic acid nucleosides		Ezomycins (XXXIV)
		Herbicidins (XXXV)

TABLE 22.1—*Continued*

Streptomyces albus var. *pathocidicus*[1~3], *Streptomyces moro-okaensis*[4,5]
Bacillus megaterium[6]
Streptomyces griseochromogenes[7,8]

Helminthosporium sp.[9], *Cordyceps militaris*[10], *Aspergillus nidulans*[11]
Cordyceps militaris[12], *A. nidulans*[13]
Streptomyces antibioticus[14]
Streptomyces hygroscopicus[15~18]
Streptomyces hygroscopicus[15~18]
Streptomyces citricolor[19]
Streptomyces clavus[20~22]
Aerobacter sp.[23]

Streptomyces tubercidicus[24,25], *Streptomyces sparsogenes* var. *sparsogenes*[26]
Streptomyces toyocaensis[27], *Streptomyces* sp.[28]
Streptomyces rimosus[29,30]
Streptomyces ladakanus[31]
Agaricus (*Clitocybe*) *neburalis*[32], *Streptomyces yokosukaensis*[33,34]
Eupenicillium brefeldianum[35]

Nocardia interforma[33], *Streptomyces lavendulae*[37]
N. interforma[38], *St. lavendulae*[37]
Streptomyces candidus[39]
Streptomyces showdoensis[40]
Streptomyces hygroscopicus[41], *Streptomyces tanesashiensis*[42]

Cordyceps militaris[43]
C. militaris[44]
Streptomyces alboniger[45]
Streptomyces gougerotii[46]
Streptomyces griseochromogenes[47]
Streptomyces cacaoi var. *asoensis*[48,49]

Streptomyces venaceus[50], *Streptomyces fasciculatis*[51], *Streptomyces plicatus*[52]
Streptomyces plicatus[52]
Streptomyces plicatus[52]
Arthrobacter oxamicetus sp.[53,54]
Streptomyces sp.[55]
Streptomyces kitazawaensis[56]
Streptomyces sp.[57]

Fig 22.1

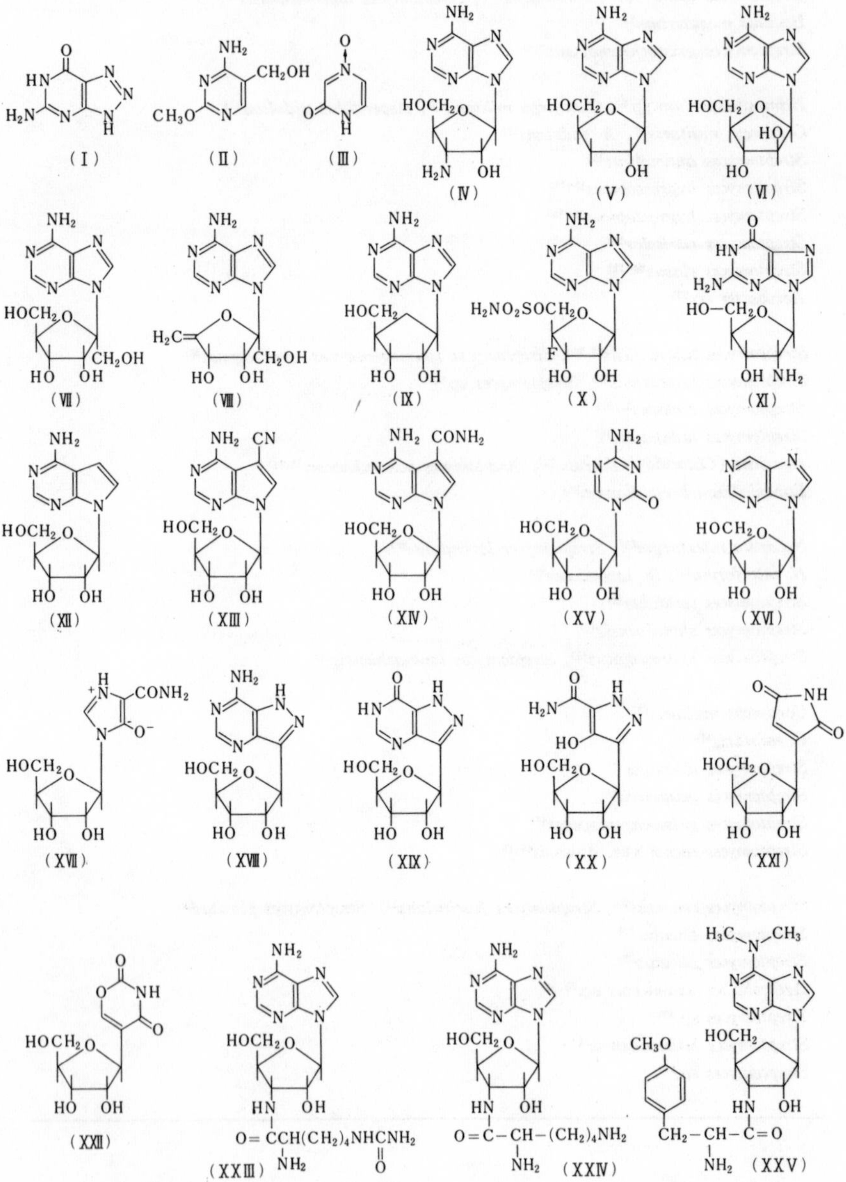

(I) (II) (III) (IV) (V) (VI)

(VII) (VIII) (IX) (X) (XI)

(XII) (XIII) (XIV) (XV) (XVI)

(XVII) (XVIII) (XIX) (XX) (XXI)

(XXII) (XXIII) (XXIV) (XXV)

Fig. 22.1—Continued

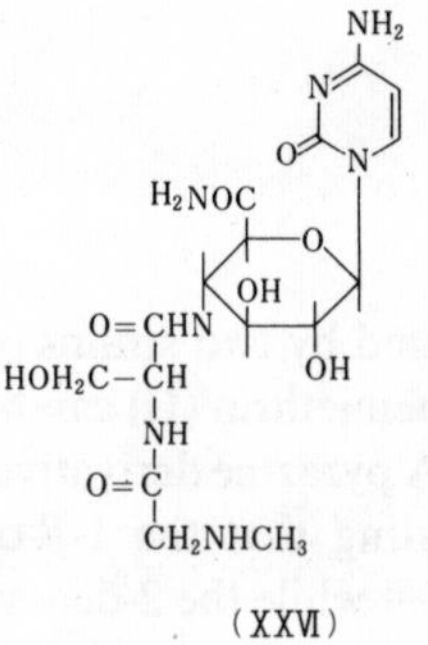

(XXVI) 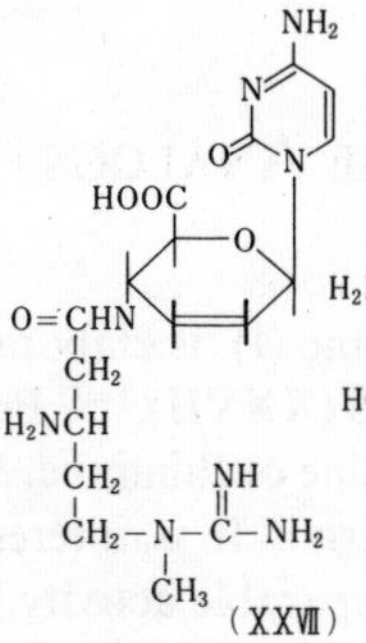 (XXVII) 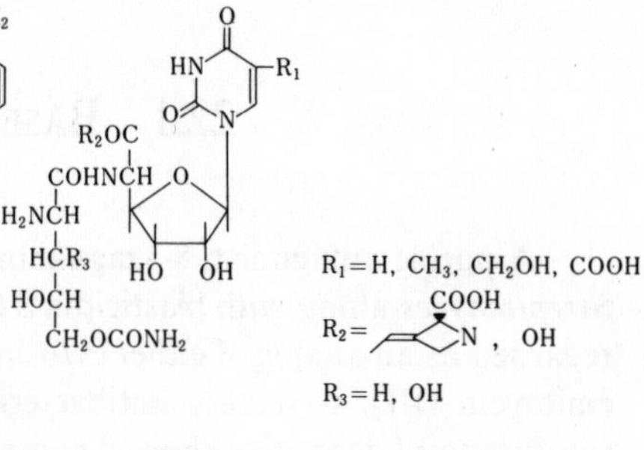 (XXVIII)

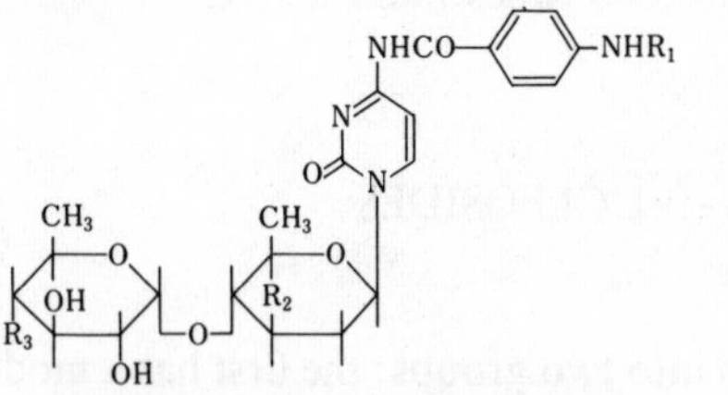

$R_1 = H, CH_3, CH_2OH, COOH$

$R_2 = $ [azetidine-2-carboxylic acid residue], OH

$R_3 = H, OH$

(XXIX) $R_1 = \alpha$-Methylseryl, $R_2 = H$, $R_3 = N(CH_3)_2$

(XXX) $R_1 = \alpha$-Methylseryl, $R_2 = H$, $R_3 = NHCH_3$

(XXXI) $R_1 = H$, $R_2 = H$, $R_3 = N(CH_3)_2$

(XXXII) $R_1 = \alpha$-Methylseryl, $R_2 = OH$, $R_2 = N(CH_3)_2$

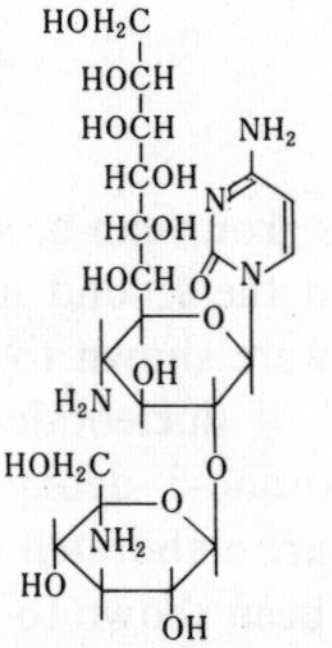

(XXXIII)

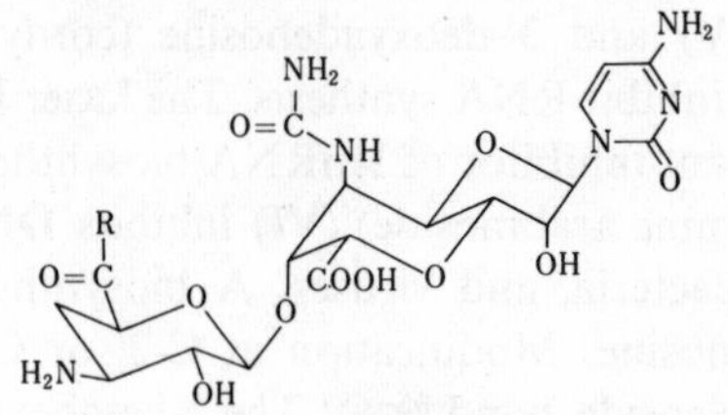

(XXIV) R = OH, L-Cystathionyl

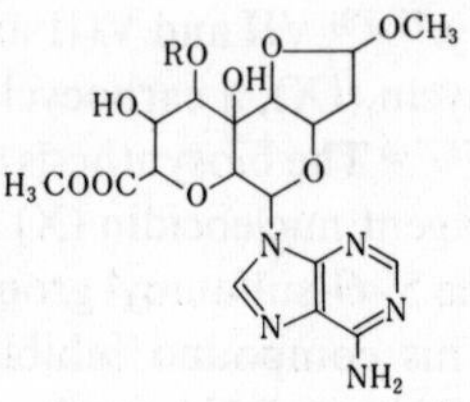

(XXXV)

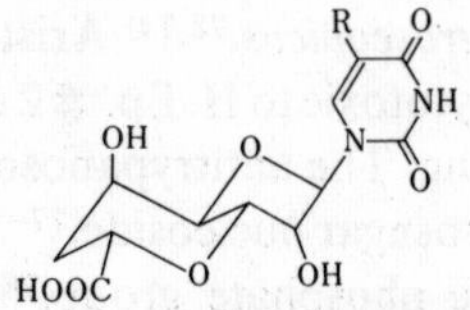

$R = CH_2OH, COOH$

(XXXVI)

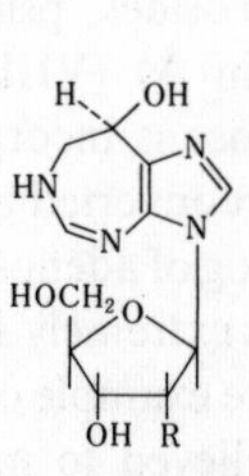

(XXXVII) R = OH

(XXXVIII) R = H

can be divided into six groups, as shown in Table 22.1. Interestingly, the mode of action and the biosynthesis are related to this classification, as will be discussed later. The chemical structures are shown in Fig. 22.1.

22.1 Base Analogs

A purine antagonist, 8-azaguanine (I), is elaborated by two strains of *Streptomyces* along with blasticidin S (XXVII).[1–5] Bacimethrin (II) can be regarded as an analog of either cytosine or thiamine. A pyrazine derivative, emimycin (III), is weakly antibacterial. It is interesting that the 1-β-D-ribofuranosyl derivative showed comparable activity,[58] while the 2-deoxyribofuranoside derivative was 1,000,000 times more active.[59]

22.2 *N*-Nucleosides

This group can be subdivided into two groups; the first has a modified base, and the second has a modified sugar moiety. Many of these compounds were shown to be phosphorylated *in vivo* and were inhibitory at the level of nucleotide metabolism. Inhibition is often multifunctional.

3′-Amino-3′-deoxyadenosine (IV) and 3′-deoxyadenosine (cordycepin) (V) are elaborated by fungi and inhibit RNA synthesis. The latter has recently been shown to be an important inhibitor of HnRNA biosynthesis in eucaryotic cells.[60–64] Ara A (adenine arabinoside) (VI) inhibits DNA synthesis in cultured animal cells, bacteria, and viruses. A biosynthetic precursor of IV, V, and VI is adenosine. Modification at C-2′ or C-3′ proceeds without cleavage of the nucleoside bond.[65–68] The 2-ketohexose nucleosides, psicofuranine (angustmycin C) (VII) and decoyinine (angustmycin A) (VIII), inhibit XMP aminase.[69–71] The carbon skeleton of glucose is incorporated intact into the sugar moiety.[72,73] VII and VIII are interconverted by *S. hygroscopicus*.[73,74] Aristeromycin (IX), a carbocyclic analog of adenosine, is cytotoxic to H. Ep. #2 cells.[75,76] The biosynthesis of IX is extremely interesting. The antitrypanosomal agent nucleocidin (X) is a rare example of a fluorosugar nucleoside.[77–79] The 5′-*O*-sulfamoyl group is believed to mimic the phosphate group.[78,79] This compound inhibits protein synthesis,[80] in contrast to the inhibition of DNA or RNA synthesis shown by other members of this group. 2′-Amino-2′-deoxyguanosine

(XI), produced by *Aerobacter* sp., is toxic to HeLa cells, Sarcoma 180, and *E. coli* KY 8323.[81)]

The pyrrolopyrimidine nucleosides, tubercidin (XII), toyocamycin (XIII), and sangivamycin (XIV), are elaborated by several strains of *Streptomyces*. This group of adenosine antagonists is highly cytotoxic to vertebrate cell lines is culture. They are phosphorylated *in vivo* and often replace the corresponding adenosine phosphates. GTP is believed to be a biosynthetic precursor for this group of antibiotics. Like pterines, the imidazole ring is cleaved by GTP-formylhydrolase and the carbon skeleton of ribose serves as the carbon source for the pyrrole ring.[82-84)] Toyocamycin nitrilhydrolase was partially purified; it converts XIII to XIV.[85)] 5-Azacytidine (XV) is effective against transplanted leukemia of AKR mice. Neburaline (XVI), obtained from a mushroom,[32)] was later shown to be elaborated by *Streptomyces* sp.[33,34)] A fungal metabolite, bredinin (XVII), is considered to be an AICAR analog and shows immunosuppressive activity. An aglycone added to the medium was ribosylated to give XVII.[86)]

22.3 *C*-Nucleosides

The pyrazolopyrimidine nucleosides, formycin (XVIII) and formycin B (XIX), were the first examples of C-nucleosides. An adenosine analog, XVIII shows a wide range of toxicity, but an inosine analog, XIX was selectively active against *Xanthomonas oryzae*. XVIII and XII are important biochemical reagents. Coformycin (XXXVII), produced by *Nocardia interforma* together with formycins, is a potent inhibitor of adenosine deaminase. It inhibits the reaction XVIII → XIX, thus exerting synergistic action with XVIII.[87,88)] It is interesting that *S. antibioticus*, which produces ara A (VI), also elaborates 2′-deoxycoformycin (XXXVIII).[89)] Pyrazomycin (XX), which is structurally related to formycins, shows antiviral activity. The 5′-phosphate is a potent inhibitor of orotidylate decarboxylase.[90)] Covalent bond formation between the maleimide ring of showdomycin (XXI) and the sulfhydryl group of specific enzymes is believed to be the basis of the selective activity of XXI.[91)] C-2 through C-5 of α-ketoglutarate are asymmetrically incorporated into the maleimide ring of XXI.[92-94)] The carbon skeleton of the oxazine ring and C-riboside of minimycin (XXII) is suggested to come from a C_7 sugar such as sedoheptulose.[95)] Thus, the biosynthesis of C-nucleosides is considered not to be related to nucleic acid metabolism.

22.4 AMINOACYLNUCLEOSIDES

One or two amino acids are amide-linked to the amino group of the amino sugar moiety in this group of antibiotics. The site of action is related to protein synthesis, except for the polyoxins (XXVIII). Inhibition of peptidyltransferase by puromycin (XXV) (the puromycin reaction) is an important reaction in studies of the mechanism of protein synthesis.[96,97] It is interesting to note that homocitrullyladenosine (XXIII) inhibits protein synthesis, whereas the nucleoside moiety, 3′-amino-3′-deoxyadenosine (IV), shows inhibition of RNA synthesis. Blasticidin S (XXVII) shows a broad antibacterial spectrum by inhibiting protein synthesis. It has been used for the prevention of rice blast disease. The polyoxins (XXVIII) represent a group of nucleoside peptides. Thirteen members are known.[98] Structural resemblance to UDP-*N*-acetylglucosamine is the basis of their action as competitive inhibitors of chitin synthetase for the fungal cell wall.[99] They have excellent selective toxicity and are widely used as an agricultural fungicide for sheath blight disease of rice plants, *Alternaria* diseases of fruit trees, etc. 5-Substituted uracils are biosynthesized from uracil and C-3 of serine. Such synthesis is independent of thymidylate synthetase.[100] It is particularly interesting that the feeding of 5-fluorouracil resulted in the formation of 5-fluoropolyoxins (XXVIII, $R_1 = F$).[101]

22.5 AMINOGLYCOSYLNUCLEOSIDES

Amicetin (XXIX) has a 4-amino-4-deoxy sugar glycosyl moiety bonded to the cytosine nucleoside. It inhibits protein synthesis. Bacimethrin (XXX), plicacetin (XXXI), and oxamicetin (XXXII) belong to this group. Hikizimycin (XXXIII) has a 3-amino-3-deoxy sugar glycosyl moiety bonded to a 4-amino-4-deoxy sugar nucleoside. It also inhibits protein synthesis.[102] The common structural feature of aminoacyl- and aminoglycosyl-pyrimidine nucleosides which cause inhibition of protein synthesis is the presence of (1) cytosine and (2) a 4-amino-4-deoxy sugar.

22.6 ANHYDROURONIC ACID NUCLEOSIDES

Ezomycins (XXXV), which are inhibitory to a very narrow range of fungi, belong to the aminoglycosylnucleosides. In addition, they have an anhydrooctose uronic acid nucleoside structure.[104] Some members have recently been shown to possess a pseudouridine-type 5-uracil nucleoside structure.[106] The same carbon skeleton appears in octosyl acids (XXXVI), metabolites of the polyoxin producing *Streptomyces*.[105] Herbicidins A and B (XXXV) are weakly antifungal and show herbicidal activity.[57,94] They are adenosine nucleosides having an anhydro-sugar uronic acid moiety.[107] These three groups of nucleosides constitute a new class of anhydrouronic acid nucleosides.

References

Reviews: the following reviews cover nucleoside antibiotics discovered before 1969:
(a) R. J. Suhadolnik, *Nucleoside Antibiotics*, Interscience, 1970.
(b) P. Roy-Burman, *Analogues of Nucleic Acid Components, Mechanism of Action, Recent Results in Cancer Research* 25, Springer-Verlag, 1970.

1. J. Nagatsu, K. Anzai and S. Suzuki, *J. Antibiotics*, **A15**, 103 (1962).
2. K. Anzai and S. Suzuki, *ibid.*, **A14**, 253 (1961).
3. K. Anzai, S. Suzuki and J. Nagatsu, *ibid.*, **A14**, 340 (1961).
4. T. Niida, K. Hamamoto, T. Tsuruoka and T. Hara, *Meika Kenkyu Nenpo* (Japanese), **6**, 27 (1963).
5. H. Tsuruoka and T. Niida, *ibid.*, **6**, 23 (1963).
6. H. Yonehara, H. Umezawa and Y. Sumiki, *J. Antibiotics*, **A16**, 161 (1961).
7. M. Terao, K. Karasawa, N. Tanaka, H. Yonehara and H. Umezawa, *ibid.*, **A13**, 401 (1960).
8. M. Terao, *ibid.*, **A16**, 182 (1963).
9. C. A. Ammann and R. S. Safferman, *Antibiot. Chemother.* **8**, 1 (1958).
10. A. J. Guarino and N. M. Kredich, *Biochim. Biophys. Acta*, **68**, 317 (1963).
11. R. J. Suhadolnik, B. M. Chassy and G. R. Walter, ~~*Biochim. Biophys. Acta*,~~ **179**, 258 (1969).
12. K. G. Cunningham, S. A. Hutchinson, W. Manson, and F. S. Spring, *J. Chem. Soc.*, **1951**, 2299.
13. E. A. Kaczka, E. L. Dulancy, C. O. Gitterman, H. B. Woodruff and K. Folkers, *Biochem. Biophys. Res. Commn.*, **14**, 452 (1964).
14. Park, Davis and Co. *Belgian Patent* No. 671557 (1967).
15. H. Sakai, H. Yüntsen and F. Ishikawa, *J. Antibiotics*, **A7**, 116 (1954).
16. H. Yüntzen, H. Yonehara and H. Ui, *ibid.*, **A7**, 113 (1954).
17. J. J. Vavra, A. Dietz, B. W. Churchill, P. Siminoff, and H. J. Kaepsell, *Antibiot. Chemother*, **9**, 427 (1959).
18. H. Hoeksema, G. Slomp and E. E. van Tamelen, *Tetr. Lett.*, **1964**, 1787.
19. T. Kusaka, H. Yamamoto, M. Shibata, M. Muroi, T. Kishi and K. Mizuno, *J. Antibiotics*, **A21**, 255 (1968).
20. E. J. Backus, H. D. Tresner and T. H. Campobell, *Antibiot. Chemother.*, **7**, 532 (1957).
21. S. O. Thomas, V. L. Singleton, J. A. Lowery, R. W. Sharpe, L. M. Pruess, J. N. Porter, J. H. Mowat and N. Bohonos, *Antibiot. Ann.*, **716**, 1956–57.
22. R. I. Hewitt, A. R. Gumble, L. H. Taylor and W. S. Wallace, *ibid.*, **722**, 1956–57.
23. T. Nakanishi, F. Tomita and T. Suzuki, *Agr. Biol. Chem.*, **38**, 2465 (1974).
24. K. Anzai and S. Suzuki, *J. Antibiotics*, **A10**, 20 (1957).
25. G. Nakamura, *ibid.*, **A14**, 90 (1961).
26. W. J. Wechter and A. R. Hanze, *U. S. Patent* No. 3336289 (1967).

27. H. Nishimura, K. Katagiri, K. Sato M. Mayama and N. Shimaoka, *J. Antibiotics*, **A9**, 60 (1956).
28. K. Ohkuma, *ibid.*, **A13**, 361 (1960).
29. K. V. Rao and D. W. Renn, *Antimicrob. Agents Chemother.*, **1963**, 77.
30. K. V. Rao, W. S. Marsh and D. W. Renn, *U.S. Patent* No. 3423398 (1969); *Chem. Abstr.*, **70**, 86268y (1969).
31. L. J. Hanka, J. S. Evans, D. J. Mason and A. Dietz, *Antimicrob. Agents Chemother.*, **1966**, 619.
32. L. Ehrenberg, H. Hedstrom, N. Löfgren and B. Takman, *Svensk Kem. Tidsskr.*, **58**, 269 (1946).
33. K. Isono and S. Suzuki, *J. Antibiotics*, **A13**, 270 (1960).
34. G. Nakamura, *ibid.*, **A14**, 94 (1961).
35. K. Mizuno, M. Tsujino, M. Takada, M. Hayashi, K. Atsumi, K. Asano and T. Matsuda, *ibid.*, **27**, 775 (1974).
36. M. Hori, E. Ito, T. Takita, G. Koyama, T. Takeuchi and H. Umezawa, *ibid.*, **A17**, 96 (1964).
37. S. Aizawa, T. Hidaka, N. Otake, H. Yonehara, K. Isono, N. Igarashi and S. Suzuki, *Agr. Biol. Chem.*, **29**, 375 (1965).
38. G. Koyama, K. Maeda, H. Umezawa and Y. Iitaka, *Tetr. Lett.*, **1966**, 597.
39. R. H. Williams, K.Gerzon, M. Hoehn and D. C. DeLong, 158th Nat. Mtg. Am. Chem. Soc., 1969. Abstracts, Micro. 38.
40. H. Nishimura, M. Mayama, Y. Komatsu, H. Kato, N. Shimaoka and Y. Tanaka, *J. Antibiotics*, **A17**, 148 (1964).
41. Y. Kusakabe, J. Nagatsu, N. Shibuya, O. Kawaguchi, C. Hirose and S. Shirato, *ibid.*, **25**, 44 (1972).
42. T. Haneishi, T. Okazaki, T. Hata, C. Tamura, M. Nomura, A. Naito, I. Seki and M. Arai, *ibid.*, **24**, 797 (1971).
43. A. J. Guarino, M. L. Ibershof and R. Swain, *Biochim. Biophys. Acta*, **72**, 62 (1963).
44. A. J. Guarino and N. M. Kredich, *Fed. Proc.*, **23**, 371 (1964).
45. J. N. Porter, R. I. Hewitt, C. W. Hesseltine, G. Krupka, J. A. Lowery, W. S. Wallace, N. Bohonos and J. H. Williams, *Antibiot. Chemother.* **2**, 409 (1952).
46. T. Kanazaki, E. Higashide, H. Yamamoto, M. Shibata, K. Nakazawa, H. Iwasaki, Takewaka and A. Miyake, *J. Antibiotics*, **A15**, 93 (1962).
47. S. Takeuchi, K. Hirayama, K. Ueda, H. Sakai and H. Yonehara, *ibid.*, **A11**, 1 (1958).
48. K. Isono, J. Nagatsu, Y. Kawashima and S. Suzuki, *Agr. Biol. Chem.*, **29**, 848 (1965).
49. K. Isono, J. Nagatsu, K. Kobinata, K. Sasaki and S. Suzuki, *ibid.*, **31**, 190 (1967).
50. C. Deboer, E. L. Caron and J. W. Hinman, *J. Am. Chem. Soc.*, **75**, 499 (1953).
51. E. H. Flynn, J. W. Hinman, E. L. Caron and D. O. Woolf, Jr., *ibid.*, **75**, 5867 (1953).
52. T. H. Haskell, A. Ryder, R. P. Frohardt, S. A. Fusari, Z. L. Jakubowski and Q. R. Bartz, *ibid.*, **80**, 743 (1958).
53. M. Konishi, M. Kimeda, H. Tsukiura, H. Yamamoto, T. Hoshiya, T. Miyaki, K. Fujisawa, H. Koshiyama and H. Kawaguchi, *J. Antibiotics*, **26**, 752 (1973).
54. K. Tomita, Y. Uenoyama, K. Fuzisawa and H. Kawaguchi, *ibid.*, **26**, 765 (1973).
55. K. Uchida, T. Ichikawa, Y. Shimauchi, T. Ishikura and A. Ozaki, *ibid.*, **24**, 259 (1971).
56. K. Takaoka, T. Kuwayama and A. Aoki, *Japanese Patent* No. 46–615332 (1971).
57. M. Arai, T. Haneishi, A. Terahara, N. Kitahara, H. Kashimori and K. Kawakubo, *Japanese Patent* No. 49–101523 (1974).
58. M. Bobek and A. Bloch, *J. Med. Chem.*, **15**, 164 (1972).
59. P.T. Berkowitz, T. J. Bardos and A. Bloch, *ibid.*, **16**, 183 (1973).
60. J. E. Darnell, L. Philipson, R. Wall and M. Adesnik, *Science*, **174**, 507 (1971).
61. J. Mendecki, S. Y. Lee and G. Brawerman, *Biochemistry*, **11**, 637 (1972).
62. H. Nakazato, M. Edmonds and D. W. Kopp, *Proc. Natl. Acad. Sci. U.S.A.*, **71**, 200 (1974).
63. R. A. Weinberg, *Ann. Rev. Biochem.*, **42**, 329 (1973).

64. G. Maale, G. Stein and R. Mans, *Nature*, **255**, 81 (1975).
65. R. J. Suhadolnik, G. Weinbaum and P. Meloche, *J. Am. Chem. Soc.*, **86**, 948 (1964).
66. B. M. Chassy and R. J. Suhadolnik, *Biochim. Biophys. Acta*, **182**, 316 (1969).
67. P. B. Farmer and R. J. Suhadolnik, *Biochemistry*, **11**, 911 (1972).
68. P. B. Farmer, T. Uematsu, H. Hogenkamp and R. J. Suhadolnik, *J. Biol. Chem.*, **247**, 1844 (1973).
69. H. S. Moyed, *Cold Spring Harbor Symp. Quant. Biol.*, **26**, 323 (1961).
70. T. T. Fukuyama and H. S. Moyed, *Biochemistry*, **3**, 1488 (1964).
71. T. T. Fukuyama, *J. Biol. Chem.*, **241**, 4745 (1964).
72. T. Sugimori and R. J. Suhadolnik, *J. Am. Chem. Soc.*, **87**, 1136 (1965).
73. B. M. Chassy, T. Sugimori, and R. J. Suhadolnik, *Biochim. Biophys. Acta*, **130**, 12 (1966).
74. H. Hoeksema, G. Slomp, and E. E. van Tamelen, *Tetr. Lett.*, **1964**, 1787.
75. P. W. Allan, D. I. Hill and L. L. Bennett, Jr., *Fed. Proc.*, **26**, 730 (1967).
76. L. L. Bennett, Jr., H. P. Schnebli, M. H. Vail, P. W. Allan and J. A. Montgomery, *Mol. Pharmacol.*, **2**, 432 (1966).
77. G. O. Morton, J. E. Lancaster, G. E. van Lear, W. Fulmor and W. E. Meyer, *J. Am. Chem. Soc.*, **91**, 1535 (1969).
78. D. A. Shuman, R. K. Robins and M.J. Robins, *ibid.*, **91**, 3391 (1969).
79. D. A. Shuman and M. J. Robins, *ibid.*, **91**, 3434 (1970).
80. J. R. Florini, H. H. Bird and P. H. Bell, *J. Biol. Chem.*, **241**, 1091 (1966).
81. T. Nakanishi, F. Tomita and T. Suzuki, *Agr. Biol. Chem.*, **38**, 2465 (1974).
82. T. Uematsu and R. J. Suhadolnik, *Biochemistry*, **9**, 1260 (1970).
83. R. J. Suhadolnik and T. Uematsu, *J. Biol. Chem.*, **245**, 4365 (1970).
84. E. F. Elstner and R. J. Suhadolnik, *ibid.*, **246**, 6973 (1971).
85. T. Uematsu and R. J. Suhadolnik, *Arch. Biochem. Biophys.*, **162**, 614 (1974).
86. K. Mizuno, M. Takada, S. Atsumi, T. Ando, and T. Matsuda, 189th Mtg. Japan Antibiot. Res. Ass., 1973.
87. R. J. Suhadolnik, *Nucleoside Antibiotics*, p. 366, Interscience, 1970.
88. H. Nakamura, G. Koyama, Y. Iitaka, M. Ohno, N. Yagisawa, S. Kondo, K. Maeda and H. Umezawa, *J. Am. Chem. Soc.*, **96**, 4327 (1974).
89. P. W. K. Woo and H. W. Dikon, *J. Hetero. Chem.*, **11**, 641 (1974).
90. R. J. Suhadolnik, *Nucleoside Antibiotics*, p. 390, Interscience, 1970.
91. p. Roy-Burman, *Analogues of Nucleic Acid Components, Mechanisms of Action, Recent Results in Cancer Research* 25, p. 81, Springer-Verlag, 1970.
92. E. F. Elstner and R. J. Suhadolnik, *Biochemistry*, **10**, 3608 (1971).
93. E. F. Elstner and R. J. Suhadolnik, ~~*Biochemistry*,~~ **11**, 2578 (1972).
94. E. F. Elstner, R. J. Suhadolnik, and A. Allerhand, *J. Biol. Chem.*, **247**, 3732 (1973).
95. K. Isono and R. J. Suhadolnik, *Ann. N. Y. Acad. Sci.*, **255**, 390 (1975).
96. R. J. Suhadolnik, *Nucleoside Antibiotics*, p. 8, Interscience, 1970.
97. E. F. Gale E. Cundliffe, P. E. Reynolds, M. H. Richmond and M. J. Waring, *The Molecular Basis of Antibiotic Action*, p. 283, Interscience, 1972.
98. K. Isono, K. Asahi and S. Suzuki, *J. Am. Chem. Soc.*, **91**, 7490 (1969).
99. A. Endo, K. Kakiki and T. Misato, *J. Bact.*, **104**, 189 (1970).
100. K. Isono and R. J. Suhadolnik, *Arch. Biochem. Biophys.* (1976), in press.
101. K. Isono, P. F. Crain, T. J. Odiorne, J. A. McCloskey and R. J. Suhadolnik, *J. Am. Chem. Soc.*, **95**, 5788 (1973).
102. K. Uchida and H. Wolf, *J. Antibiotics*, **27**, 783 (1974).
103. K. Sakata, A. Sakurai, and S. Tamura, *Tetr. Lett.*, **1974**, 1533.
104. K. Sakata, A. Sakurai and S. Tamura, *ibid.*, **1974**, 4327.
105. K. Isono, P. F. Crain and J. A. McCloskey, *J. Am. Chem. Soc.*, **97**, 943 (1975).
106. K. Sakata, A. Sakurai and S. Tamura, *Tetr. Lett.*, **1975**, 3191.
107. A. Terahara, T. Haneishi, M. Arai and H. Kuwano, 18th Symp. Chemistry of Natural Products, 1974. Abstracts, p. 286.

64. G. Mayer, [illegible] and E. Manz, [illegible], 286, 91 (1975).
65. R. J. Suhadolnik, G. Weinbaum, and H. P. Meloche, J. Am. Chem. Soc., 86, 948 (1964).
66. R. M. Glazer and R. J. Suhadolnik, [illegible], 184, 316 (1969).
67. T. R. Farmer and R. J. Suhadolnik, Biochemistry, 11, 911 (1972).
68. T. E. Farmer, [illegible], and R. J. Suhadolnik, J. [illegible] Chem., 247, [illegible] (1972).
69. H. S. Moyed, Cold Spring Harbor Symp. Quant. Biol., 26, 323 (1961).
70. [illegible] and H. S. Moyed, Biochemistry, 3, 1585 (1964).
71. T. S. [illegible], J. Biol. Chem., 241, 4265 (1966).
72. [illegible] and R. J. Suhadolnik, J. Am. Chem. Soc., 87, 1150 (1965).
73. R. W. Chassy, T. Sugimori, and R. J. Suhadolnik, Biochim. Biophys. Acta, 130, 12 (1966).
74. H. [illegible], G. Sinrock, and E. L. van Tamelen, Tetr. Lett., 1964, 1785.
75. [illegible], Proc. Nat. Acad. Sci. [illegible] (1964).
76. L. L. Bennett, [illegible], M. H. Vail, [illegible], and J. A. Montgomery, Mol. Pharmacol., 2, [illegible] (1966).
77. G. O. Morton, J. E. Lancaster, G. E. Van Lear, W. Fulmor, and W. E. Meyer, J. Am. Chem. Soc., 91, 1535 (1969).
78. D. A. Shuman, [illegible] and M. J. Robins, [illegible], 91, 3391 (1969).
79. D. A. Shuman and M. J. Robins, [illegible], 92, 3434 (1970).
80. J. R. Tolman, H. D. Hill, and [illegible], J. Biol. Chem., 243, 3389 (1968).
81. [illegible], J. Am. Chem. Soc., 96, 3463 (1974).
82. [illegible] and R. J. Suhadolnik, Biochemistry, [illegible] (1970).
83. M. Elstner and R. J. Suhadolnik, J. Biol. Chem., 245, 4853 (1970).
84. T. F. Elstner and R. J. Suhadolnik, ibid., 246, 6973 (1971).
85. [illegible] and R. J. Suhadolnik, [illegible], 102, 616 (1974).
86. K. [illegible], M. [illegible], S. [illegible], T. [illegible], and [illegible], [illegible], Agric. [illegible], 35, [illegible] (1971).
87. R. J. Suhadolnik, Nucleoside Antibiotics, Wiley-Interscience, 1970.
88. H. Nakamura, [illegible], Y. [illegible], H. [illegible], S. Kondo, K. Maeda, and H. Umezawa, J. Am. Chem. Soc., 96, 4327 (1974).
89. [illegible] W. K. Woo and H. W. Dion, J. Heterocycl. Chem., 11, 641 (1974).
90. R. J. Suhadolnik, Nucleoside Antibiotics, Wiley-Interscience, 1970.
91. [illegible], [illegible] Chemistry, Mechanisms of Action, [illegible], 1979.
92. [illegible] Elstner and R. J. Suhadolnik, [illegible], 10 Suppl. (1971).
93. [illegible] Elstner and R. J. Suhadolnik, Biochemistry, 11, 2578 (1972).
94. [illegible] Elstner, R. J. Suhadolnik, and [illegible], J. Biol. Chem., 248, 5385 (1973).
95. M. Boyd and R. J. Suhadolnik, [illegible], 264, 130 (1975).
96. R. J. Suhadolnik, Nucleoside Antibiotics, Wiley-Interscience, 1970.
97. [illegible], [illegible] M. H. Richmond and M. J. Waring, The Molecular Basis of Antibiotic Action, [illegible] Wiley-Interscience, 1972.
98. [illegible], S. Asahi and [illegible], J. Am. Chem. Soc., 91, 7490 (1969).
99. [illegible], K. [illegible], Nippon [illegible], 184, [illegible] (1969).
100. K. Isono and R. J. Suhadolnik, Arch. Biochem. Biophys. (1975), in press.
101. K. Isono, P. F. Crain, T. J. Odiorne, J. A. McCloskey, and R. J. Suhadolnik, J. Am. Chem. Soc., 95, 5788 (1973).
102. K. [illegible] and [illegible], [illegible], 22, [illegible] (1968).
103. K. [illegible], A. [illegible], and [illegible], Tetrahedron Lett., 1974, 1355.
104. [illegible], Tetrahedron Lett., 1974, 4152.
105. K. Isono, P. F. Crain, and J. A. McCloskey, J. Am. Chem. Soc., 97, 943 (1975).
106. K. [illegible], and [illegible], Tetrahedron Lett., 1972, 1991.
107. [illegible], [illegible], M. [illegible] and H. [illegible], [illegible], Chemistry of Natural Products, 1974, Abstracts, p. [illegible].

Index

L, M

N

O

P